TOPICS IN GRAPH THEORY

ANNALS OF THE NEW YORK ACADEMY OF SCIENCES
Volume 328

TOPICS IN GRAPH THEORY

Edited by Frank Harary

The New York Academy of Sciences
New York, New York
1979

Library of Congress Cataloging in Publication Data

Main entry under title:

Topics in graph theory.

(Annals of the New York Academy of Sciences; v. 328)
"This series of papers is the result of the Academy's scientist-in-residence program honoring Professor Harary on May 2–6, 1977."
1. Graph theory—Congresses. 2. Harary, Frank. I. Harary, Frank. II. Series: New York Academy of Sciences. Annals; v. 328.
Q11.N5 vol. 328 [QA166] 508'.1s [511'.5] 79–15781

S/CCP
Printed in the United States of America
ISBN 0-89766-028-5

ANNALS OF THE NEW YORK ACADEMY OF SCIENCES

VOLUME 328

June 20, 1979

TOPICS IN GRAPH THEORY*

Editor
FRANK HARARY

Scientist-in-Residence Program Coordinators
ALLAN GEWIRTZ AND LOUIS V. QUINTAS

CONTENTS

*This series of papers is the result of The Academy's Scientist-in-Residence Program honoring Professor Harary on May 2–6, 1977.

FOREWORD

If you look up "graph theory" in the latest (1976) edition of the *Mathematics Dictionary*, edited by Glenn and Robert James, you are in for a shock. There is no such entry! There is not even a definition of "graph" as graph theorists use the term. I take this as a double tribute: to the rapidity with which graph theory has become a major discipline of modern mathematics and to the degree to which the James and James dictionary is hopelessly out of date.

As few as ten years ago mathematicians were still inclined to regard graph theory as a minor aspect of either topology or combinatorics. Today it is a flourishing field of research with its own textbooks, its own journal, its own eminent specialists—above all, its increasingly useful applications to almost every branch of science and technology. It is only a matter of time, I am persuaded, until children in the primary grades will be introduced to the basic concepts of graph theory and until it becomes an essential part of high school instruction.

Young students need all the help they can get in visualizing mathematical structures, in grasping them intuitively before they take up rigorous proofs. Graph theory, although as rigorous as any other branch of mathematics, is also a marvelous form of hand waving. It is astonishing how many deep theorems of a combinatorial nature, their binary relations hidden under complex algebraic notation, become suddenly clear when the graph theorist waves his graphs like a magician waving his hands to produce a rabbit. The child's imagination is stimulated by the mystery of mathematical isomorphism, and the teaching task is made easier. Moreover, educators are slowly discovering the value of recreational mathematics in combating dullness. Because recreational mathematics is rich in graph theory applications (especially in the use of directed graphs), we have another reason to expect an increasing emphasis on graph theory at all levels of precollege teaching.

The papers in this valuable collection are the outgrowth of a week devoted to graph theory at The New York Academy of Sciences, May 2–6, 1977. About two dozen persons, classified as either "Scientist-in-Residence" or a "Participating Scholar," attended. The conference leader, Frank Harary, asked each of them to write a survey article on a topic of their choice, with the view of gathering these articles into a book. The papers were reviewed by an anonymous referee who was, I am told on good authority, "Mr. Graph Theory" himself.

It was Seymour Schuster, in an anonymous telegraphic book review in the *American Mathematical Monthly*, who first applied this phrase to Frank Harary, and with good reason. For several decades Professor Harary has been graph theory's number one promoter, as evidenced by his "city list" appended below.* (Gabriel Dirac once introduced him as "the Ambassador of Graph Theory.") His *Graph Theory* has become the standard world textbook. His papers on graph theory, written alone or in collaboration, now approach the 300 mark. His humor, creativity, and enthusiasm—you'll find all three reflected in this volume—are legendary.

Let me reflect some of it myself by closing with a bit of Hararian word play. Professor Harary has a habit of saying, as he walks to the blackboard to pick up the chalk, "Let's consider an arbitrary, chosen-at-random graph." He then invariably draws the following:

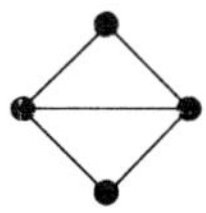

* These are the cities outside the United States where Professor Harary has been invited to give lectures at universities or at conferences. The alphabet was completed on June 6, 1978, when he spoke at a conference in Xanten-on-the-Rhine.

Gardner: Foreword

This simple, incomplete graph on four points has a name, though it is not likely to appear in the next revision of James and James. Among the legion of students who learned graph theory from Professor Harary at Ann Arbor or from his book, his supposedly "random graph" is now widely known as the "anarbortrary graph."

Martin Gardner
Hastings-on-Hudson, New York

City List

Aachen, Aarhus, Aberdeen, Abidjan, Accra, Adelaide, Amsterdam, Auckland, Balatonfüred, Bangkok, Beograd, BeerSheva, Berlin, Birmingham, Bochum, Bonn, Bradford, Bratislava, Braunschweig, Bremen, Bridgetown, Brisbane, Bruxelles, Bucuresti, Budapest, Buenos Aires, Bundoora, Burnaby, Calgary, Cambridge, Canberra, Caracas, Charlotte Amalie, Christchurch, Clayton, Dakar, Delft, Dublin, Duisburg, Dunedin, Edinburgh, Edmonton, Eindhoven, Erlangen, Firenze, Frascati, Freising, Genève, Glasgow, Haifa, Hamburg, Hamilton, Hannover, Helsingør, Helsinki, Hobart, Hong Kong, Ibadan, Ife, Ilmenau, Johannesburg, Katmandu, Kensington, Kingston, Klagenfurt, København, Köln, Kraków, Kuala Lumpur, Lagos, Leicester, Leiden, Lima, Lisbon, London, London (Canada), Lund, Lyngby, Melbourne, Montpellier, Montréal, Mülheim, München, Nairobi, Nathan, Newcastle, Newcastle (Australia), Niamey, North Ryde, Nottingham, Oberwolfach, Ottawa, Oslo, Oulu, Oxford, Paris, Parkville, Pavia, Penang, Perth, Praha, Pretoria, Punta Arenas, Québec, Rangoon, Reading, Rehovot, Rio de Janeiro, Roma, St. Andrews, St. Lucia, San Juan, Santiago, Singapore, Stockholm, Sydney, Taipei, Tehran, Tel Aviv, Tierra del Fuego, Tihany, Tokyo, Toronto, Townsville, Udine, Utrecht, Vancouver, Varenna, Victoria, Warszawa, Waterloo, Wellington, Winnipeg, Wollongong, Xanten, York, Zürich.

PREFACE

During the week of 2–6 May 1977, The New York Academy of Sciences (NYAS) hosted a meeting on graph theory in the framework of the Scientist-in-Residence (SIR) program. There had been just two previous SIRs: the celebrated mathematician-at-large Paul Erdös and the Nobel Laureate physicist Julian Schwinger. Thus I felt particularly honored to be designated as a SIR.

The meeting was flawlessly organized by my friends and colleagues Allan Gewirtz and Louis Quintas. There were also "participating scholars" ranging in academic age from graduate student to full professor. The presentations were scheduled in a quite civilized manner: two hours were allowed for lunch, a full hour was taken for both the morning coffee break and the afternoon tea break, Wednesday afternoon was free, and there were no evening meetings (except for the Thursday evening dinner, after which I spoke on "Independent discoveries in graph theory"). Each participating scholar had the opportunity to describe his own research to the others and to benefit from their comments. The resulting contributions to this book range from definitive surveys of a topic in graph theory to interesting new observations of concepts and theorems. My thanks go to all of them.

There are many other people I must thank and it is a pleasure to do so. Mark Batell, a mathematical psychologist, helped by setting up a filing system for the articles. My doctoral students Jerry Kabell and Zevi Miller were most helpful with the refereeing of the papers, as was my colleague Andreas Blass.

In accordance with the current practice of several combinatorial journals, there were two referees for each paper and I was one of them. This explains one of the sentences in the eloquent Foreword by Martin Gardner. I could not agree with him more in his prediction that, in time, graph theory will eventually find its way not only into the high school curriculum but even into elementary education. Furthermore, it is my conviction that games played on graphs will also become an essential ingredient in the educational process.

The first week of May 1977 at the NYAS was not at all a conventional scientific conference. Rather it was a stimulating research, teaching, and learning experience. There was enough time for extended discussions in which all the participating scholars participated. The ambience was intellectually active, but relaxed rather than hectic. The pleasant physical surroundings of the NYAS and the comfortable decor of the meeting room considerably enhanced the occasion. We were all delighted with the smooth operation at the NYAS, which included gourmet meals prepared by the talented chef. Finally, I am pleased that Bill Boland handled the publication of this volume. I hope it will prove to be useful to those who discover theorems in graph theory and to those who use graph theory as a mathematical model for application to other subjects.

Frank Harary
Ann Arbor, Michigan

INDEPENDENT DISCOVERIES IN GRAPH THEORY*

Frank Harary

Department of Mathematics
University of Michigan
Ann Arbor, Michigan 48109

I am very honored to have been designated the "Scientist-in-Residence" of the New York Academy of Sciences for the current week. This is an honor comparable to any of those which I have already enjoyed.

By definition, an *independent discovery in graph theory* occurs when two people or groups of people working independently discover essentially the same result. Usually these discoveries occur at about the same time, but they can be separated by periods of many years, as you will see from some of our examples. I should make it clear at the outset that I really mean independent discoveries *in* graph theory, not independent discoveries *of* graph theory. The latter, of course, refers to discoveries of the entire field, which I shall mention briefly before moving on.

The first recorded discovery of graph theory was made by the great Swiss mathematician Leonhard Euler [7] when he developed a theorem on graphs to solve the famous problem of the Seven Bridges of Königsberg in 1736. This theorem is now known as the characterization of eulerian graphs. The discovery of graph theory which has had the greatest impact on modern life was by Georg Kirchhoff in 1847 in his study [16] of electrical networks. Cayley, in 1857, while considering the problem of changes of variables in differential calculus, discovered trees [5], and some seventeen years later found in them the ideal tool for counting chemical isomers [6]. The great Irish mathematician and physicist, Sir William Hamilton, was led into graph theory as a result of a game he invented in 1859. The game, which essentially required the finding of a spanning cycle in the skeleton of a dodecahedron, was never a commercial success (except to Hamilton, who sold the idea for £25). However, it resulted in Hamilton's name being attached to one of the more recalcitrant problems in graph theory, that of characterizing hamiltonian graphs, which have a cycle containing all the points. Another famous problem which has led many people into graph theory is the Four Color Conjecture, only recently proved by Haken *et al.* [10] (with a computer). This by no means exhausts the list of discoverers of graph theory. In very many cases and in disciplines in the physical sciences, the social sciences, computer science, and the humanities, graphs frequently occur as a natural, useful, and intuitive mathematical model. The consequence is that those investigators who were not aware of the existence of graph theory as a study in its own right were led to rediscover it in order to apply it.

One of the most fundamental theorems is that of Kasimierz Kuratowski characterizing planar graphs [18]. When it appeared in 1930, two American mathematicians, Orrin Frink and Paul A. Smith, had already submitted (independently, and independently of each other) papers containing precisely the same theorem, which they promptly withdrew.

Now I would like to share with you a number of cases of independent discoveries *in* graph theory. These are classified as win, lose, or draw according to

* Adapted by J. A. Kabell from the recording of the address delivered at the New York Academy of Sciences on 5 May 1977.

0077-8923/79/0328-0001$1.75/1 © 1979, NYAS

chronological priority. A former student of mine, Lowell Beineke, published [1] a technique for partioning complete graphs into line-disjoint paths. It was only later that we learned that the great French number theorist and collector of puzzles, E. Lucas, had published exactly the same construction in 1887 [19]. Another loser was the characterization of planar graphs in terms of contractions (dualizing Kuratowski's theorem) which Bill Tutte and I published in 1965 [15], only to find later that Klaus Wagner had obtained the same result in 1937 [22].

Beineke and I, in 1965, calculated and published [2] the genus of the n-cube. Unfortunately for us, Gerhard Ringel [20] had published the same result ten years earlier. Another example of a loser is a manuscript I wrote with my first doctoral student, Bob Norman. In this note we showed that the sum of the point independence and covering numbers, and likewise the sum of the line independence and covering numbers, is always just the number of points in the graph ($\alpha_0 + \beta_0 = \alpha_1 + \beta_1 = p$). The only problem was that Tibor Gallai [9] had done precisely the same work a few years earlier. Fortunately, the referee to whom our paper was sent by Paul Halmos was familiar with Gallai's result, so Bob and I were saved some embarrassment.

Now that we have seen some losers, let me tell you about one *bona fide* draw. Pavol Hell and I, at an AMS meeting in Las Vegas, happened to hit on the idea of the ramsey number of a directed graph. Eventually, we wrote up and published [14] our results. The very same month in which our paper appeared, another journal had a paper by Jean-Claude Bermond [4], defining exactly the same concept and proving many of the same results! That can truly be called a draw.

Here are two more draw stories both involving the Japanese graph theorist, Jin Akiyama. By way of background, I met in person both Dragos Cvetković and his doctoral student S. Simic on arrival at the Belgrade airport in 1974. At about the same time, Akiyama and his student K. Kaneko were also deriving graph equations for line graphs and nth power graphs. Friendly correspondence between Tokyo and Belgrade led to a triply joint paper which is to appear in 1980.

Akiyama was also doing the research in another Ph.D. thesis elsewhere in the world when he derived a forbidden subgraph characterization of iterated line graphs. The referee recommended that Jin write to D. G. Akka of Shahabad, India, who had also obtained the same result, with the suggestion that they get together and write one joint article, and Jin has agreed to do so.

Now let us look at a few winners. In 1958, Claude Berge published the second book ever written [3] on graph theory (the first was, of course, König [17], in 1936), and *Mathematical Reviews* (MR) sent me a copy to review. It concluded with a section of fifteen unsolved problems. The last four were disguised versions of Four Color Conjecture which happily has since been proved, as mentioned above. Problem 11, however, was simple enough for me to solve at once. It asked for the maximum connectivity of a graph with a given number of points and lines, and I included the result in my review in MR. Shortly thereafter, I received several papers to referee in which it was solved. Therefore I decided to publish this straightforward result [13].

Another winner I would like to present tonight actually concerns a real world application of graph theory rather than a new theorem. The idea of signed graphs and balance, which I introduced in 1953 [11], grew naturally out of attempts to describe both positive and negative psychological and social interactions in graphical terms. As a war involves essentially the same types of interactions, but between countries rather than individuals, I analyzed the Mideast conflict of 1956 using signed graphs and the tendency toward balance. The resulting paper [12] was published in the *Journal of Conflict Resolution* in 1962. In 1969, I received a frantic long-distance telephone call from Greece, from a man named George. First he spoke to me in Greek and I spoke to him in English, with neither of us understanding a word of the other language. Then he tried Turkish and I tried Spanish with the same result, and we finally settled on very fractured French. He was calling to tell me that he had just found out that his Ph.D. thesis in political science was essentially subsumed by this article of mine and needed to have my written permission to receive his doctorate from the University of Athens. Naturally I was so delighted to learn that my approach was worthy of a Ph.D. in political science that I immediately gave my blessing.

Here is an example of multiple independent discovery. It is a well-established result that any planar graph can be drawn in the plane in such a way that all its edges are straight line segments. This was published by Fáry [8] in 1948 and is often called Fáry's Theorem. It was rediscovered by Sherman Stein [21] in 1951 and by several others since then. Once again, however, Klaus Wagner [23] anticipated everyone by publishing the same theorem in 1936.

An amusing incident concerns two papers received not long ago by the Managing Editor of the *Journal of Graph Theory*, my good friend and colleague Gary Chartrand. The result, which was submitted nearly simultaneously as a joint note by two mathematicians in one university and also by one chap in another university, was a novel observation which stated that if a graph satisfied three very simple sounding conditions, it didn't exist. Of course, Gary wrote to these three gentlemen and suggested that they submit one joint paper which we would be glad to publish. The sole author agreed, but the two coauthors did not, so unfortunately we had to decline both notes.

These anecdotes do not by any means exhaust the examples of independent discovery in graph theory. In fact, independent discovery is probably more the rule than the exception, particularly in a field which is growing as rapidly as graph theory. One of the reasons is that we now have results on problems of interest circulated widely among people of varying backgrounds. It is not at all unreasonable that more than one person should have a similar insight at about the same time.

References

1. Beineke, L. W. 1964. Decompositions of complete graphs into forests. Magyar Tud. Akad. Mat. Kutato Int. Kozl. **9:** 589–594.
2. Beineke, L. W. & F. Harary. 1965. The genus of the n-cube. Canad. J. Math. **17:** 494–496.
3. Berge, C. 1958. Theorie des Graphes et ses Applications. Dunod, Paris.
4. Bermond, J. C. 1974. Some ramsey numbers for directed graphs. Discrete Math. **9:** 313–321.
5. Cayley, A. 1857. On the theory of the analytical forms called trees. Philos. Mag. **13:** 19–30.
6. Cayley, A. 1874. On the mathematical theory of isomers. Philos. Mag. **67:** 444–446.
7. Euler, L. 1736. Solutio problematis ad geometriam situs pertinentis. Comment. Acad. Sci. I. Petropolitanae **8:** 128–140.
8. Fáry, I. 1948. On straight line representation of planar graphs. Acta Sci. Math. Szeged. **11:** 229–233.
9. Gallai, T. 1959. Über extreme Punkt- und Kantenmengen. Ann. Univ. Sci. Budapest, Eötvös Sect. Math. **2:** 133–138.
10. Haken, W., K. Appel & J. Koch. 1977. Every planar map is four-colorable. Illinois J. Math. **21:** 429–567.
11. Harary, F. 1953. On the notion of balance of a signed graph. Michigan Math. J. **2:** 143–146.

12. HARARY, F. 1961. A structural analysis of the situation in the Middle East in 1956. J. Conflict Resolution. **5:** 167–178.
13. HARARY, F. 1962. The maximum connectivity of a graph. Proc. Nat. Acad. Sci. U.S.A. **48:** 1142–1146.
14. HARARY, F. & P. HELL. 1975. Generalized ramsey theory for graphs V: The ramsey number of a digraph. Bull. London Math. Soc. **6:** 175–182; **7:** 87–88.
15. HARARY, F. & W. T. TUTTE. 1965. A dual form of Kuratowski's theorem. Canad. Math. Bull. **8:** 17–20, 373.
16. KIRCHHOFF, G. 1847. Über die Auflosung der Gleichungen, auf welche der linearen Verteilung galvanischer Strömme geführt wird. Ann. Phys. Chem. **72:** 497–508.
17. KÖNIG, D. 1936. Theorie der endlichen und unendlichen Graphen. Akademische Verlagsgesellschaft M.B.H. Leipzig. Reprinted Chelsea, New York (1950).
18. KURATOWSKI, K. 1930. Sur le problème des courbes gauches en topologie. Fund. Math. **15:** 271–283.
19. LUCAS, E. Récréations Mathématiques, I-IV, Paris, 1882–1894.
20. RINGEL, G. 1955. Über drei kombinatorische Probleme am *n*-dimensionalen Würfel und Würfelgitter. Abh. Math. Sem. Univ. Hamburg. **20:** 10–19.
21. STEIN, S. K. 1951. Convex maps. Proc. Amer. Math. Soc. **2:** 464–466.
22. WAGNER, K., 1936. Bemerkungen zum Vierfarbenproblem. Jber. Deutsch. Math.-Verein. **46:** 26–32.
23. WAGNER, K. 1937. Über eine Eigenschaft der ebenen Komplexe. Math. Ann. **114:** 570–590.

THE EXPLOSIVE GROWTH OF GRAPH THEORY

Frank Harary

Department of Mathematics
University of Michigan
Ann Arbor, Michigan 48109

The preparation of an article that documents the remarkable growth of research activity in the theory of graphs during the past few decades was suggested to me by the important commentary of May [12] on the growth and quality of mathematical literature.

Books and Conference Books

The first book on graph theory as a separate subject in mathematics was written by König [11] in 1936. He wrote in German because he correctly decided that if it were written in his native Hungarian, very few people would be able to read it. This book was reprinted in the United States in 1950. When I first became interested in graph theory in September 1950, this was the only book available on the subject. There were some theorems on graphs in the pioneering book on topology by Veblen [15] whose first chapter takes up "linear graphs." There were also some booklets by Lucas and Sainté-Lague, referenced in König's book, on the use of graphs in solving puzzles and in recreational mathematics. There were in 1950 literally only about a dozen mathematicians in the world actively doing research in graph theory!

The second book on graph theory, by Berge [1], appeared in French in 1958. I had begun working on a book on graph theory since 1950, virtually the moment I was exposed to this fascinating aesthetic art form. As I was writing each chapter, so many tractable open questions suggested themselves that research impeded the process of writing. For this reason, my book [10] did not appear until 1969. Chart 1 displays the number of books which dealt with either graph theory as a separate subject, some special topic in graph theory, or a particular application of graph

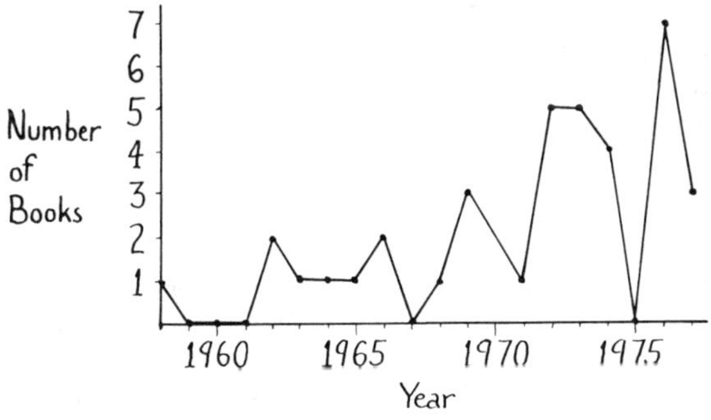

Chart 1. Books devoted wholly or in part to Graph Theory.

0077-8923/79/0328-0005$1.75/1

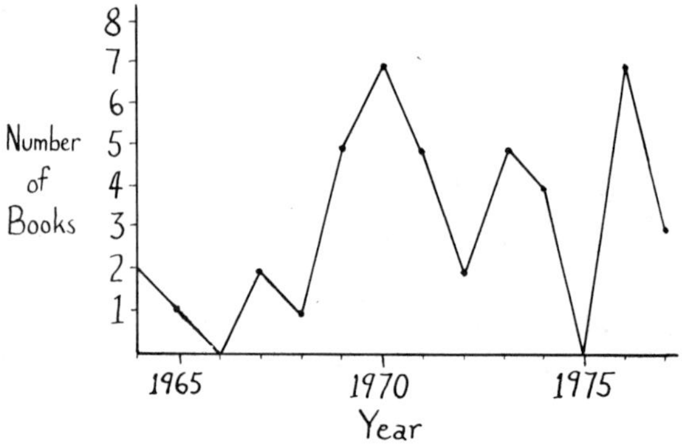

CHART 2. Proceedings of conferences devoted wholly or in part to Graph Theory.

theory to another science: physical, social, or engineering, beginning with 1958 (it is called a chart because I cannot call it a graph).

There is another important outlet for research in graph theory in book form. These are the "conference books," shown in CHART 2, which report the proceedings of symposia on graph theory, combinatorial theory, and their applications. The first of these conference books, to my knowledge, was the book [8] of the Proceedings of the First Czechoslovakian Symposium on Graph Theory held in Smolenice in 1963.

In constructing CHARTS 1 and 2, it was a matter of subjective judgment as to how much graph theory should be adequate for inclusion. When a book has been translated, only the original was included. When a conference book appeared in more than one volume, it was still counted as just one book. When there was more than one printing of a book, it was included as just one book. A list of books and conference books which includes most of those reported in CHARTS 1 and 2 appeared in the Graph Theory Newsletter [9].

PUBLISHED PAPERS ON GRAPH THEORY

In ancient times, articles on graph theory were reviewed in *Mathematical Reviews* (MR) in one of the following categories:

(1) Foundations, Theory of Sets, Logic
(2) Algebraic Topology
(3) Geometry
(4) Numerical Analysis
(5) Probability
(6) Statistics
(7) Physical Applications, in the subsection on Optics, Electromagnetic Theory, Circuits
(8) Other Applications, in the subsections on Economics, Management Science, Biology and Sociology, Information and Communication Theory.

Even though most papers on the subject were reviewed under Algebraic Topology (at the end of the section on Topology), in general the papers were neither algebraic nor topological, but combinatorial or as I would prefer to express it, *graphical*.

The first mention of Combinatorial Analysis as an area of mathematics in MR was in Volume 20, Number 1, January 1959. But *Graph Theory* (GT) was not yet mentioned. Papers on graph theory were reviewed in this new category as well as in the eight sections listed above. It wasn't until 1968 that MR in Volume 35 stated Category 05 as "Combinatorial Theory, Graph Theory." Mysteriously, the beautiful words "Graph Theory" then disappeared from the front cover of MR beginning with the July 1973 issue, Volume 46, when the name of 05 was changed to "Combinatorics." It would be splendid if this could be altered again so that 05 reads "Combinatorics, Graph Theory" and this would improve the front cover of MR.

Since its establishment, the number of papers in Category 05 has been increasing exponentially, far more rapidly than the number of papers in all branches of mathematics. There have been so many papers in Category 05 in recent years that one could even use the *number of pages* in MR as an indication of the growth rate of publication. Graph Theory is now the subcategory 05C.

The number of graphical papers reviewed in Volume 20, 1959, of MR was only seven. In fact, my review of Berge's book appeared in MR, Volume 21, under Algebraic Topology.

Beginning in 1973, MR launched in Volume 45 the extremely useful inclusion in its Index issues of the names of authors within the subject index. A remarkably convincing CHART 3 which displays the recent exponential growth in 05C can be constructed by estimating accurately, to the nearest fourth of a column, the number of columns in the author list in the index issue. This is shown using solid lines.

The subcategory 94A20 includes papers on applications of graph theory. When the number of columns from this source is added to the number for 05C, the result is shown in CHART 3 as dotted lines; it also has an exponential appearance. Unfortunately, MR decided to change this format in Volume 53 so that CHART 3 only reports Volumes 45–52.

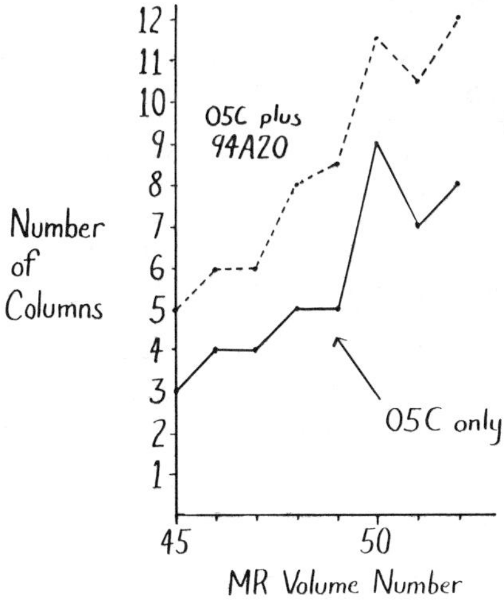

CHART 3. Number of columns devoted to Graph Theory in MR Subject Index.

Journals

The first mathematical journal devoted to combinatorics and graph theory began publication with Volume 1 in 1966. This was the *Journal of Combinatorial Theory* (JCT) which was founded by G.-C. Rota and myself as a result of exploratory initial discussions in Frascati during July 1964 over much Italian wine and pasta. It began as a quarterly and almost immediately acquired a three-year backlog of papers. In 1970, it was separated into two parts which are easily described in reverse alphabetical order: Series B comprises Graph Theory and Matroids; Series A includes all other areas of combinatorics. Since then, JCT has been a monthly.

The second such journal was *Discrete Mathematics* (DM) which was founded by Peter Hammer. It also began as a quarterly with Volume 1 having its first three issues in 1971 and its fourth in early 1972. With four issues comprising each volume, DM in 1973 became a monthly periodical having three volumes per year, with only one exception in 1974. That year DM had four volumes as a result of a special report on coding theory. Its plans are to remain a monthly indefinitely. However, in order to cope with its backlog, it will soon increase the number of pages per issue.

The journal DM has recently given birth to a series of books titled *Annals of Discrete Mathematics*. These resemble somewhat the Springer Lecture Notes in Mathematics, of which Volumes 110, 186, 303, 406, 411, and 642 report proceedings of combinatorial conferences with a primary emphasis on graph theory. In addition DM has another spin-off: a new journal, *Discrete Applied Mathematics*, with Volume 1 in 1979.

There are other publications involving graph theory which merit mention here. The informal and informative, but important, *Graph Theory Newsletter* was founded at Western Michigan University largely influenced by Gary Chartrand. It began in 1971 and has as its basic purpose the rapid, relatively inexpensive dissemination of interesting information to graph theorists all over the world. It includes abstracts of results, announcements of forthcoming meetings, lists of recently published papers by title, and lists of books. More recently it has been presenting short expository articles.

In the field of applications, the journal *Networks* was launched in 1972, founded by I. T. Frisch and A. B. Frank. By definition, a *network* consists of a graph (or a digraph, a multigraph, or other generalization) together with values assigned to its edges or its vertices, or both). Of course networks are directly involved with real world problems involving time and money, while graph theory constitutes the basic mathematical model underlying networks.

Other active applications using GT are in the fields of electrical networks, operations research and computer science. The journals in these fields often publish articles on either so-called "pure" graph theory or its applications to the respective fields. These journals include, amongst many others:

> *Transactions of the IEEE* (Institute for Electrical and Electronic Engineering) on *CAS* (Circuits and Systems), *Operations Research*, *Journal of the ACM* and *Communications of the ACM* (Association for Computing Machinery), *SIAM* (Society for Industrial and Applied Mathematics) *Journal on Computing.*

In fact, "graph theory" is a subject in the classification schema of both *Physics Abstracts* and *Chemical Abstracts*.

In the applications of graph theory to the social sciences, relevant articles have appeared in the journals:

Behaviorial Science, Journal of Mathematical Sociology, Psychometrika, Social Networks, Sociometry.

The year 1976 saw a noticeable increase in combinatorial journals as it was the first year for both *Ars Combinatoria* and the *Journal of Combinatorics, Information and System Sciences.*

The *Journal of Graph Theory* (JGT) began in 1977. Of the 10,000 new journals in the professional and scientific fields which were launched worldwide that year, JGT was voted by the Association of American Publishers (Technical, Scientific and Medical Division) as being the most outstanding new journal of 1977.

Bibliographies of Graph Theory

The first really comprehensive bibliography of GT was the list of references up to 1936 in the book by König. Very recently, the first book on the history of graph theory by Biggs *et al.* [6] pointed out several additional appropriate references dating before 1936. Together, these two sources provide an admirable bibliography of early works on GT.

The next few listings appeared in books on GT including those by Berge [1], myself [10], and many others (see [9]). But these cannot be considered comprehensive because they were not, as they reflected the authors' subjective criteria for inclusion.

The genuine bibliographies of graph theory which have appeared to date are those compiled by Turner [13], Deo [7], and most recently, by Berman [2, 3, 5] of works through the middle of 1973. The latter proved particularly useful as the articles in it enabled Berman [4] to develop citation indexes for papers and books on GT.

Turner and Katz [14] compiled an informative annotated bibliography of research articles on graph theory in the Soviet Union to 1969. There has been a notable further increase in graph theory and its applications in the Russian literature during the past decade and it is now time to recommend that a sequel to [14] be continued on a regular basis. It would appear quite safe to predict that we shall soon see an awakening to the importance of graph theory for applications to the real world in the Chinese research literature.

Together, these bibliographies provide dramatic evidence of the explosive growth of graph theory.

Symposia

There have been many national and international symposia on graph theory, combinatorial theory, and their applications. These have taken place intermittently in Canada, Czechoslovakia, France, Germany, Great Britain, Hungary, Italy, and the United States. The author has had occasion to speak at conferences on the applications of graph theory to the following topics: electrical networks, management sciences, social sciences, architecture, linguistics, theoretical physics, operational research, linear programming, sparse matrices in numerical analysis, information retrieval, biology, design problems in engineering, didactics, social networks, peace science, theoretical chemistry, and theoretical computer science. Whenever such an invitation materializes, I take the point of view that the organizers of the conference are aware of the fact that the only subject I know in some detail is GT. Hence

I ask them what aspects of GT are the most relevant and applicable to their fields and prepare my lectures accordingly.

During my world-wide travels, I have lectured not only at colloquia in departments of mathematics at various universities but also in the following other departments:

> Applied mathematics, statistics, operations research, civil engineering, architecture, psychology, physics, chemistry, electrical engineering, computer science, combinatorics and optimization, anthropology, sociology, and Australian environmental studies.

This remarkable variety of subjects attests vividly to the essentially universal applicability of the theory of graphs as a useful and valid mathematical model.

Conclusion and Prediction

Graph Theory is a very dynamic subject: some areas are slowing down (attempts to *prove* the Four Color Conjecture), others are growing (ramsey theory, reconstruction conjecture), and new ones are being born (isomorphic factorizations, multiple interval graphs, coloring digraphs). The most established scientific areas in which GT is regularly used as a mathematical model are electrical engineering and theoretical chemistry. In fact, a series of papers titled "The graph-like state of matter" is being written by M. Gordon *et al.* In the social sciences, the most recent application areas are anthropology, sociology, and political science. Several underdeveloped subjects are also beginning to use GT.

The evidence is uncontrovertible that the present growth status of graph theory and its applications is genuinely explosive. My prediction is that we are now witnessing only the early beginnings of this phenomenon and that in time the importance of discrete mathematics including the fields of matroids and networks, and of GT in particular, will become comparable to the present scope of the more traditional applied mathematics.

Summary

The growth of the literature in graph theory is described in several ways: (1) by graphing the number of books on the subject per year, (2) by analyzing the number of published papers as reported in *Mathematical Reviews*, (3) by displaying the growth of journals whose primary emphasis is on graph theory and combinatorial theory, and (4) by noting various bibliographies on graph theory. The observations lead to the conclusion that the literature of graph theory is growing explosively.

References

1. Berge, C. 1958. Theorie des Graphes et ses Applications. Dunod, Paris.
2. Berman, G. 1975. Forward Citations in Graph Theory. University of Waterloo. Waterloo, Canada.
3. Berman, G. 1977. Applied Graph Theory Bibliography with Forward Citations. University of Waterloo, Waterloo, Canada.

4. BERMAN, G. 1977. Frequently cited publications in pure graph theory. J. Graph Theory **1:** 175–180.
5. BERMAN, G. & D. M. JACKSON. 1976. Forward Citations in Enumerations. University of Waterloo, Waterloo, Canada.
6. BIGGS, N., K. LLOYD & R. WILSON. 1976. Graph Theory 1736–1936. Oxford University Press, Oxford.
7. DEO, N. 1969. An extensive English language bibliography on graph theory and its applications. Jet Propulsion Laboratory, Pasadena, Technical Report (October): Supplement 1 (April 15, 1971).
8. FIEDLER, M., EDITOR. 1964. Theory of Graphs and its Applications. Academia, Prague.
9. Graph Theory Newsletter, Supplement: A List of Books in Graph Theory and Related Areas 2(197), No. 4.
10. HARARY, F. 1969. Graph Theory, Addison-Wesley, Reading, Mass.
11. KÖNIG, D. 1936. Theorie der endlichen und unendlichen Graphen. Leipzig; Reprinted Chelsea, New York (1950).
12. MAY, K. O. 1968/69. Growth and quality of the mathematical literature. Isis. **59:** 363–371.
13. TURNER, J. 1969. Key-word index bibliography of graph theory. Proof Techniques in Graph Theory. F. Harary, Ed. Academic Press, New York, N.Y. pp. 189–330.
14. TURNER, J. & W. H. KATZ. 1970. A survey of progress in graph theory in the Soviet Union. SIAM Review (Supplement). **12:** 1–68.
15. VEBLEN, O. 1922. Analysis Situs. Amer. Math. Soc. Colloq. Publ. Vol. 5, Cambridge. (2d edit. New York. 1931).

GRAPHS AND PERMUTATION GROUPS WHOSE TRANSITIVE CONSTITUENTS HAVE PRIME DEGREE*

Brian Alspach

Department of Mathematics
Simon Fraser University
Burnaby, British Columbia V5A 1S6
Canada

Introduction

In [10] the structure of vertex-transitive graphs of prime order was determined and they were enumerated. In [1, 3, 4] the structure of their automorphism groups was completely determined while in [1, 5] the number of graphs and digraphs of order p, p a prime, having a specified transitive permutation group $G \leq S^p$ as their automorphism group was calculated. In [2] the number of tournaments of order p having G as their automorphism group was given. In this paper we use the above ideas to investigate the relationship between graphs and digraphs and permutation groups for which each transitive constituent has prime degree, that is, each orbit has prime length.

We use [6] for graph theoretic terminology and [11] for permutation group terminology. We denote by Aut(X) the automorphism group of the graph or digraph X. We say X is vertex-transitive if Aut(X) acts transitively on the vertices of X. If G is a permutation group each of whose orbits has prime length, let $\Delta(p)$ denote the union of all the orbits of fixed prime length p and let $G^{\Delta(p)}$ denote the restriction of G to $\Delta(p)$. If $\sigma \in G$ is a fixed permutation and S denotes a union of orbits of G, then σ^S denotes the restriction of σ to S.

Graphical Groups

We use $\leq$ and $<$ to denote subgroup inclusion and proper inclusion, respectively. If $G \leq S^n$, the symmetric group of degree n, let $\eta(G)$ and $\mu(G)$ denote the number of graphs and digraphs, respectively, of order n that have G as their automorphism group. We say that G is graphical if $\eta(G) > 0$. In the following we shall use $\equiv$ to denote isomorphism as permutation groups, whereas $\cong$ shall denote isomorphism as abstract groups.

Lemma 1. Let $G \leq S^n$ be graphical and every orbit of G have prime length. If $p_1, p_2, \ldots, p_k$ are the distinct primes occurring for orbit lengths of G, then

$$G \equiv G^{\Delta(p_1)} \times \cdots \times G^{\Delta(p_k)} \tag{1}$$

the direct product of the $G^{\Delta(p_i)}$'s.

Proof: Let X be a graph of order n such that Aut$(X) \equiv G$. If every orbit has the same prime length, the result is true since $G = G^{\Delta(p)}$ in this case. So suppose there are two or more distinct primes occurring as orbit lengths of Aut(X) and let

* This research was supported by the National Research Council of Canada.

0077-8923/79/0328-0012$1.75/1 © 1979, NYAS

Δ_i and Δ_j be any two orbits of Aut(X) of different prime lengths p_i and p_j. Obviously, there exists $\sigma \in \text{Aut}(X)$ such that σ^{Δ_i} is a cycle of length p_i. Since $p_j \neq p_i$, some power of σ will fix some $u \in \Delta_j$ but still be a cycle of length p_i when restricted to Δ_i. Hence, either u is adjacent to every vertex of Δ_i or adjacent to no vertex of Δ_i. The same is then true for every vertex of Δ_j. Therefore, given two orbits of different prime length, either no edges connect pairs of vertices one from each orbit or all such edges are in the graph. The result now follows. □

COROLLARY 1.1. Let $G \leq S^n$ satisfy $\mu(G) > 0$ and the other hypotheses of LEMMA 1. Then

$$G \equiv G^{\Delta(p_1)} \times \cdots \times G^{\Delta(p_k)} \tag{2}$$

Proof: The same proof used for LEMMA 1 works for COROLLARY 1.1 except that between the vertices of two orbits Δ and Δ' of different prime lengths there are either no arcs, all arcs from Δ to Δ', all arcs from Δ' to Δ, or all arcs between Δ and Δ'. □

LEMMA 1 and its corollary allow us to restrict our attention to the situation in which all orbits have the same prime length. If G is a permutation group in which each orbit has the same prime length p and if G is graphical, let X be a graph with $\text{Aut}(X) \equiv G$. Define a new graph based on X in the following way: The vertices of the graph correspond to the orbits of Aut(X) and there is an edge between two vertices of the graph if and only if there are edges between vertices of the corresponding orbits but not all such edges are in X. This graph is called the *orbital graph* of X and is denoted Orb(X).

Let $C_1, C_2, \ldots, C_k$ be the components of Orb(X) and let Ω_i, $i = 1, 2, \ldots, k$, be the union of the orbits corresponding to the vertices of the component C_i. Consider Aut(X) restricted to any Ω_i. Since the vertices in any orbit that comprises part of Ω_i are either all adjacent to every vertex or all adjacent to no vertex of any given orbit that is not part of Ω_i, it is clear that every automorphism of $\text{Aut}(X)^{\Omega_i}$ extends to an automorphism of X that fixes all other vertices of X. In other words, we have proved that Aut(X) is a direct product of its restrictions to each of the Ω_i's. This proves the following lemma.

LEMMA 2. If G is a permutation group each of whose orbits has prime length p and X is a graph such that $\text{Aut}(X) \equiv G$, then

$$\text{Aut}(X) \equiv G^{\Omega_1} \times \cdots \times G^{\Omega_k} \tag{3}$$

For a directed graph X, we define its orbital graph by letting two vertices be adjacent if neither no arcs, nor all arcs in one direction and no others, nor all possible arcs go between vertices of the corresponding orbits. Notice that the orbital graph of a digraph is a graph. Again we obtain a similar result for digraphs.

COROLLARY 2.2. If G is a permutation group each of whose orbits has prime length p and X is a digraph such that $\text{Aut}(X) \equiv G$, then

$$\text{Aut}(X) \equiv G^{\Omega_1} \times \cdots \times G^{\Omega_k} \tag{4}$$

LEMMA 2 and its corollary allow us to concentrate on graphs (or digraphs) X for which $\text{Aut}(X) = G$ has orbits all of length p, p a prime, and for which the orbital graph is connected.

We now consider another product for permutation groups. If G and H are two permutation groups acting on disjoint object sets S and T, respectively, and if

$G \cong H$ via the isomorphism $f\colon G \to H$, then the *f-diagonal product*, denoted $G \overset{f}{*} H$ (the role of f is usually suppressed in which case the notation $G * H$ is used), is obtained by considering the group of permutations $\{g * f(g)\colon g \in G\}$, where $g * f(g)$ acts on $S \cup T$ and

$$g * f(g)(a) = \begin{cases} g(a), \text{ if } a \in S \\ f(g)(a), \text{ if } a \in T \end{cases} \tag{5}$$

For example, $\langle(1\ 2\ 3)\rangle * \langle(4\ 5\ 6)(7\ 8\ 9)\rangle \equiv \langle(1\ 2\ 3)(4\ 5\ 6)(7\ 8\ 9)\rangle$ under the isomorphism mapping $(1\ 2\ 3)$ to $(4\ 5\ 6)(7\ 8\ 9)$. It is then obvious how to define diagonal products for finitely many isomorphic permutation groups $G_1, G_2, \ldots, G_n$ acting on mutually disjoint object sets $S_1, S_2, \ldots, S_n$, respectively. Let $f_i\colon G_1 \to G_i$, for $i = 2, 3, \ldots, n$, be an isomorphism. Then the $(f_2, f_3, \ldots, f_n)$*-diagonal product*, denoted by $G_1 \overset{f_2}{*} G_2 \overset{f_3}{*} \cdots \overset{f_n}{*} G_n$, is the group of permutations $\{g * f_2(g) * \cdots * f_n(g)\colon g \in G\}$, where $g * f_2(g) * \cdots * f_n(g)$ acts on $S_1 \cup \cdots \cup S_n$ and

$$g * f_2(g) * \cdots * f_n(g)(a) = \begin{cases} g(a), \text{ if } a \in S_1 \\ f_i(g)(a), \text{ if } a \in S_i, \text{ for } i \geq 2 \end{cases} \tag{6}$$

Again, if there is no possibility of confusion, we shall suppress the f_i's and simply write $G_1 * G_2 * \cdots * G_n$. We shall have many occasions to consider the diagonal product of a permutation group with itself. We make this definition explicit. If G is a permutation group acting on the set S, then the diagonal product $\overset{n}{*}\, G$ acts on the Cartesian product $S \times \{1, 2, \ldots, n\}$. We have $\overset{n}{*}\, G = \{g * g * \cdots * g\colon g \in G\}$, where

$$g * g * \cdots * g(s, k) = (g(s), k) \tag{7}$$

for $1 \leq k \leq n$.

We now quote a theorem of Burnside that will be used many times in the following material. We use Z_p to denote the field of integers modulo p, Z_p^* denotes the multiplicative subgroup $Z_p - \{0\}$, and $T_{a,b}$ denotes the permutation of Z_p given by $T_{a,b}(x) = ax + b$, for $a \in Z_p^*$ and $b \in Z_p$.

THEOREM (Burnside [7, Theorem 7.8]). If G is a transitive permutation group of prime degree p, then either G is doubly transitive or

$$G \equiv \{T_{a,b}\colon a \in H < Z_p^*, b \in Z_p\} \tag{8}$$

If S is a union of orbits of the permutation group G and $T \subseteq S$, then G^S denotes the restriction of G to S and $G_T^S = \{g \in G^S\colon g(t) = t, \text{ for every } t \in T\}$. Let G be a permutation group on S. A subset B of S is called a *block* of G if for every $g \in G$ either $g(B) = B$ or $g(B)$ has no point in common with B. Clearly, S, the empty set ϕ, and any singleton $\{s\}$ is a block of G. These blocks are called *trivial blocks*. Any other block of G is called a *nontrivial block*. If B is a block of G, then for any $g \in G$ we have that $g(B)$ is also a block of G. Two such blocks are called *conjugate*. Any two conjugate blocks are either equal or disjoint and the collection of all blocks conjugate to a given block are said to form a *complete block system*. If G is transitive, then all blocks in a complete block system have the same cardinality. If G is transitive and has no nontrivial blocks, then G is said to be a *primitive* permutation group. Otherwise, G is called an *imprimitive* permutation group.

LEMMA 3. If X is a graph or digraph such that $\text{Aut}(X) = G$ has n orbits each of prime length p and if $\text{Orb}(X)$ is connected, then

$$\text{Aut}(X) \equiv G^{\Delta_1} * G^{\Delta_2} * \cdots * G^{\Delta_n} \tag{9}$$

Proof: Since the orbital graph $\text{Orb}(X)$ is connected it contains a spanning tree T. Let Δ_1 and Δ_2 be the orbits corresponding to an endvertex and its adjacent vertex in T, respectively. By definition of Orb(X) there are some edges between Δ_1 and Δ_2 (we give the proof for graphs and point out that the proof for digraphs is essentially the same) but not all edges. We want to show that $G_{\Delta_1}^{\Delta_1 \cup \Delta_2}$ is the identity group.

Suppose there exists $\sigma \in G_{\Delta_1}^{\Delta_1 \cup \Delta_2}$, with $\sigma \neq 1$. Since σ fixes each vertex of Δ_1, any two vertices of Δ_2 in the same cycle of the cyclic decomposition of σ must be adjacent to exactly the same set of vertices of Δ_1. Let U_1 denote the set of vertices in Δ_1 that are adjacent to each of the vertices of Δ_2 in a given cycle of length 2 or more in the cyclic decomposition of σ. Let U_2 be the vertices of Δ_2 that are adjacent to the vertices of U_1. Since $\sigma \neq 1$ implies there is a cycle of length 2 or more, we know $|U_2| \geqq 2$. On the other hand, $|U_2| = p$ would imply that any vertex of U_1 is adjacent to all p vertices of Δ_2. This in turn would imply that every vertex of Δ_1 is adjacent to every vertex of Δ_2 which would contradict the corresponding vertices being adjacent in Orb(X). Hence, we conclude that $2 \leqq |U_2| \leqq p - 1$. Since $G^{\Delta_1 \cup \Delta_2}$ has Δ_1 and Δ_2 for its orbits, $G(U_2)$ is a nontrivial block system for G^{Δ_2}. But the latter is impossible because any transitive group of prime degree is primitive. Therefore, we conclude that $G_{\Delta_1}^{\Delta_1 \cup \Delta_2}$ is the identity group. Hence, $|G^{\Delta_1}| = |G^{\Delta_1 \cup \Delta_2}|$. Similarly, $|G^{\Delta_2}| = |G^{\Delta_1 \cup \Delta_2}| = |G^{\Delta_1}|$.

By choosing an orbit Δ_3 corresponding to a vertex of T that is adjacent to the vertex corresponding to Δ_2 and repeating the above, we obtain $|G^{\Delta_3}| = |G^{\Delta_2}| = |G^{\Delta_1 \cup \Delta_2 \cup \Delta_3}| = |G^{\Delta_1}|$. Eventually, we obtain $|G^{\Delta_i}| = |G|$, for $i = 1, 2, \ldots, n$. Thus, if f_i is the restriction of G to Δ_i, then f_i is an isomorphism of G onto G^{Δ_i}. Furthermore, $f_i f_1^{-1}$ is an isomorphism of G^{Δ_1} onto G^{Δ_i} and each $g \in G$ is expressible as

$$h * f_2 f_1^{-1}(h) * f_3 f_1^{-1}(h) * \cdots * f_n f_1^{-1}(h), \text{ where } h = g^{\Delta_1}$$

so that G is of the form $G^{\Delta_1} * G^{\Delta_2} * \cdots * G^{\Delta_n}$. □

Corollary 3.1. Under the same hypotheses as Lemma 3, if $G^{\Delta_1} = S^p$ or if G is not doubly transitive, then $\text{Aut}(X) \equiv \overset{n}{*}\, G^{\Delta_1}$.

Proof: First consider the case that G^{Δ_1} is not doubly transitive and let Δ_1 and Δ_2 be as in the preceding proof. By Burnside's Theorem quoted earlier, we can label the vertices of Δ_1 with $u_{1,0}, u_{1,1}, \ldots, u_{1,p-1}$ so that $G^{\Delta_1} = \{T_{a,b}\colon a \in H < Z_p^*, b \in Z_p\}$, where $T_{a,b}(u_{1,i}) = u_{1,ai+b}$, with the second subscript reduced modulo p. In Lemma 3 it was shown that $|G^{\Delta_2}| = |G^{\Delta_1}|$ and, by Burnside's Theorem, we know G^{Δ_2} is interpretable as the same set of linear maps as G^{Δ_1}. However, we want to be able to label the vertices of Δ_2 with $u_{2,0}, u_{2,1}, \ldots, u_{2,p-1}$ so that every $g \in G^{\Delta_1 \cup \Delta_2}$ is of the form $T_{a,b} * T_{a,b}$ *with the same* a and b.

In the proof of Burnside's Theorem in [7] the labelling $0, 1, 2, \ldots, p-1$ is obtained by choosing successive elements in any p-cycle of the group. In other words, our labelling of the vertices of Δ_1 arose from the p-cycle $(u_{1,0}\ u_{1,1}\ \cdots\ u_{1,p-1}) \in G^{\Delta_1}$. From Burnside's Theorem, of the hp elements in G^{Δ_1}, where $h = |H|$, one element (the identity) leaves every vertex fixed, $p-1$ elements consist of a p-cycle, and the remaining $hp - p$ elements each fix exactly one vertex. Now since not every edge nor no edges are between Δ_1 and Δ_2, the unique permutation $\sigma \in G^{\Delta_2}$ satisfying $(u_{1,0}\ u_{1,1}\ \cdots\ u_{1,p-1}) * \sigma \in G^{\Delta_1 \cup \Delta_2}$ must itself be a p-cycle. We shall use σ to pick the labelling of Δ_2. Consider the permutation $\tau \in G^{\Delta_1 \cup \Delta_2}$ such that $\tau^{\Delta_1} = T_{a,0}$, where a generates H. If $a \neq 1$, that is, if $|H| > 1$, then τ^{Δ_2} must also fix a unique vertex of Δ_2. Let this vertex be $u_{2,0}$ in the p-cycle σ which then determines the rest of the labelling for Δ_2. So we have

$$(u_{1,0}\ u_{1,1}\ \cdots\ u_{1,p-1}) * (u_{2,0}\ u_{2,1}\ \cdots\ u_{2,p-1}) = T_{0,1} * T_{0,1} \in G^{\Delta_1 \cup \Delta_2}$$

where $T_{a',b}(u_{2,i}) = u_{2,a'i+b}$. Reconsider τ which is of the form $T_{a,0} * T_{a',0}$ since $u_{2,0}$ was chosen so that $\tau(u_{2,0}) = u_{2,0}$. If we show $a' = a$, we are done in the case of

$|H| > 1$ since $T_{a,0}$ and $T_{0,1}$ generate G^{Δ_1}. This implies every $\sigma \in G^{\Delta_1 \cup \Delta_2}$ is of the form $T_{a,b} * T_{a,b}$. We then continue along the spanning tree T to obtain $\mathrm{Aut}(X) \equiv \overset{n}{*}\, G^{\Delta_1}$.

Consider the unordered pair $e = \{u_{1,0}, u_{2,0}\}$. The various powers of $T_{0,1} * T_{0,1}$ take e onto $\Gamma = \{u_{1,i}, u_{2,i}\}$, $i = 0, 1, \ldots, p-1$. Now a' also must generate H or else some power of τ would satisfy $(\tau^r)^{\Delta_2} = 1$ but $(\tau^r)^{\Delta_1} \neq 1$. Thus, every automorphism of $G^{\Delta_1 \cup \Delta_2}$ that fixes $u_{1,0}$ also fixes $u_{2,0}$ so that the orbit containing e is exactly Γ. If $a \neq a'$ were true, then $T_{a,0} * T_{a',0}$ maps $\{u_{1,1}, u_{2,1}\}$ onto $\{u_{1,a}, u_{2,a'}\}$ which implies that the orbit containing e properly contains Γ. Hence, $a = a'$ and, as stated earlier, we are done in the case that $|H| > 1$.

In the case that $|H| = 1$, each G^{Δ_i} is cyclic of order p and it is easy to see that $G \equiv \overset{n}{*}\, G^{\Delta_1}$.

Now consider the case that $G^{\Delta_1} \equiv S^p$. Let the elements of Δ_1 be labelled $u_{1,0}, u_{1,1}, \ldots, u_{1,p-1}$ and consider $G_{u_{1,0}}^{\Delta_1 \cup \Delta_2}$ which has order $(p-1)!$ This group cannot act transitively on Δ_2 or else either every edge or no edges are between Δ_1 and Δ_2. Either way we contradict the fact that the vertices corresponding to Δ_1 and Δ_2 are adjacent in $\mathrm{Orb}(X)$. Thus, $G_{u_{1,0}}^{\Delta_1 \cup \Delta_2}$ has two orbits on Δ_2, and since $|G_{u_{1,0}}^{\Delta_1 \cup \Delta_2}| = (p-1)!$, one orbit must be a single vertex of Δ_2, call it $u_{2,0}$; the other orbit must be the remaining $p - 1$ vertices of Δ_2. Hence, the orbit containing the unordered pair $\{u_{1,0}, u_{2,0}\}$ consists of unordered pairs $\{u_{1,i}, u(i)\}$, where $u(i)$ is a distinct vertex of Δ_2 for each $u_{1,i} \in \Delta_1$. Consequently, if we label $u(i)$ with $u_{2,i}$, then whenever $\sigma \in G^{\Delta_1 \cup \Delta_2}$ satisfies $\sigma(u_{1,i}) = u_{1,j}$ it must also satisfy $\sigma(u_{2,i}) = u_{2,j}$. That is, $G^{\Delta_1 \cup \Delta_2} \equiv S^p * S^p$. Continuing along the spanning tree T gives the result that $\mathrm{Aut}(X) \equiv \overset{n}{*}\, S^p$. □

It should be noted that from the preceding proof it is evident that if $\mathrm{Aut}(X) \equiv \overset{n}{*}\, S^p$, then between the vertices in distinct orbits of $\mathrm{Aut}(X)$ either there are no edges, or all edges are present, or each vertex of one orbit is adjacent to exactly one vertex of the other orbit and vice versa, or, finally, each vertex of one orbit is adjacent to all but exactly one vertex of the other orbit and vice versa. Of course, the subgraphs induced on each orbit must be either complete or the complement of a complete graph.

MAIN RESULTS

We now state the previous results as one theorem and devote the remainder of the section to examining the converse.

THEOREM 4. Let X be a graph or digraph such that every orbit of $\mathrm{Aut}(X)$ has prime cardinality, let $\Omega_1, \Omega_2, \ldots, \Omega_n$ denote the connected components of $\mathrm{Orb}(X)$, and let $\Delta_{i,1}, \Delta_{i,2}, \ldots, \Delta_{i,i_k}$ denote the orbits corresponding to the vertices of Ω_i. Then

$$\mathrm{Aut}(X) \equiv \mathrm{Aut}(X)^{\Omega_1} \times \cdots \times \mathrm{Aut}(X)^{\Omega_n}$$

where

$$\mathrm{Aut}(X)^{\Omega_i} \equiv \mathrm{Aut}(X)^{\Delta_{i,1}} * \cdots * \mathrm{Aut}(X)^{\Delta_{i,i_k}}$$

Moreover, if $\mathrm{Aut}(X)^{\Delta_{i,1}}$ is either S^{p_i}, where $p_i = |\Delta_{i_1}|$, or not doubly transitive, then $\mathrm{Aut}(X)^{\Omega_i} \equiv \overset{i_k}{*}\, \mathrm{Aut}(X)^{\Delta_{i,1}}$.

The converse deals with the question: Given a permutation group each of whose orbits has prime cardinality and which is a direct product of diagonal products, when is it graphical? As a partial answer to this question we state the following result.

THEOREM 5. Let G be a permutation group each of whose transitive constituents has prime degree and such that G is a direct product of diagonal products of transitive constituents of the form

$$(\overset{n_1}{*} G_1) \times (\overset{n_2}{*} G_2) \times \cdots \times (\overset{n_r}{*} G_r)$$

Then, if each G_i, $i = 1, \ldots, r$, is either the full symmetric group or is not doubly transitive, then $\mu(G) > 0$ always holds while $\eta(G) > 0$ whenever the following three conditions are satisfied:

(a) $n_i \geqq 2$ if degree $G_i \geqq 7$ and $|G_i|$ is odd,
(b) $n_i \geqq 3$ if degree $G_i \leqq 5$ and $|G_i|$ is odd or G_i is the transitive group of order $p(p-1)/2$ and degree $p \equiv 3 (\text{mod } 4)$, and
(c) $G \neq \overset{m}{\times} H$, $m = 2, 3, 4$ or 5 if H is the transitive group of order $p(p-1)/2$ and degree $p \equiv 1 (\text{mod } 4)$.

Proof: Suppose G is a permutation group of the form given in the hypotheses. Consider $\overset{n_i}{*} G_i$. If G_i has degree p and $G_i = S^p$, let $Y_{i,1} = DK_p$, the complete directed graph on p vertices, with vertices labelled $u_{1,0}, \ldots, u_{1,p-1}$. Let each of $Y_{i,j}$, $j = 2, \ldots, n_i$ be the complement of DK_p, that is, the directed graph with p vertices and no arcs. Let the vertices of $Y_{i,j}$ be labelled $u_{j,0}, \ldots, u_{j,p-1}$. Now let the arc $(u_{i,k}, u_{i+1,k})$ be in the directed graph for $i = 1, \ldots, n_i - 1$ and $k = 0, 1, \ldots, p-1$. The automorphism group of the resulting directed graph clearly has the vertices of each $Y_{i,j}$ as an orbit and this group is $\overset{n_i}{*} G_i$, as required.

If G_i is the transitive group of degree p and order hp, $1 \leqq h < p - 1$ (there is only one to within conjugacy, by Burnside's Theorem), then let $Y_{i,1}$ have vertices $u_{1,0}, \ldots, u_{1,p-1}$ with an arc from $u_{1,i}$ to $u_{1,j}$ if and only if $j - i \in H$, where H is the unique subgroup of Z_p^* of order h. By [1, 3, 4], we know $\text{Aut}(Y_{i,1}) \equiv G_i$. Then let $Y_{i,j} = DK_p$, for $j = 2, 3, \ldots, n_i$. Again let the arc $(u_{i,j}, u_{i+1,j})$ be in the directed graph, for $i = 1, 2, \ldots, n_i - 1$ and $j = 0, 1, \ldots, p-1$. The automorphism group of the resulting directed graph clearly has the vertices of each $Y_{i,j}$ as its orbits and this group is $\overset{n_i}{*} G_i$.

Go through the above construction for each of the diagonal products and call the resulting directed graphs $Y_1, Y_2, \ldots, Y_r$. Then let X be the directed graph one obtains by letting each vertex of Y_i dominate every vertex of Y_j whenever $i < j$, for $i = 1, 2, \ldots, r-1$. It is clear that $\text{Aut}(X) \equiv \text{Aut}(Y_1) \times \text{Aut}(Y_2) \times \cdots \times \text{Aut}(Y_r)$ which is the required group. This completes the proof of $\mu(G) > 0$.

The procedure for graphs is analogous with the preceding although a little more care is needed in order to prevent additional automorphisms from occurring. If G_i has degree p and $G_i \equiv S^p$, let $Y_{i,1} = K_p$, the complete graph with vertices $u_{1,0}, \ldots, u_{1,p-1}$. Let $Y_{i,j} = \bar{K}_p$, the complement of K_p, with vertices $u_{j,0}, \ldots, u_{j,p-1}$, for $j = 2, \ldots, n_i$. In addition, put in the edges $u_{j,k}u_{j+1,k}$, for $j = 1, 2, \ldots, n_i - 1$ and $k = 0, 1, \ldots, p-1$. The resulting graph, call it Y_i, satisfies $\text{Aut}(Y_i) \equiv \overset{n_i}{*} G_i \equiv \overset{n_i}{*} S^p$ as desired. If there are other G_j's such that $G_j \equiv S^p$ and $n_j = n_i$, $i \neq j$, produce each Y_j by letting $Y_{j,1} = \bar{K}_p$ and $Y_{j,k} = K_p$, for $k = 2, \ldots, n_j$, with the edges between being $u_{i,k}u_{i+1,t}$, for $i = 1, 2, \ldots, n_j - 1$ and $k \neq t$. We claim this then guarantees that Y_j is not isomorphic to Y_i unless $p = 2$ and $n_i = 2$. Assume this is not the case and the Y_j's are labelled $Y_{j_1}, Y_{j_2}, \ldots, Y_{j_s}$. Then put in all edges between Y_i and Y_{j_1}, between Y_{j_1} and $Y_{j_2}, \ldots,$ and between $Y_{j_{s-1}}$ and Y_{j_s}. The resulting graph has automorphism group $\text{Aut}(Y_i) \times \text{Aut}(Y_{j_1}) \times \cdots \times \text{Aut}(Y_{j_s})$. In this way, we can get the direct product of $\overset{n_i}{*} S^p$ as many times as desired.

As mentioned above, the preceding construction will work for all cases with the exception of the direct product of $m \geqq 2$ copies of $S^2 * S^2$. In this case, let Y be the graph with vertex set $\{u_1, u_2, v_1, v_2\}$ and edge set $\{u_1u_2, u_1v_1, u_2v_2\}$. Then $\text{Aut}(Y) \equiv S^2 * S^2$ taking care of the case $m = 1$. Now let Y' be a copy of Y with the corresponding vertices labelled with u'. Between Y and Y' insert the edges $u_1v'_1$, $u_1v'_2$, $u_2v'_1$, and $u_2v'_2$. The automorphism group of the resulting graph is $(S^2 * S^2) \times (S^2 * S^2)$ taking care of the case $m = 2$. For $m \geqq 3$, take $m - 2$ copies of Y calling them $Y_1, Y_2, \ldots, Y_{m-2}$. Join vertices of Y and Y' as above, let every vertex of Y_1 be adjacent to every vertex of Y, and let every vertex of Y_i be adjacent to every vertex of Y_{i-1}, for $i = 2, 3, \ldots, m - 2$. The resulting graph has automorphism group $\overset{m}{\times} (S^2 * S^2)$. This completes the case when G_i is the full symmetric group.

If G_i is not doubly transitive and has even order, that is, $G_i \equiv \{T_{a,b}: a \in H < Z_p^*, b \in Z_p\}$ where $|H|$ is even, then let $Y_{i,1}$ have vertices $u_{1,0}, \ldots, u_{1,p-1}$ with $u_{1,k}u_{1,j} \in E(Y_{i,1})$ if and only if $j - k \in H$. By [1, 3, 4], we know that $\text{Aut}(Y_{i,1}) \equiv G_i$. Let $Y_{i,j} = K_p$ for $j = 2, \ldots, n_i$ and insert edges $u_{k,j}u_{k+1,j}$, for $k = 1, 2, \ldots, n_i - 1$ and $j = 0, 1, \ldots, p - 1$. The group of the resulting graph Y_i is $\overset{n_i}{*} G_i$, as required. If we need more than one $\overset{n_i}{*} G_i$ in the direct product for G, we cannot simply take copies of Y_i because there is then the danger that we may introduce automorphisms that interchange some of the Y_i's. To prevent this from happening we now introduce a device that is used several times in the remainder of the proof. In this particular case, in all other Y_j's, let $Y_{j,k} = \bar{K}_p$, for $k = 2, \ldots, n_i$ instead of K_p. Other than this change, Y_j will be the same as Y_i. It is easy to see that Y_j and Y_i are not isomorphic if $n_i \geqq 2$. Let $Y_{j_1}, Y_{j_2}, \ldots, Y_{j_t}$ be any other such Y_j's so obtained. Let every vertex of Y_i be adjacent to every vertex of Y_{j_1} and let every vertex of Y_{j_a} be adjacent to every vertex of Y_{j_a+1}, for $a = 1, 2, \ldots, t - 1$. The automorphism group of the resulting graph is the appropriate direct product of $\overset{n_i}{*} G_i$.

The preceding works whenever $n_i \geqq 2$. Now consider the case in which we want $G_i \times G_i \times \cdots \times G_i$, m times, when $|G_i|$ is even. Suppose G_i is not the transitive group of order $p(p - 1)/2$ when $p \equiv 1 (\text{mod } 4)$. So $G_i = \{T_{a,b}: a \in H < Z_p^*, b \in Z_p\}$. If Y_1 is the graph with vertices $u_{1,0}, u_{1,1}, \ldots, u_{1,p-1}$ such that $u_{1,i}u_{1,j} \in E(Y_1)$ if and only if $j - i \in H$, then by [1, 3, 4], we have $\text{Aut}(Y_1) \equiv G_i$. Now let Y_2 be the graph with vertices $u_{2,0}, \ldots, u_{2,p-1}$ and $u_{2,i}u_{2,j} \in E(Y_2)$ if and only if $j - i \in H \cup Hx$, where Hx is the coset of H such that $Hx \neq H$ and $H \cup Hx$ is not a subgroup of Z_p^* (this is possible because $|H| < (p - 1)/2$). Then, by [1, Theorem 2], we have $\text{Aut}(Y_2) \equiv G_i$ and $Y_1 \not\equiv Y_2$. Then let $Y_3, \ldots, Y_m$ all be copies of Y_1. Insert all edges between Y_i and Y_{i+1}, for $i = 1, 2, \ldots, m - 1$. The automorphism group of the resulting graph is $\overset{m}{\times} G_i$.

If $G_i = \{T_{a,b}: a \in Q_p, b \in Z_p\}$, $p \equiv 1 (\text{mod } 4)$, where Q_p is the group of quadratic residues, then there is only one graph of order p whose group is G_i, namely, the *quadratic residue graph* $Y(Q_p)$ with vertices $u_0, u_1, \ldots, u_{p-1}$ and $u_iu_j \in E(Y(Q_p))$ if and only if $j - i \in Q_p$ (see [3, 4]). If the group G has at least one other transitive constituent $G_j \not\equiv G_i$, then take m copies $Y_1, Y_2, \ldots, Y_m$ of $Y(Q_p)$ and join every vertex of Y_i to every vertex of Y_{i+1}, for $i = 1, 2, \ldots, m - 1$. In addition, let every vertex of the subgraph corresponding to G_j be adjacent to every vertex of Y_1. Then the automorphism group restricted to $Y_1 \cup \cdots \cup Y_m$ is indeed $\overset{m}{\times} G_i$. The present argument works even if there is another factor in the direct product which is not G_i provided it is graphical. In the case that $G = \overset{m}{\times} G_i$, the idea is to take the wreath product of any identity graph on m vertices with $Y(Q_p)$. (An identity graph is a graph whose automorphism group has order 1). The wreath product is obtained by replacing each vertex of the identity graph with a copy of $Y(Q_p)$ and then putting all edges between two copies if and only if there was an edge between the corresponding vertices in the identity graph. The automorphism group of the resulting graph is $\overset{m}{\times} G_i$ [9] and since identity graphs exist for $m = 1$ and $m \geqq 6$, we see why condition (c) of the theorem is present.

Finally, we consider the case when $|G_i|$ is odd. If G_i is of degree 3, then G_i must be C_3 the cyclic group of order 3 generated by a 3-cycle. As is well known [8], the smallest graph whose automorphism group is abstractly isomorphic to C_3 has nine vertices and its automorphism group has three orbits of length 3 and is of the form $\overset{3}{*} C_3$. To obtain $\overset{m}{*} C_3$, $m \geqq 4$, simply take $m - 3$ copies of $\bar{K}_3$ with vertices $u_{i,0}$, $u_{i,1}$, and $u_{i,2}$, $i = 4, \ldots, m$ and join them to the graph Z of [8, p. 127, case (2)] by $u_{4,j}$, $j = 0, 1, 2$, are each adjacent to one (distinct) vertex of degree 5 in Z and $u_{i,j}$ is adjacent to $u_{i+1,j}$, for $j = 0, 1, 2$ and $i = 4, 5, \ldots, m - 1$. The group of the resulting graph is $\overset{m}{*} C_3$.

The only odd order transitive group of degree 5 is C_5. A similar approach as used for C_3 produces a graph whose automorphism group is $\overset{m}{*} C_5$, for $m \geqq 3$. The order 15 graph we build on is that of [8].

Let $G_i = \{T_{a,b}\colon a \in H < Z_p^*,\ b \in Z_p, \text{ and } |H| \text{ is odd}\}$, where $p \geqq 7$. There is no graph of order p with automorphism group G_i [1] so that $n_i \geqq 2$. If $G_i = C_p$, in [8] graphs Y with $\mathrm{Aut}(Y) \equiv C_p * C_p$ were constructed. By attaching as many copies of $\bar{K}_p$ as necessary similar to the case of C_3 above, we can produce a graph whose automorphism group is $\overset{n_i}{*} C_p$. If $G_i \neq C_p$, then $|H| > 1$ and, additionally, assume $|H| < (p - 1)/2$. Since Z_p^* is cyclic of order $p - 1$, there exists a subgroup $H' < Z_p^*$ with $|H'| = 2|H|$. Let $G_i' = \{T_{a,b}\colon a \in H',\ b \in Z_p\}$ so that $G_i < G_i'$ and $[G_i'\colon G_i] = 2$. Let Y_1 be the graph with vertices $u_{1,0}, u_{1,1}, \ldots, u_{1,p-1}$ and $u_{1,i}u_{1,j} \in Y_1$ if and only if $j - i \in H'$. Let $Y_2 = \bar{Y}_1$ and have vertices $u_{2,0}, \ldots, u_{2,p-1}$. Notice $Y_1 \not\cong Y_2$ because $|H| < (p - 1)/2$. Then insert the edges $u_{1,i}u_{2,j}$ if $i = j$ or $j - i \in H$. Any permutation of the form $T_{a,b} * T_{a,b}$ for $a \in H$ and $b \in Z_p$ is an automorphism of the resulting graph Y.

Hence, $\mathrm{Aut}(Y) \geq G_1 * G_1$. Since $\mathrm{Aut}(Y_1) \equiv \mathrm{Aut}(Y_2) \equiv G_i'$, we know $\mathrm{Aut}(Y) \leq G_i' * G_i'$. But $T_{-1,0} * T_{-1,0}$ maps $u_{1,0}u_{2,1} \in E(Y)$ to $u_{1,0}u_{2,-1} \notin E(Y)$ and thus $\mathrm{Aut}(Y) \equiv G_i * G_i$. To obtain $\overset{m}{*} G_i$, for arbitrary $m \geqq 3$, we attach $m - 2$ copies of Y_1 as we have done several times.

The last case we must consider is $G_i \equiv \{T_{a,b}\colon a \in Q_p,\ b \in Z_p,\ p \equiv 3 \pmod 4\}$. There is no graph on p vertices whose automorphism group is G_i [3, 4] so that $n_i \geqq 2$. The case of $n_i = 2$ is not settled. For $n_i \geqq 3$, let Y_i be the graph with vertices $u_{i,0}, u_{i,1}, \ldots, u_{i,p-1}$, for $i = 1, 2, 3$, where $Y_1 = K_p$ and $Y_2 = Y_3 = \bar{K}_p$. Insert the edges $u_{1,i}u_{2,j}$ if and only if $j - i \in Q_p$. Also insert the edges $u_{i,k}u_{3,k}$ and $u_{2,k}u_{3,k}$, for $k = 0, 1, \ldots, p - 1$. Call the resulting graph Y. Clearly, $\mathrm{Aut}(Y)$ has three orbits, namely, Y_1, Y_2, and Y_3 and if $\sigma(u_{1,i}) = u_{1,j}$, for some $\sigma \in \mathrm{Aut}(Y)$, then $\sigma(u_{2,i}) = u_{2,j}$ and $\sigma(u_{3,i}) = u_{3,j}$. Because vertices in Y_1, Y_2, and Y_3 with the same second coordinate all move together, the automorphisms are controlled by the edges between Y_1 and Y_2. Thus, if we identify vertex $u_{1,i}$ and $u_{2,i}$, we see that the group $\mathrm{Aut}(Y)$ acts on $Y_1 \cup Y_2$ the same as $\mathrm{Aut}(D)$, where D is the directed graph with vertices $w_0, w_1, \ldots, w_{p-1}$, where there is an arc from w_i to w_j if and only if $j - i \in Q_p$. But $\mathrm{Aut}(D) \equiv G_i$ [1] and thus $\mathrm{Aut}(Y) \equiv G_i * G_i * G_i$. For $n_i \geqq 4$ we attach the appropriate number of copies of Y_3 and connect them to Y as done several times earlier in this proof.

Thus, we have produced the diagonal products, as asserted. If some of them are isomorphic, we can prevent them from getting mapped to each other by using the device used earlier in the proof. To get the direct product of the diagonal products, we simply take the union of the graphs constructed for the diagonal products or direct product of isomorphic diagonal products. This completes the proof. □

Conclusion

Theorem 5 is not a complete answer to when is $\eta(G) > 0$. The problem arises when the induced subgraph on each orbit is either K_p or $\bar{K}_p$. It is not immediately apparent how to determine the automorphism group in this case. For example, let $Y_1 = K_7$ have vertices $u_{1,0}, \ldots, u_{1,6}$, let $Y_2 = \bar{K}_7$ have vertices $u_{2,0}, \ldots, u_{2,6}$, and let $u_{1,i}u_{2,j}$ be an edge if and only if $j - i \in Q_7$. The automorphism group of this graph has order $4 \cdot 7 \cdot 6$ and, by Burnside's Theorem and Theorem 4, each transitive constituent is double transitive. The preceding graph is one of the graphs to try in an attempt to find a graph whose group is $G * G$, where $G \equiv \{T_{a,b}\colon a \in Q_7,\ b \in Z_7\}$. The general question is one of finding $\mathrm{Aut}(X)$ when X has been constructed according to the techniques in the proof of Theorem 5 when the induced subgraph for each orbit is complete or the complement of a complete graph. A second question is what is the value of $\mu(G)$ and $\eta(G)$ when the former or both are known to be positive?

References

1. Alspach, Brian. 1973. Point-symmetric graphs and digraphs of prime order and transitive permutation groups of prime degree. J. Comb. Theory (B). **15:** 12–17.
2. Astie, Annie. 1972. Groupes d'automorphismes des tournois sommet-symmétriques d'ordre et dénombrement de ces tournois. C. R. Acad. Sci. Ser. A. **275:** 167–169.
3. Berggren, J. L. 1972. An algebraic characterization of symmetric graphs with a prime number of vertices. Bull. Austral. Math. Soc. **7:** 131–134.
4. Chao, Chung-Yun. 1971. On the classification of symmetric graphs with a prime number of vertices. Trans. Amer. Math. Soc. **158:** 247–256.
5. Chao, Chung-Yun & Jacqueline G. Wells. 1973. A class of vertex-transitive digraphs. J. Comb. Theory (B). **14:** 246–255.
6. Harary, Frank. 1969. Graph Theory. Addison-Wesley, Reading, Mass.
7. Passman, D. S. 1968. Permutation Groups. W. A. Benjamin, New York, N.Y.
8. Sabidussi, G. 1959. On the minimum order of graphs with given automorphism group. Monatsh. Math. **63:** 124–127.
9. Sabidussi, G. 1959. The composition of graphs. Duke Math. J. **26:** 693–696.
10. Turner, James. 1967. Point-symmetric graphs with a prime number of points. J. Comb. Theory **3:** 136–145.
11. Wielandt, H. 1964. Finite Permutation Groups. Academic Press, New York, N.Y.

A COMBINATORIAL APPROACH TO GRAPHICAL POLYNOMIALS AND SPANNING SUBGRAPHS

Ruth A. Bari

Department of Mathematics
George Washington University
Washington, D.C. 20052

1. INTRODUCTION

Let G be a connected (p, q)-graph, that is, a graph with p points and q lines. Let $T(G; x, y)$ be the Tutte polynomial of G, and let $P(G, \lambda)$ be the chromatic polynomial of G.

We consider a certain $2p \times 2p$ matrix B, whose nonzero entries are binomial coefficients $\binom{n}{r}$; $n = p-1, p-2, \ldots, 1, 0$; $r = 0, 1, 2, \ldots, p-2, p-1$, and a vector v whose entries are the coefficients of $T(G; x, 0)$ followed by the coefficients of $P(G, \lambda)$. Using a combinatorial argument, we prove that the vector v is a unit eigenvector of the matrix B.

The first part of this paper is based on results obtained in [1]. The really new part is THEOREM 3.2.

2. BASIC DEFINITIONS AND THEOREMS

Let G be a connected (p, q)-graph and label the lines of G with integers $1, 2, \ldots, q$. Then G is a *line-labelled graph.*

In this paper, we consider only line-labelled graphs. A line will be referred to by its label.

DEFINITION 2.1. Let Z be a cycle of G, m the largest integer assigned to the lines of Z. The path $P = Z - m$ is the *broken circuit* associated with Z. The line m is the *break* in the broken circuit P.

THEOREM 2.1 (Whitney-Broken Circuits [6]). Let G be a line-labelled (p, q)-graph and let $P(G, \lambda)$ be the chromatic polynomial of G. Then

$$P(G, \lambda) = \sum_{i=0}^{p-1} (-1)^i m_i \lambda^{p-i}$$

where m_i is the number of spanning subgraphs of G that have i lines and contain no broken circuits.

DEFINITION 2.2. Let S be a spanning subgraph of G with i lines that contains no broken circuits. Then S is called an i-line *Whitney subgraph* of G.

From THEOREM 2.1, it is clear that the coefficient m_i counts the i-line Whitney subgraphs of G.

In his paper [5], W. T. Tutte gives the following recursive definition of the *dichromatic* or *Tutte polynomial* $T(G, xy)$.

0077-8923/79/0328-0021$1.75/1 © 1979, NYAS

DEFINITION 2.3. With any graph G, there is associated a polynomial $T(G; x, y)$ in two variables x and y having the following two properties:

I. If G has i bridges, ℓ loops, and no other lines, then $T(G; x, y) = x^i y^\ell$

II. If e is a line of G that is neither a bridge nor a loop, then

$$T(G; x, y) = T(G'_e; x, y) + T(G''_e; x, y)$$

where G'_e is obtained from G by deleting line e, and G''_e is the graph obtained from G by contracting e, with its two ends, to a single point.

THEOREM 2.2. Let G be connected. Then $T(G; 1, 1)$ counts the number of spanning trees of G.

Proof: Let $\tau(G)$ denote the number of spanning trees of G.

I. If G has i bridges and ℓ loops, and no other lines, then G has exactly one spanning tree, whose lines are the bridges of G. Thus $\tau(G) = 1$. Since, in this case, $T(G; x, y) = x^i y^\ell$, $T(G; 1, 1) = 1^i 1^\ell = 1 = \tau(G)$.

II. If G has a line e that is neither a bridge nor a loop and t is a spanning tree of G, then either t does not contain line e or t contains line e.

If t does not contain line e, then t is a spanning tree of $G'_e = G - e$. Thus $\tau(G'_e) = T(G'_e; 1, 1)$.

If t contains line e, there is a spanning tree t''_e in G''_e that corresponds to t, where t can be derived from t''_e by restoring line e. Thus $\tau(G''_e) = T(G''_e; 1, 1)$.

Thus it follows that

$$\tau(G) = \tau(G'_e) + \tau(G''_e) = T(G'_e; 1, 1) + T(G''_e; 1, 1) = T(G; 1, 1). \quad \square$$

For any connected line-labelled graph G, the recursive formula for $\tau(G) = T(G; 1, 1)$ can be used to enumerate the spanning trees of G and to determine which of these trees have no broken circuits.

To accomplish this, we delete and contract the lines of G in increasing numerical order of their labels, always removing the line with the smallest label that is neither a bridge nor a loop.

If C_n is a cycle of G, the process of deleting a line of C_n and then contracting that line will, after the contraction, replace C_n by C_{n-1}. After $n - 1$ deletions and contractions, we obtain a graph in which C_n has been replaced by a loop. Since the lines of C_n were removed in increasing order, the loop represents the line of C_n with the largest label, that is, the break in the broken circuit.

The recursion is complete when we obtain a set $\mathscr{H}$ of graphs, all of whose lines are bridges or loops. Each spanning tree of G will be represented by a graph in $\mathscr{H}$ and each broken circuit of G will be represented by a loop in an element of $\mathscr{H}$.

DEFINITION 2.4. Let t be a spanning tree of G that is represented by a graph h in $\mathscr{H}$ with i bridges and j loops. Then t is said to have *internal activity* i and *external activity* j.

Note: DEFINITION 2.4 is not Tutte's original definition, but is equivalent to it. This is proved in [1].

Using DEFINITION 2.4, we may restate DEFINITION 2.3 as follows:

DEFINITION 2.3a. Let G be a connected line-labelled graph. The Tutte polynomial, or dichromate of G is the function

$$T(G; x, y) = \sum_{i,j} a_{i,j} x^i y^j$$

where $a_{i,j}$ is the number of spanning subgraphs of G with internal activity i and external activity j.

3. COUNTING THE WHITNEY SUBGRAPHS OF G

Since a maximal acyclic subgraph of G is a spanning tree, if G' is a subgraph of G with more than $p - 1$ lines, G' contains at least one cycle, and hence a broken circuit. Therefore, the Whitney subgraphs of G are subgraphs of the spanning trees of G that contain no broken circuits, that is, the trees with external activity zero.

To enumerate the spanning trees of G, we use the recursion formula for $\tau(G) = T(G; 1, 1)$. At each step of the recursion, we partition the spanning trees of G into those that do not contain a specified line e, and are therefore spanning trees of a graph in which e has been deleted, and those that contain line e, and therefore correspond to the spanning trees of a graph in which line e has been contracted to a single point. Since line e does not appear in the latter graph, we indicate the fact that we are enumerating trees that *contain* line e by using a circled ⓔ.

EXAMPLE 3.1. Spanning trees of $K_4 - e$.

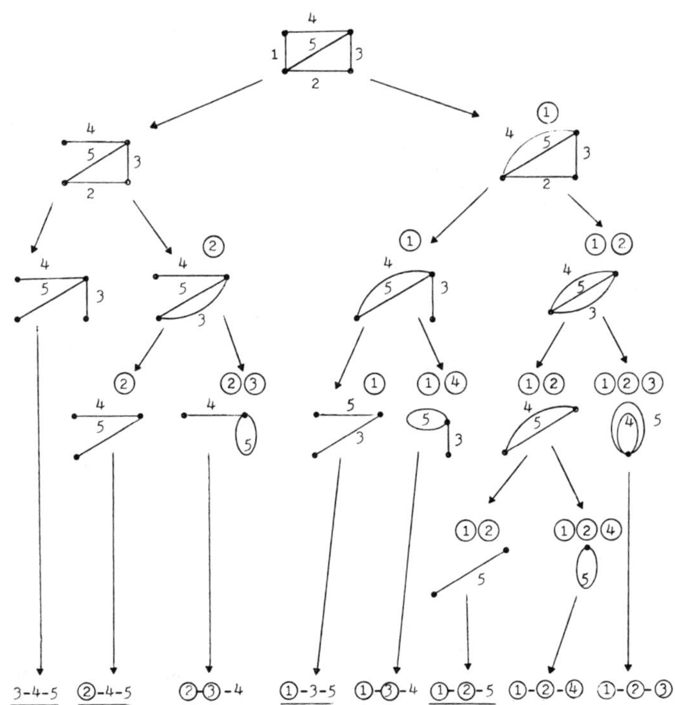

The last line is the list of spanning trees. The trees that are underlined have no broken circuits. We see that $a_{3,0} = 1$, $a_{2,0} = 2$, $a_{1,0} = 1$, $a_{0,0} = 0$. Also $a_{0,2} = 1$, $a_{0,1} = 1$, and $a_{1,1} = 2$.

DEFINITION 3.1. Let G be a line-labelled connected graph, and let $\mathscr{H}$ be the set of graphs that represent the spanning trees of G at the end of the recursive process for finding $\tau(G) = T(G; 1, 1)$. If t is a spanning tree of G and t is represented by a graph h_t in $\mathscr{H}$, then a line e of G is *internally passive*, or simply *passive* in t if its label appears encircled near h_t.

To count the Whitney subgraphs of G, we partition them as follows: Two Whitney subgraphs S_1 and S_2 are in the same class if they are chosen from the same tree t with external activity zero. Each passive line of t *must* belong to every Whitney subgraph chosen from t.

EXAMPLE 3.2. In EXAMPLE 3.1, the spanning trees 3-4-5 and ②-4-5 have external activity zero. The spanning tree 4-5 *may* be chosen from 3-4-5, but may *not* be chosen from ②-4-5, since every spanning tree of ②-4-5 *must* contain line 2. Also, every Whitney subgraph of G chosen from the spanning tree ①-②-5 has at least two lines, as it *must* contain lines 1 and 2.

EXAMPLE 3.3 (The Whitney subgraphs of $K_4 - e$). From EXAMPLE 3.1, we list the spanning trees of $K_4 - e$ with external activity zero, their internal activities, and their internal passivities, where the *internal passivity of* t is the number of internally passive lines in t.

TABLE

Tree	Internal Activity	Internal Passivity
t_1 = 3-4-5	3	0
t_2 = ①-3-5	2	1
t_3 = ②-4-5	2	1
t_4 = ①-②-5	1	2

These trees partition the Whitney subgraphs of G into four classes, C_1, C_2, C_3, and C_4, where, if S is a Whitney subgraph,

$S \in C_1$, if S contains no passive lines.

$S \in C_2$, if S contains ① but not ②.

$S \in C_3$, if S contains ② but not ①.

$S \in C_4$, if S contains both ① and ②.

Thus every Whitney subgraph chosen from a tree with passive lines *must* contain the passive lines. We may choose the subgraph 4-5 from t_1 = 3-4-5, but *not* from t_3 = ②-4-5, since every Whitney subgraph chosen from t_3 *must* contain line ②.

If $P(G, \lambda) = m_0 \lambda^4 - m_1 \lambda^3 + m_2 \lambda^2 - m_3 \lambda$, where m_i is the number of i-line Whitney subgraphs of G, we may use the above partition to evaluate m_i.

$$m_0 = \binom{3}{0} = 1$$

$$m_1 = \binom{3}{1} + \binom{2}{0} + \binom{2}{0} = \binom{3}{1} + 2\binom{2}{0} = 5$$

$$m_2 = \binom{3}{2} + \binom{2}{1} + \binom{2}{1} + \binom{1}{0} = \binom{3}{2} + 2\binom{2}{1} + \binom{1}{0} = 8$$

$$m_3 = \binom{3}{3} + \binom{2}{2} + \binom{2}{2} + \binom{1}{1} = \binom{3}{3} + 2\binom{2}{2} + \binom{1}{1} = 4$$

In each case, if a tree t has r passive lines and s active lines, we choose from t every Whitney subgraph of G with $r + \ell$ lines, $0 \leq \ell \leq s$, containing all of the r passive lines, and any selection of ℓ lines of t chosen from the s active lines.

THEOREM 3.1. Let G be a connected (p, q)-graph, and let $P(G, \lambda)$ be the chromatic polynomial of G. Then

$$P(G, \lambda) = \sum_{s=1}^{p-1} (-1)^r a_{s,0} \lambda(\lambda - 1)^s$$

where $a_{s,0}$ is the number of spanning trees of G with internal activity s and external activity zero, where $r = p - 1 - s$ is the number of internally passive lines in a tree with s internally active lines.

Proof: By Whitney's theorem on broken circuits,

$$P(G, \lambda) = \sum_{i=1}^{p-1} (-1)^i m_i \lambda^{p-i}$$

where m_i is the number of Whitney subgraphs with i lines.

If t is a spanning tree of G with external activity zero, then t contains no broken circuits.

Suppose t has r internally passive lines and s internally active lines. The contribution of t to $P(G, \lambda)$ is the number of subgraphs of t that contain the r passive lines and any selection of the s active lines, that is,

$$(-1)^r \left[\binom{s}{0} \lambda^{p-r} - \binom{s}{1} \lambda^{p-r-1} + \cdots + (-1)^s \binom{s}{s} \lambda^{p-r-s}\right]$$

Since t has $p - 1$ lines, $r + s = p - 1$, so $p - r = s + 1$, and $p - r - s = 1$. Hence t contributes

$$(-1)^r \left[\binom{s}{0} \lambda^{s+1} - \binom{s}{1} \lambda^{s} + \cdots + (-1)^s \binom{s}{s} \lambda\right] = (-1)^r \lambda(\lambda - 1)^s$$

Every spanning subgraph of G with no broken circuits can be chosen from the spanning trees of G with external activity zero. If a tree has external activity zero and internal activity s, then $1 \leq s \leq p - 1$. Therefore,

$$P(G, \lambda) = \sum_{s=1}^{p-1} (-1)^r a_{s,0} \lambda(\lambda - 1)^s \quad \square$$

Note: This theorem is proved by a different method in [2].

DEFINITION. The *choose-matrix* C_p is the following matrix whose nonzero entries are binomial coefficients $\binom{n}{r}$ (read "n-choose r").

$$C_p = \text{Diag. } [1, -1, 1, -1, \ldots, (-1)^{p-1}] \times$$

$$\begin{bmatrix} \binom{p-1}{0} & 0 & 0 & \cdots & 0 & 0 & 0 \\ \binom{p-1}{1} & \binom{p-2}{0} & 0 & \cdots & 0 & 0 & 0 \\ \binom{p-1}{2} & \binom{p-2}{1} & \binom{p-3}{0} & \cdots & 0 & 0 & 0 \\ \vdots & \vdots & \vdots & \vdots & \vdots & \vdots & \vdots \\ \binom{p-1}{p-3} & \binom{p-2}{p-4} & \binom{p-3}{p-5} & \cdots & \binom{2}{0} & 0 & 0 \\ \binom{p-1}{p-2} & \binom{p-2}{p-3} & \binom{p-3}{p-4} & \cdots & \binom{2}{1} & \binom{1}{0} & 0 \\ \binom{p-1}{p-1} & \binom{p-2}{p-2} & \binom{p-3}{p-3} & \cdots & \binom{2}{2} & \binom{1}{1} & \binom{0}{0} \end{bmatrix}$$

COROLLARY. Let $d^T = [a_{p-1,0}, a_{p-2,0}, \ldots, a_{2,0}, a_{1,0}, a_{0,0}]$ and let $e^T = [m_0, -m_1, m_2, -m_3, \ldots, (-1)^{p-1}m_{p-1}]$, where $a_{r,0}$ is the number of spanning trees of G with internal activity r and external activity 0, and m_r is the number of r-line Whitney subgraphs. Then $C_p d = e$.

EXAMPLE 3.4. From EXAMPLE 3.3, it follows that

$$\begin{bmatrix} m_0 \\ -m_1 \\ m_2 \\ -m_3 \end{bmatrix} = \begin{bmatrix} \binom{3}{0} & 0 & 0 & 0 \\ -\binom{3}{1} & -\binom{2}{0} & 0 & 0 \\ \binom{3}{2} & \binom{2}{1} & \binom{1}{0} & 0 \\ -\binom{3}{3} & -\binom{2}{2} & -\binom{1}{1} & \binom{0}{0} \end{bmatrix} \begin{bmatrix} 1 \\ 2 \\ 1 \\ 0 \end{bmatrix}$$

$$= \begin{bmatrix} \binom{3}{0} & 0 & 0 \\ -\binom{3}{1} & -2\binom{2}{0} & 0 \\ \binom{3}{2} & +2\binom{2}{1} & +\binom{1}{0} \\ -\binom{3}{3} & -2\binom{2}{2} & -\binom{1}{1} \end{bmatrix} = \begin{bmatrix} 1 \\ -5 \\ 8 \\ -4 \end{bmatrix}$$

THEOREM 3.2. Let G be a connected (p, q)-graph and let $a_{s,0}$ be the number of spanning trees of G with internal activity s and external activity 0.

Let $r = p - 1 - s$ be the number of internally passive lines in any spanning tree of G with s internally active lines and let m_i be the number of i-line Whitney subgraphs of G, $i = 0, 1, \ldots, p - 1$. Then

$$a_{s,0} = \binom{s}{0} m_r - \binom{s+1}{1} m_{r-1} + \binom{s+2}{2} m_{r-2} - \cdots + (-1)^r \binom{s+r}{r} m_0$$

Proof: Since $a_{s,0}$ counts only spanning trees with external activity zero, we may refer to internally active or passive lines simply as active or passive lines.

If t is a spanning tree of G with s active lines and r passive lines, there is a unique r-line Whitney subgraph that may be chosen from t by our counting scheme, the one which contains all of the r passive lines. Hence $a_{s,0}$ is simply the number of r-line Whitney subgraph that contains r passive lines and no active lines. We must therefore count the number of Whitney subgraphs that have no passive lines.

Let N be the number of r-line Whitney subgraphs of G.

Let b_i be the property of an r-line Whitney subgraph that it has i active lines, and let b_i' be the property of this subgraph that it does not have i active lines.

Let $N(b_i)$ be the number of r-line Whitney subgraphs with i active lines.

Then $a_{s,0} = N(b_1' \, b_2' \cdots b_r')$, the number of r-line Whitney subgraphs with no active lines.

This formulation suggests that we may apply the principle of inclusion and exclusion:

$$\begin{aligned} N(b_1' \, b_2' \cdots b_r') = N &- [N(b_1) + N(b_2) + \cdots + N(b_r)] \\ &+ [N(b_1 b_2) + N(b_1 b_3) + \cdots + N(b_{r-1} b_r)] \\ &- \cdots + (-1)^r N(b_1 b_2 \cdots b_r) \end{aligned} \tag{1}$$

If $i_1 < i_2 < \cdots < i_k$, then $N(b_{i_1}, b_{i_2}, \ldots, b_{i_k}) = N(b_{i_k})$ since every Whitney subgraph with i_k active lines also has i_j active lines whenever $i_j < i_k$.

Thus $N(b_1) + N(b_2) + \cdots + N(b_r)$ is the number of r-line Whitney subgraphs with at least one active line $= N(b_1)$.

$N(b_1 b_2) + N(b_1 b_3) + \cdots + N(b_{r-1} b_r)$ is the number of r-line Whitney subgraphs with at least two active lines $= N(b_2)$.

. . . .

$N(b_1 b_2 \cdots b_k) + N(b_1 b_2 \cdots b_{k-1} b_{k+1}) + \cdots + N(b_{r-k+1}, b_{r-k+2}, \ldots, b_r)$ is the number of r-line Whitney subgraphs with at least k active lines $= N(b_k)$.

. . . .

$N(b_1 b_2, \ldots, b_r)$ is the number of r-line Whitney subgraphs with r active lines $= N(b_r)$. □

LEMMA. $N(b_k) = \binom{s+k}{k} m_{r-k}$.

Proof: $N(b_k)$ is the number of r-line Whitney subgraphs that contain k active lines.

There are m_{r-k} Whitney subgraphs with $r - k$ lines, each chosen from a spanning tree t of G with at most $r - k$ passive lines. The remaining $r + s - (r - k) = s + k$ lines of t are therefore active lines.

To form an r-line Whitney subgraph from an $(r-k)$-line subgraph H, we need merely augment the latter with k additional lines. These must be chosen from the $s+k$ active lines of t that are not in H. This may be done in $\binom{s+k}{k}$ ways.

Since there are $\binom{s+k}{k}$ r-line subgraphs for each of the m_r $(r-k)$-line subgraphs,

$$N(b_k) = \binom{s+k}{k} m_{r-k} \quad \square$$

Applying this information to **(1)**, we get

$$a_{s,0} = \binom{s}{0} m_r - \binom{s+1}{1} m_{r-1} + \binom{s+2}{2} m_{r-2} - \cdots + (-1)^r \binom{s+r}{r} m_r$$

COROLLARY. Let G be a connected (p, q) graph, $d^T = [a_{p-1,0}, a_{p-2,0}, \ldots, a_{2,0}, a_{1,0}, a_{0,0}]$ and let $e^T = [m_0, -m_1, m_2, -m_3, \ldots, (-1)^{p-1} m_{p-1}]$, where $a_{r,0}$ is the number of spanning trees of G with internal activity r and external activity zero and m_r is the number of r-line spanning subgraphs of G. Then $C_p e = d$.

DEFINITION. The matrix B_n is the matrix

$$B_p = \begin{bmatrix} 0 & C_p \\ C_p & 0 \end{bmatrix}$$

COROLLARY. Let $v^T = [a_{p-1,0}, a_{p-2,0}, \ldots, a_{0,0}, m_0, -m_1, \ldots, (-1)^r m_r]$. Then v is a unit eigenvector of B_p, that is, an eigenvector with eigenvalue 1.

EXAMPLE 3.5.

(a) The vectors $v_1^T = [1, 2, 1, 0, 1, -5, 8, -4]$
$v_2^T = [1, 3, 2, 0, 1, -6, 11, -6]$

are two unit eigenvectors of B_4. The first is derived from $K_4 - e$, the second from K_4.

(b) The vectors $v_1^T = [1, 2, 2, 1, 0, 1, -6, 14, -15, 6]$ and
$v_2^T = [1, 6, 11, 6, 0, 1, -10, 35, -50, 24]$

are two unit eigenvectors of B_5. The first is derived from a 4-cycle and a triangle with one common line. The second is derived from K_5.

SUMMARY

We define a certain $2p \times 2p$ matrix B whose nonzero entries are binomial coefficients. Using a combinatorial argument, we prove that, for each connected (p, q)-graph G, there is a unit eigenvector v of B. The components of v are the coefficients of the Tutte polynomial $T(G; x, 0)$ followed by the coefficients of the chromatic polynomial $P(G, \lambda)$.

REFERENCES

1. BARI, RUTH A. 1979. Chromatic polynomials and the internal and external activities of Tutte. Proc. Int. Conf. Graph Theory Comb. University of Waterloo, July 1977. *In* Graph Theory and Related Topics. A. Bondy & U. S. R. Murty, Eds. Academic Press, New York, N.Y.

2. BIGGS, NORMAN. 1974. Algebraic Graph Theory. Cambridge Tracts in Mathematics, 67. Cambridge University Press, Cambridge, England.
3. TUTTE, W. T. 1953. A contribution to the theory of chromatic polynomials. Can. J. Math. **6:** 80–91.
4. TUTTE, W. T. 1967. On dichromatic polynomials. J. Comb. Theory. **2:** 301–320.
5. TUTTE, W. T. 1974. Codichromatic graphs. J. Comb. Theory **16:** 168–174.
6. WHITNEY, H. 1932. A logical expansion in mathematics. Bull. Am. Math. Soc. **38:** 572, 579.

ON REGULAR LINE GRAPHS

Douglas Bauer

Department of Pure and Applied Mathematics
Stevens Institute of Technology
Hoboken, New Jersey 07030

The purpose of this note is to determine the values of p and k for which there is a line graph on p points that is regular of degree k. This question arises from the study of a more general question; namely, when is a given degree sequence realizable as a line graph? This problem was suggested by F. T. Boesch and inspired by a recent survey paper by Hakimi and Schmeichel [1] in which a "degree" is associated with each edge of a graph. The solution for the particular case of regular graphs is obtained by combining (1) known conditions for the existence of k-regular graphs and of (c, b)-biregular bipartite graphs, and (2) the fact that $L(G)$ is regular if and only if G is regular or bipartite biregular. The solution is summarized in the following theorem. Terminology and notation will be standard, as in Harary [2].

THEOREM. There is a connected regular line graph of degree k on p points if and only if one of the following holds:

(1) $p = k + 1$.
(2) For some integer $c \geq 2$, $k = 2c - 2$, $2p \geq (c + 1)c$, and $2p \equiv 0 \mod c$.
(3) For some integers $c, b \geq 2$, $k = c + b - 2$, $p \geq cb$, and $p \equiv 0 \mod lcm\ (c, b)$.

Proof: We first establish the sufficiency of each of the above conditions by constructing in each case a graph G such that $H = L(G)$ is regular of degree k and has p points. In case (1) we may let G be the star $K_{1,p}$. For case (2), since c divides $2p$ and $2p \geq (c + 1)c$, there is a number $p' \geq c + 1$ such that $2p = p'c$. Since $p'c$ is even and $p' \geq c + 1$, there is a connected regular graph G of degree c on p' points. Hence $H = L(G)$ is a connected regular graph of degree $k = 2c - 2$ and has $p = p'c/2$ points.

Finally, we consider case (3). Since p is congruent to 0 modulo the least common multiple of c and b, $p = mb = nc$ for integers m and n, and since $p \geq cb$, we have $m \geq c$ and $n \geq b$. Let $U = \{u_1, \ldots, u_n\}$ and $V = \{v_1, \ldots, v_m\}$ be disjoint sets. We shall construct a (c, b)-biregular bipartitie graph G with parts U and V such that each u_i has degree c and each v_j has degree b in G. We define u_i and v_j to be adjacent in G if and only if j is congruent mod m to some l satisfying $(i - 1)c + 1 \leq l \leq ic$. Since $(i - 1)c + 1, \ldots, ic$ is a sequence of consecutive integers of length c and $c \leq m$, there are exactly c integers j satisfying this condition, so u_i has degree c in G. It is easily seen that each v_j is adjacent to exactly b vertices u_i, since an analogous condition prevails. If G is connected, $H = L(G)$ is the desired line graph. If G fails to be connected, we make a simple modification: if G_1 and G_2 are components of G, choose edges $e = \{u, v\}$ in G_1, $e' = \{u', v'\}$ in G_2 so that $u, u' \in U$ and $v, v' \in V$. Modifying G by deleting e and e' and adding edges $f = \{u, v'\}$ and $f' = \{u', v\}$, we reduce the number of components of G while leaving degrees unchanged. Continuing in this manner we reach the connected case and we are done.

We now prove the necessity. Let H be a connected regular line graph of degree k having p points and let G be the graph such that $H = L(G)$. Choose a point u of G and let c be its degree; then any point v adjacent to u must have degree $b = k + 2 - c$, since the line joining u and v corresponds to a point of $H = L(G)$ having degree k. Similarly, every point adjacent to v must have degree c. Since G is connected, every point of G is accessible by a path from u and the degrees of the points on such a path will alternate between c and b.

If either c or b is 1, then it follows immediately that G is $K_{1,p}$ and hence $p = k + 1$. If $c = b \geq 2$, it follows that G is regular of degree c and $k = 2c - 2$. If G has p' points, then $p' \geq c + 1$ and $2p = p'c$ so that $2p$ is congruent to 0 mod c. Thus condition (2) holds. If $c \neq b$, then let U be the points of degree c and V those of degree b; by the above, G is bipartite with parts U and V. If U has n points and V has m points, then $n \geq b$ and $m \geq c$, and G has $mb = nc$ lines. Thus $p = mb = nc$ is divisible by $lcm(c, b)$ and satisfies $p \geq cb$. □

References

1. Hakimi, S. L. & E. F. Schmeichel. 1978. Graphs and their degree sequences: A survey. *In* Lecture Notes in Mathematics, 642. Theory and Applications of Graphs (Proceedings, Michigan, 1976). Springer-Verlag, New York, N.Y.
2. Harary, F. 1969. Graph Theory. Addison-Wesley, Reading, Mass.

A CHRONOLOGY OF THE RINGEL-KOTZIG CONJECTURE AND THE CONTINUING QUEST TO CALL ALL TREES GRACEFUL*

Gary S. Bloom

Department of Computer Science
City College
City University of New York
New York, New York 10031

Introduction

This is a self-contained presentation of the status in 1978 of the 1963 conjecture of Ringel concerning the decomposition of K_n into isomorphs of an arbitrary tree. This conjecture is explained and traced through its modification by Kotzig to the series of attacks intent on proving that trees are graceful. This latter condition for trees is shown to be one of a series of conditions whose truth is sufficient to prove the original conjecture. The current knowledge of graceful trees is surveyed.

A recent approach to graceful trees based on the properties of "completely augmented" adjency matrices for graceful graphs is introduced. Several theorems are proved including ones characterizing all graceful numberings by their adjacency matrices, characterizing graceful numberings of caterpillars, and counting all caterpillars by using their graceful adjacency matrices. Techniques that allow taking graceful adjacency matrices and creating new graceful graphs are mentioned.

Graceful Graphs and Other Preliminaries

Let *graph* denote a nonempty collection of nodes and undirected edges joining them. Neither self-loops nor multiple edges are permitted. Such a graph becomes a *numbered graph* when each node v is assigned a value $\psi(v)$ and each edge (u, v) is assigned the value $\psi(u, v) = |\psi(u) - \psi(v)|$.

A graph G is *gracefully numbered* if the following conditions hold:

(a) All n nodes in G are numbered with distinct values from the set $\{0, 1, 2, \ldots, e\}$, where $e =$ the number of edges in G.

(b) The edge numbering of G is an assignment of all e values in $\{1, 2, \ldots, e\}$ to distinct edges.

Some gracefully numbered graphs are illustrated in Figure 1.

A graph G is a *graceful graph* if there exists a graceful numbering of G. There are many *nongraceful* graphs. For example, the three nongraceful graphs with five nodes are illustrated in Figure 2.

In the second part of this paper, a summary is given of the efforts of many people to determine if all trees are graceful. Prior to discussing that work, we shall trace the history of how this problem arose and clarify its place as the strongest unresolved conjecture in a sequence of conjectures.

* This research was supported in part by NSF Research Grant ENG 76-18569.

0077-8923/79/0328-0032$1.75/1 © 1979, NYAS

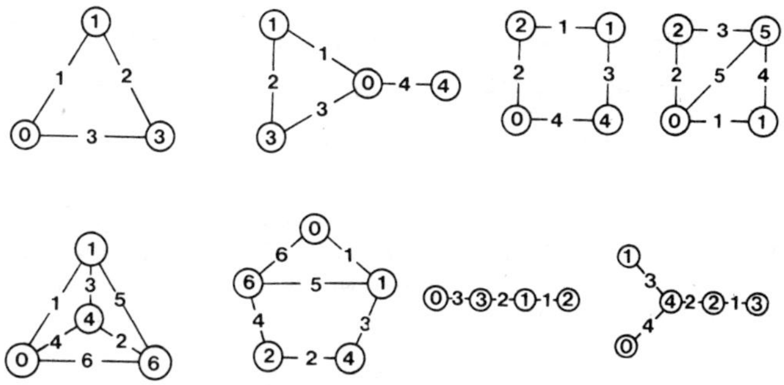

FIGURE 1. Some gracefully numbered graphs.

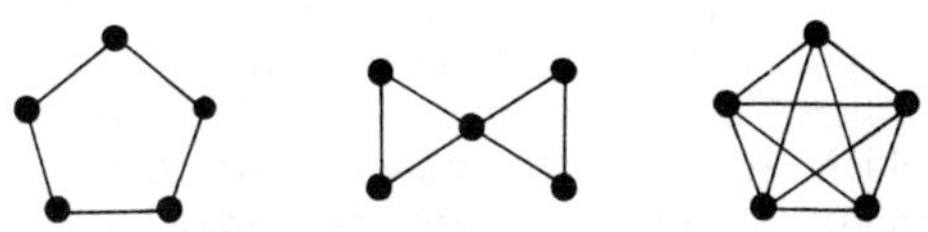

FIGURE 2. Nongraceful graphs with five nodes.

A SEQUENCE OF CONJECTURES

In 1963 Ringel [1] conjectured that the complete graph on an odd number of nodes K_{2n+1} could be decomposed into $2n+1$ subgraphs isomorphic to any given tree with n edges. For example, FIGURE 3 illustrates the decomposition of K_5 into five isomorphs of the tree with two edges.

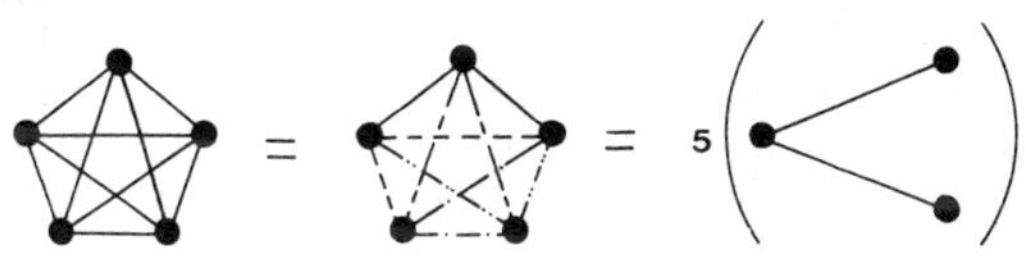

FIGURE 3. Decomposition of K_5.

In 1966 Rosa [2] reported Kotzig's suggestion that a stronger constraint would not falsify Ringel's conjecture. The stronger *Ringel-Kotzig Conjecture* asserts that: K_{2n+1} can be cyclically decomposed into $2n+1$ subgraphs isomorphic to a given tree with n edges.

Intuitively, such a cyclic decomposition is accomplished by (a) arbitrarily choosing T_n, an n-edge tree, (b) identifying the edges of T_n with a suitable set of edges in K_{2n+1}, and then (c) "rotating" each node and edge of T_n $2n$ times from its original position. Following the final rotation, every edge of K_{2n+1} will have been covered exactly once in one of the $2n+1$ positions of T_n. FIGURE 4 shows one such cyclic decomposition of K_5.

More specifically, denoting the nodes by the set of integers $\{0, 1, 2, \ldots, 2n\}$ permits one to define a *unit rotation* R of an edge (u, v) by

$$R(u, v) = (u+1, v+1) \mod(2n+1)$$

For example, $R(4, 1) = (0, 2)$ for $n = 2$.

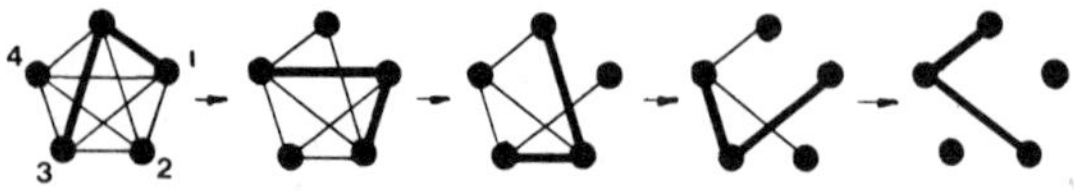

FIGURE 4. A cyclic decomposition of K_5.

Consequently, any edge (u, v) may be thought of as a representative of the equivalence class of edges of K_n which are generated by all rotations of (u, v), i.e., $R^i(u, v)$, where $0 \le i \le 2n$. An edge equivalence class so obtained may be characterized by the common "length" of the edges which it contains. The *length* of edge (u, v) is designated $L(u, v)$ and

$$L(u, v) = \begin{cases} |u - v|, & \text{if } |u - v| \le n \\ (2n + 1) - |u - v|, & \text{if } |u - v| > n \end{cases}$$

For example, in K_5, $n = 2$; $\{(0, 1), (1, 2), (2, 3), (3, 4), (4, 0)\}$ is the set of length-1 edges and $\{(0, 2), (1, 3), (2, 4), (3, 0), (4, 1), (5, 2)\}$ is the set of length-2 edges; lengths of edges in K_{2n+1} range, of course, from 1 to n. The *complement of length L* is $\bar{L} = (2n + 1) - L$ which ranges, correspondingly, from $2n$ to $n + 1$.

The cyclic decomposition of a complete graph may, thus, be described in two steps: (a) Embed an n-edge tree T into the $\binom{2n+1}{2}$ edges of K_{2n+1} in such a way that one representative of each length class is chosen; (b) apply $R(u, v)$ to each edge (u, v) of T $n - 1$ times to generate the elements of its fixed-length edge equivalence class, while preserving the tree structure of T as a unit with each rotation of all of its edges.

The stumbling block to establishing the Ringel-Kotzig Conjecture is demonstrating that every tree with $n + 1$ nodes may have consecutive distinct lengths attributed to its n edges.

Rosa [2] was first to study *valuations* of trees. He considered various classes of node numbers comprised of restricted subsets of the consecutive integers $0, 1, \ldots, 2n$ which label the nodes of K_{2n+1}. Edge labels were chosen generally as edge lengths (or, in some cases, as the complements of edge lengths) in K_{2n+1}. In particular, Rosa studied four "nested" valuation classes called (in order of increasing strength) ρ, σ, β, and α-valuations. Any n-edge graph (not *just* tree) for which a valuation exists can be used to cyclically decompose K_{2n+1}.

ρ-Valuation. The nodes of graph G are assigned n distinct integers from the set $\{0, 1, \ldots, 2n\}$ and the induced edge value for each edge (u, v) is chosen from the set $\{|u - v|, (2n + 1) - |u - v|\}$. The numbering of G must be accomplished so that all n of these two-element sets are distinct.

For example, consider the numbering of tree T_0 in FIGURE 5a. Here $n = 2$, the vertex number set $V = \{0, 1, 3\} \subseteq \{0, 1, \ldots, 2n\}$, and we choose edge $e_1 = |3 - 0| = 3$, edge $e_2 = (2 \cdot 2 + 1) - |0 - 1| = 5 - 1 = 4$. The complete graph $K_{2n+1} = K_5$ can then be considered to contain T_0 and four rotated copies of it. From FIGURE 5b it

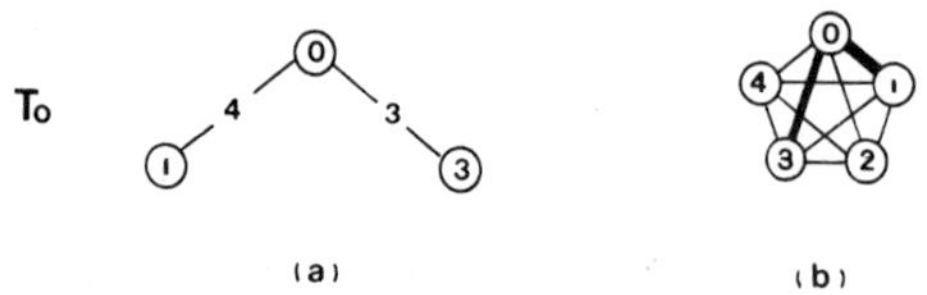

FIGURE 5. A ρ-valuation of tree T_0 and its embedding in K_5.

is clear that each rotation R of the tree T_0 contains one edge labelled 3 (of length 2 in K_5) and one labelled 4 (of length 1 in K_5). Moreover, each application of R preserves the structure of T_0, and four such rotations from the original position exhaust the edges of K_5.

Thus, if a ρ-valuation exists for any n-node graph (or tree, in particular) it may be used to decompose K_{2n+1} cyclically. As was previously stated, if all trees could be shown to have ρ-valuations, then the Ringel-Kotzig Conjecture would be proved.

σ-Valuation. In this special case of the ρ-valuation, the vertices of graph G are still assigned values from the set $\{0, 1, \ldots, 2n\}$, but the n edge values must assume the n distinct values of the set $\{1, 2, \ldots, n\}$; that is, the values in $\{n + 1, \ldots, 2n\}$ cannot appear.

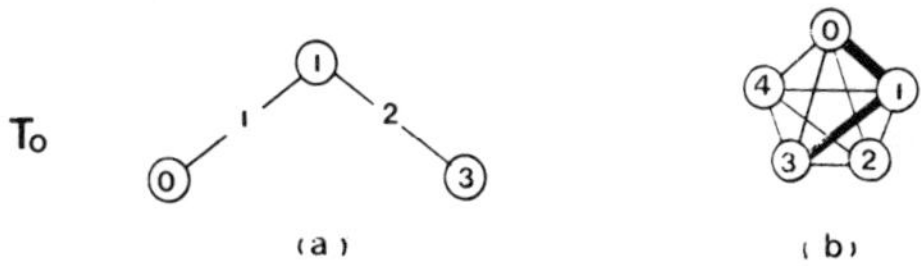

FIGURE 6. A σ-valuation of tree T_0 and its embedding in K_5.

The numbering of T_0 in FIGURE 5a is a ρ-valuation but not a σ-valuation. The numbering of T_0 in FIGURE 6a is both. In FIGURE 6b it is clearly shown that this numbering will also produce a cyclic decomposition of K_5. Each of the five isomorphs of T_0 contains one edge of length 1 and one of length 2, and the unit rotations guarantee that each edge is covered.

β-Valuation. This is a further refinement of the previous valuations in which the requirement that $E = \{1, 2, \ldots, n\}$ is retained, but, on the other hand, the set of eligible vertex numbers is almost halved, i.e., $V \subseteq \{0, 1, \ldots, n\}$. Recalling the introduction of this paper, one notes that β-valuation is simply another name for graceful numbering. Consequently, *if all trees are graceful, the Ringel-Kotzig Conjecture holds.* Although the previous two numberings of T_0 are not graceful, the numbering in FIGURE 7 is. In embedding the gracefully numbered T_0 into K_5, one notes that T_0 is in the "right half" of K_5. This "right-half" property is universal for decompositions using graceful numberings.

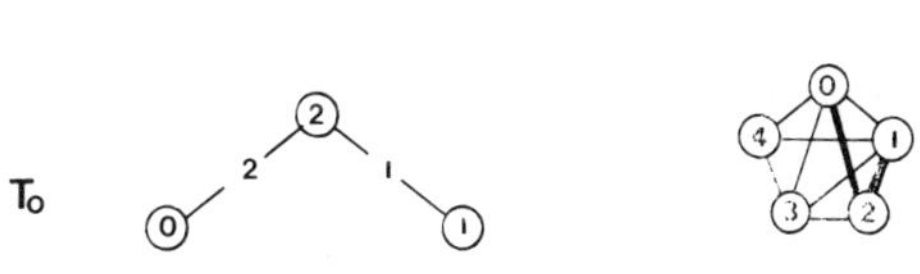

FIGURE 7. A β-valuation (graceful numbering) of tree T_0 and its embedding in K_5.

α-Valuations. This valuation occurs only if a graceful graph is *balanced* (terminology of Sheppard [3]), that is, if each edge has a "large" number at one end and a "small" number at the other. A large value is at least as great as $n/2$ and a small value is less than $n/2$. In the graceful numbering of T_0 just shown, one edge (1, 2) has both ends "large" (i.e., at least $2/2 = 1$) and so that numbering is not

balanced. The numbering of T_0 in FIGURE 8 satisfies the conditions for all four valuation classes that we have described.

Rosa knew that if all trees had even ρ-valuations, the decomposition conjecture would be resolved. It seemed most expedient to determine first if all trees admitted the most constrained numberings. Consequently, he initially asked if all trees were α-valuable.

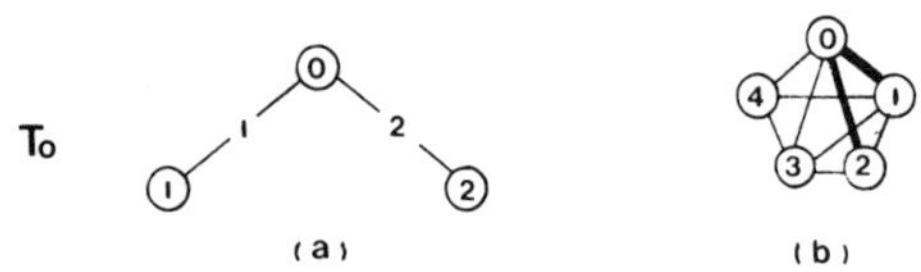

FIGURE 8. An α-valuation (balanced numbering) of tree T_0 and its embedding in K_5.

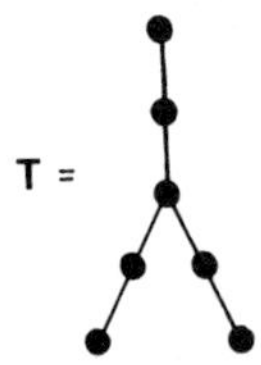

FIGURE 9. A tree with no balanced numbering.

It is easy to demonstrate that not all trees are balanced using the simple counter-example shown in FIGURE 9. (a) if tree T were balanced, each edge would be numbered with one end ≤ 3 and the other > 3. Thus the "middle structure" of T must be one of those shown in FIGURE 10. (b) Since all numbers 1, 2, ..., 6 must be assigned in an edge numbering, one notes, in particular, that the numbering of edges 5 and 6 can only occur by the two node relationships shown in FIGURE 11. (c) FIGURE 12 shows the only two ways that the balanced middle structure can occur with edges 5 and 6 in T. FIGURE 13 indicates the unique way that an edge

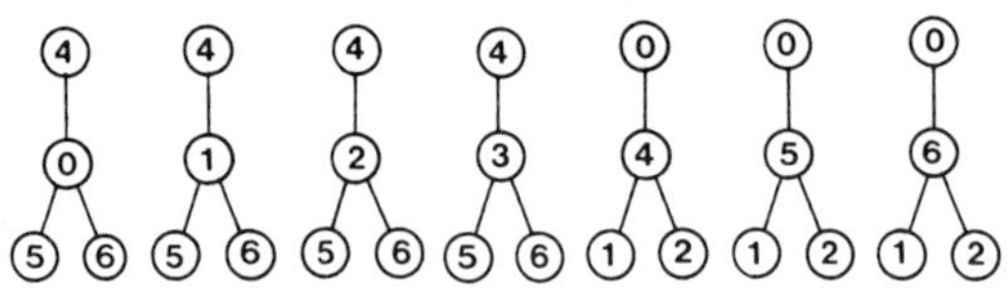

FIGURE 10. Balanced "middle structures" for T.

FIGURE 11. The only node relations that can generate edges 5 and 6 in T.

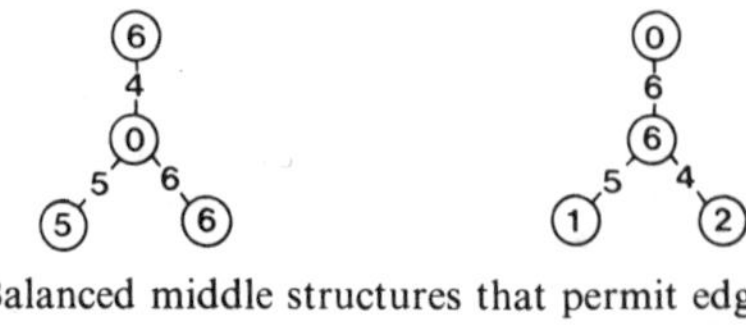

FIGURE 12. Balanced middle structures that permit edges 5 and 6 in T.

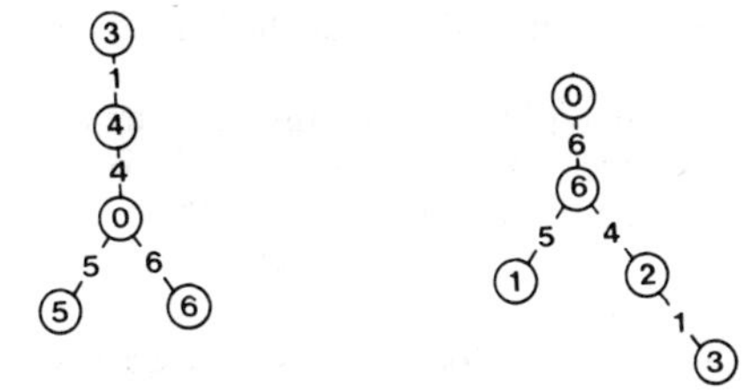

FIGURE 13. Unique augmentation of the middle structure of T by an edge labelled 1.

numbered 1 could be generated with these numberings of T. (d) Neither of the remaining numbered nodes (0 and 1 in the first case; 5 and 6 in the second) can complete the structure of T without duplicating an edge number already present (and omitting those numbers that are not). Consequently, T is unbalanced: Since not all trees are balanced the Ringel-Kotzig Conjecture remains unresolved. Just as pertinent to the conjecture and only slightly less restrictive is the following question: Are all trees graceful? This 14-year-old question remains unanswered.

The difficulty in resolving this question seems to be lack of knowledge of a proper "horticulture." That is, no *general method* is currently known which allows one to take a tree that is known to be graceful and to augment it in an arbitrary position with a twig or to graft another general tree onto it to create a new tree with a graceful numbering. For example, it would be easy to number all trees gracefully *if* the node numbered zero could be placed in *all* positions, i.e., if the graph were "zero-rotatable" [4, 5]. The tree T in FIGURE 14a is graceful and can easily be used to generate graceful numberings of the trees in FIGURE 14b because T can be gracefully numbered so that any of the dark nodes is numbered with zero. The graceful numbering of the new trees comes from adding a node labelled 6 and adjoining it to the zero-node by an edge with the induced value 6. The tree T is not zero-rotatable, however. FIGURE 15a shows where zero cannot be placed in a graceful numbering. On the other hand, the tree obtained from adjoining a node to the zero-forbidden node is graceful as FIGURE 15b shows; it just

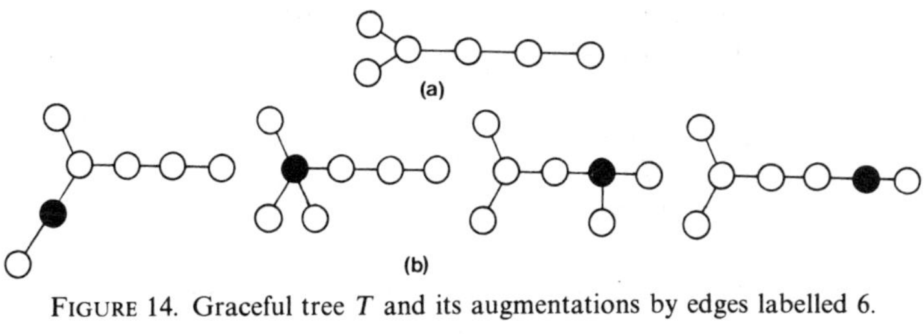

FIGURE 14. Graceful tree T and its augmentations by edges labelled 6.

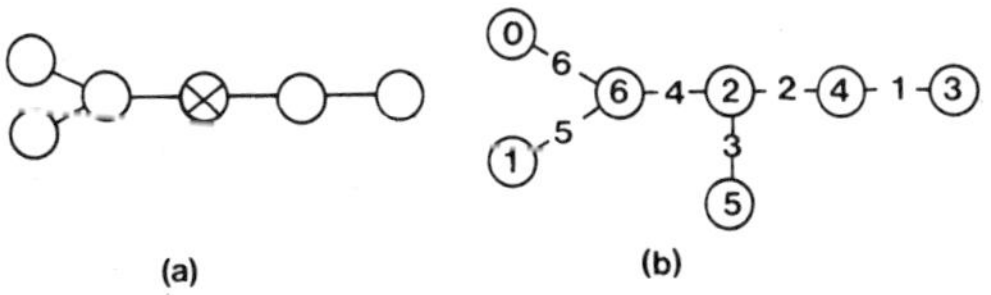

FIGURE 15. A zero-forbidden node and a graceful numbering of T with an edge adjoined at that node.

cannot be obtained by simply adding a twig labelled six to a graceful labelling of its edge-deleted subgraph.

Although no general method of systematically growing all new trees from old is currently known, various species of trees have been found to be graceful, and methods for generating some graceful numbering of these classes are known. In the next section we review a history of these discoveries.

The Ever-Expanding Set of Trees Known to Be Graceful

Early Results

The most extensive collection of graceful tree types was assembled in Rosa's 1967 paper [2] delineating the decomposition problem. His work included the following results:

(i) All trees with sixteen or fewer nodes are graceful. Hence, K_{31} and all smaller complete graphs with an odd number of vertices can be cyclically decomposed into isomorphs of an arbitrary tree giving support to the Ringel-Kotzig Conjecture.

(ii) Any tree with fewer than five leaves is graceful.

(iii) Any caterpillar is graceful. (A *caterpillar* is a tree with the property that removal of its endpoints leaves a path called its "backbone.") Rosa showed that any caterpillar may be numbered by the bipartite technique illustrated here. Start with any caterpillar, for example, C in Figure 16a. Bicolor the nodes as shown in Figure 16b. Then starting at one end of the caterpillar, consecutively number all nodes of that color, as shown in Figure 16b. Finally, as shown in Figure 16c, return to the starting end from the far end by continuing the consecutive numbering with the nodes of the second color.

(iv) F-trees are graceful. This class is made up of trees on which all branches but one must be isomorphic caterpillars. The possible exceptional branch for this class must also be a caterpillar.

(v) Two related classes of trees and their union are graceful. The U_4 trees have four arbitrary length paths growing from a root; U_3 trees have three branches growing from their root. Two are arbitrary paths, and the third is either a path or a caterpillar.

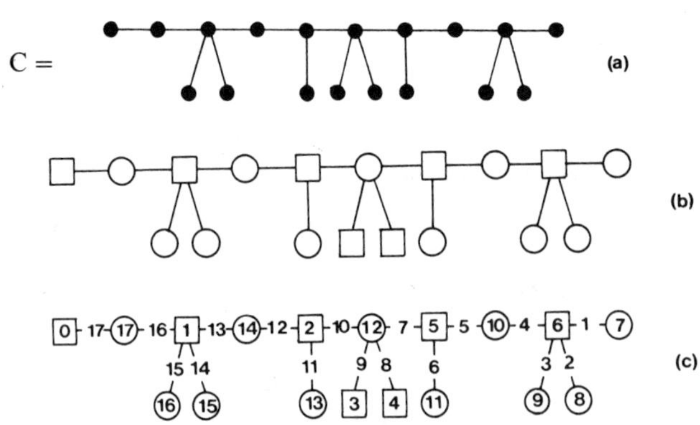

Figure 16. Labelling caterpillar C gracefully.

(vi) Rosa also noted in 1966 the problem of zero-rotatability in graceful graphs and announced his result that paths may have any of their vertices labelled zero in a graceful numbering. He recently published the complete proof of this result [4].

In 1969 at an international conference, Golomb [6] introduced the more general problem of determining which of all graphs were graceful and reiterated the unresolved graceful tree conjecture. He also demonstrated the necessary condition for all trees to be graceful (first proved by A. Lempel) that every n-node tree must have $\lfloor n/2 \rfloor$ of its nodes labelled with odd integers and $\lceil n/2 \rceil$ with even integers so that half of the tree's edges (i.e., $\lceil (n-1)/2 \rceil$) join even labelled nodes to odd labelled nodes.

In 1969 in the Research Problems section of the American Mathematics Monthly, Duke [7] discussed the conjecture and noted that there are certain nodes on some trees where the zero may not be placed.

Some Balanced Tree Results

In 1973 Kotzig [8] extended Rosa's work on balanced trees. Since balanced trees are also graceful, it is interesting to cite certain of his results while realizing that "graceful" may be substituted for "balanced" in them.

(i) He found methods to "grapht" together certain balanced trees to produce new balanced trees. (By the "graphting" procedure we mean that two trees are joined by the addition of an appropriate edge between them or by identifying nodes common to both.)

(ii) If a long enough path is graphted to any leaf of an arbitrary tree, the result is a balanced tree.

(iii) If a long enough path replaces an arbitrary edge in an arbitrary tree, the result is balanced.

Based on these results Kotzig conjectured a generalization to the above. His proposal is that only a finite number of trees homeomorphic to any given tree are not balanced. The truth of his conjecture would be equivalent to saying that almost all trees are balanced (and hence graceful).

A Powerful Construction Method

Also in 1973 Stanton and Zarnke [9] announced that they had found two (related) constructions that would create families of gracefully labelled trees from an arbitrary pair of gracefully numbered trees.

The general technique of both methods can be briefly cited and then explicated by a few simple examples.

General Method. Trees S and T with n_S and n_T nodes, respectively, are known to be graceful. The result of the algorithm is that replicas of T will be attached to either n_S or $n_S - 1$ of the nodes of S.

Steps. (1) Label n_S (or $n_S - 1$) replicas of T by "properly" adding multiples of n_T to the node numbers used in the graceful numbering of T. (2) Relabel the nodes of S by multiplying all the graceful labels by n_T and adding an arbitrary c, where $0 \leq c < n_T$. (3) Grapht together these nodes in the relabelled S and the replicas of T which have identical node numbers.

As an example of the first type of numbering, let S and T be defined and gracefully numbered as shown in FIGURE 17. In this example $n_T = 6$ and $n_S = 2$. The two replicas of T, called T_0 and T_1 are "properly" labelled with $0 \cdot 6 = 0$ or $1 \cdot 6 = 6$ added to the labels of T as shown in FIGURE 18a.

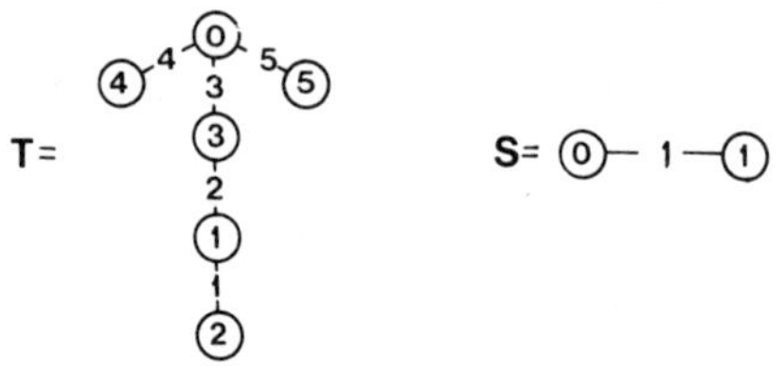

FIGURE 17. Two arbitrary trees S and with graceful labellings.

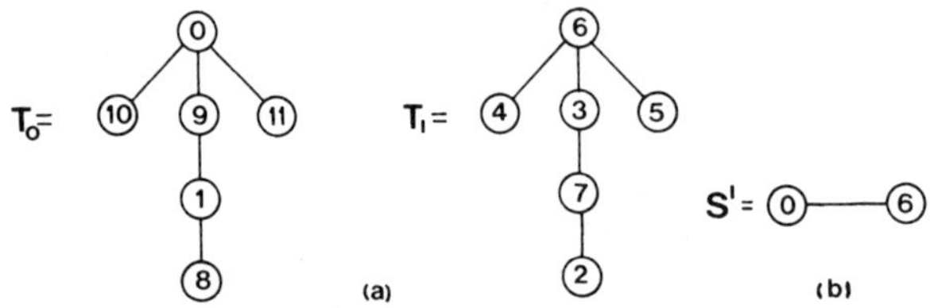

FIGURE 18. Renumbered isomorphs of T and S.

More precisely, T_0 is assigned the labels $\phi(v)$ of T on its nodes at even distances from an arbitrarily specified root. Those same nodes in replica T_i are assigned $\phi(v) + in_T$. The nodes at odd distance from the root are labelled in the "opposite direction," that is, the odd-distance nodes in replica $T_{n_S - 1}$ are given the original valuations $\phi(v)$ and those in replica T_i are assigned $\phi(v) + (n_S - 1 - i)n_T$.

On the other hand, the single replica S' of original tree S is labelled with $n_T\phi(v) + c$, where $\phi(v)$ was the original graceful number and $0 \leq c < n_T$. For the preceding example, if $c = 0$, then S' is relabelled in FIGURE 18b, and the result of the graphting operation is shown in FIGURE 19.

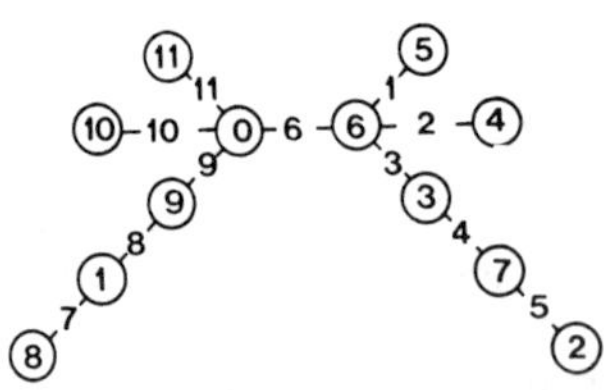

FIGURE 19. The graceful result of graphting the numbered graphs in FIGURE 18.

In this same example, if we choose $c = 4$ then the S' relabelling is shown in FIGURE 20a, and the considerably different result of graphting that S' with T_0 and T_1 is shown in FIGURE 20b. Other values of c cause T_0 and T_1 to be joined at other nodes, but always by an edge labelled 6.

In this construction of Stanton and Zarnke's, each node of S has a replica of T affixed to it. In their closely related second construction method the *root* of S does not receive a replica. Thus, for example, from trees T and S in FIGURE 21a, they obtain tree T' shown in FIGURE 21b. They can then use $T = T'$ and the same S to obtain the tree in FIGURE 21c.

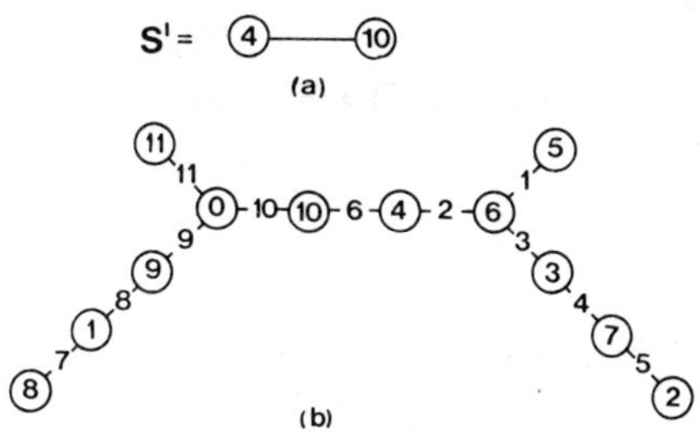

FIGURE 20. Alternative relabelling of S' and the result of using it in a graphting operation.

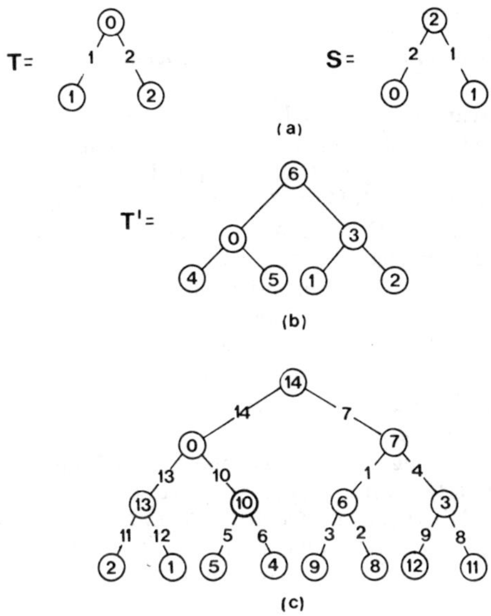

FIGURE 21. Complete binary tree construction as a special case.

Their procedure can be repeated ad infinitum with T being a complete binary tree of depth n and S being one of depth 2, giving as result a graceful complete binary tree of depth $n + 1$. Consequently, all complete binary trees are graceful, and we know how to number them.

Stanton and Zarnke describe another numbering construction in which an r-node subtree is pruned from an n-node tree and is replaced by a different r-node tree to create a new n-node tree. The required precursors of this procedure are having graceful numberings of both the original n-node and r-node trees. Also the r-node subtree which is pruned must have consecutively numbered edges. This procedure, illustrated in FIGURE 22, prunes four nodes from tree T_1 and replaces them with tree T_2 with each node value increased by 2 to give T_3. Parenthetically, one should note that although Stanton and Zarnke do not mention it, their constructions may be extended from trees to general graphs.

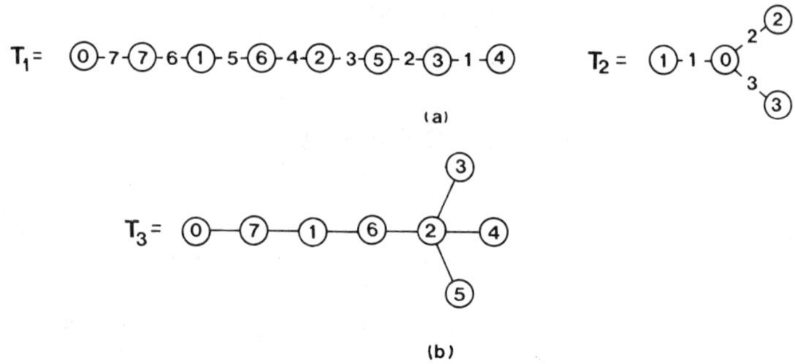

FIGURE 22. Pruning and graphting construction to generate graceful trees.

Other Items in the Chronology of Graceful Tree Numberings

Several other researchers have independently discovered some of the previous results or slight extensions of them. We briefly cite a chronology of this activity:

1971, 1973, 1975, 1977. Guy and Klee [10] and Guy [11–13] published biennial updates on the status of Duke's [7] statement of the problem.

1975. Cahit and Cahit [14] independently found algorithms for labelling caterpillars and n k-snakes attached at a single vertex.

1975. Gabow [15] made an independent discovery of one class of the Stanton and Zarnke construction which gracefully joins isomorphs of a gracefully numbered tree to the exterior nodes of an n-star. He also found a related numbering of "Fibonacci trees."

1976. Sheppard [3] studied balanced graphs and again cited the unsolved problem of all trees being graceful.

1976. Cahit [16], among others, was unaware of Stanton and Zarnke's work and again stimulated interest in the graceful tree question by asking in the Monthly Research Problems (American Mathematical Monthly) if all complete binary trees were graceful. Guy (private communication) reported that as of June 1976 affirmative answers to this question had been arrived at by Barnes, Taylor, Owens, and by Gyárfás and Lehel. Brualdi [17] in December 1976 and later Guy [13] promulgated the resolution of this problem. Although the work of all these people provided some extensions in tree labellings, no significantly new results applicable to all trees were achieved.

RECENT RESULTS

Some headway in broadening the class of trees known to be graceful has been made recently by examining the adjacency matrices of graceful trees. The initial independent observations of the character of these matrices were found by Haggard and McWha [18] and Bloom [19].

Preliminary Observations

Some fundamental observations on the nature of graceful graphs are useful.

LEMMA 1. A graceful graph G with n edges has no more than $n + 1$ nodes.

Proof: The n edges of G are distinctly labelled with the n integers 1, 2, ..., n and the nodes are distinctly labelled with elements taken from the set $\{0, 1, \ldots, n\}$. Since $\|\{0, 1, \ldots, n\}\| = n + 1$ there can be at most $n + 1$ nodes in $V(G)$. □

Thus, $K_3 \cup \bar{K}_2$ is not graceful, even though K_3 and $K_3 \cup K_1$ are, and, even though all trees may be graceful, all forests with more than one tree are not.

LEMMA 2. Every graceful graph G with n edges and v nodes can be embedded in a graceful graph with n edges and $n + 1$ nodes.

Proof: Give G any graceful labelling. Then adjoin $d = n + 1 - |v|$ isolated nodes to G and assign them the integers in $\{0, 1, \ldots, n\}$ which have not been assigned to nodes of the graph G. □

For example, K_4, which can be gracefully labelled as shown in FIGURE 23a, has $d = 3$ and can be augmented by $\bar{K}_3$ as shown in FIGURE 23b to give a graceful multicomponent graph. Clearly, the addition of d isolated nodes does not change the number of edges or their labelling from the original labelling.

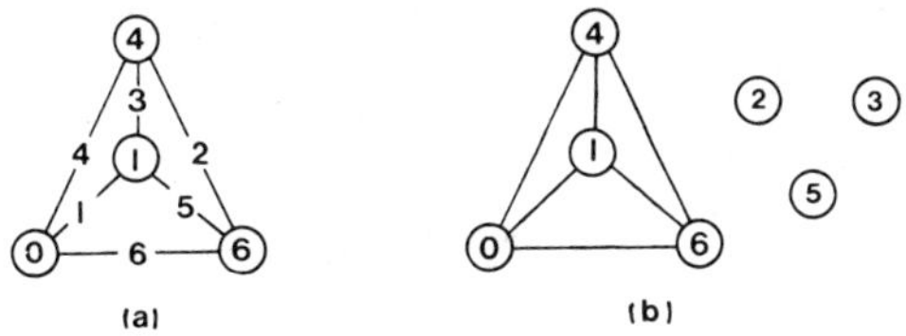

FIGURE 23. Augmenting K_4 to give a graceful, fully augmented graph.

Parameter d is the *node deficit* of the original graph G and this deficit is reduced to zero in the new graph $G' = G \cup \bar{K}_d$. Every graph with exactly n edges and $n + 1$ nodes is considered *fully augmented*. The adjacency matrices of such graphs can be easily characterized when they are canonically written with their nodes listed in the counting order of their associated node numbers.

THEOREM 1. Let $A = (a_{ij})$ be a canonical adjacency matrix of a fully augmented, gracefully numbered, graceful graph. Define the *upper kth diagonal* as $\{a_{ij} \mid j - i = k, j > i\}$ and the *lower kth diagonal* similarly. Then, each diagonal of A contains exactly one "1" entry. Conversely, any symmetric adjacency matrix with exactly one "1" on each diagonal denotes a graceful numbering of a fully augmented graph.

Proof: In a fully augmented, gracefully numbered graph all node numbers from 0 to n are present, and the condition that $a_{ij} = 1$ implies that the node numbered i is connected to the node numbered j. The edge joining these nodes is numbered $|j - i| = k$.

Because the numbering is graceful, only one edge can join any of the pairs of numbers with $|j - i| = k$, thus the kth diagonal can have only one "1" entry. Moreover, since each value of $k \in \{1, \ldots, n\}$ must be assigned to some edge, every diagonal must have a nonzero entry.

The converse observation is self-evident, since this matrix contains the appropriate number of nodes and edges, and has exactly one edge with each of the requisite values. □

To exemplify the content of the theorem, FIGURE 24 gives a catalog of all graceful numberings of 3-edge, 4-node graphs. COROLLARY 1 gives a counting of these numberings.

COROLLARY 1. There are exactly $n!$ graceful numberings of graphs with n edges and $n + 1$ nodes.

Proof: On diagonal 1 there are n possible choices for the "1" entry; on diagonal 2 there are $n - 1$; ... ; on diagonal n there is the single (corner) square possible. □

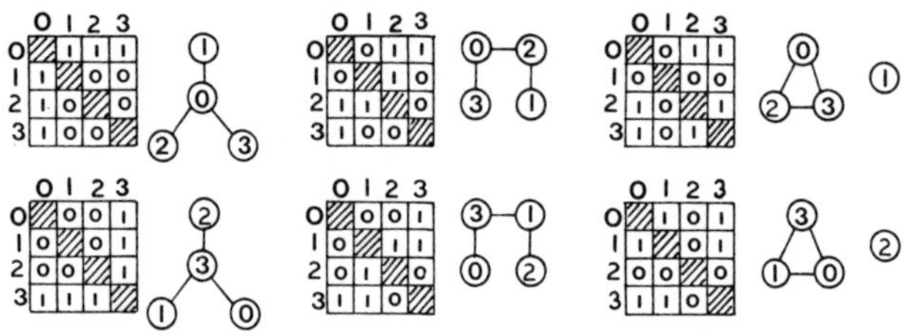

FIGURE 24. All graceful numberings of 3-edge, 4-node graphs.

Adjacency Matrices of Graceful Trees

It is useful to be able to distinguish the adjacency matrix of a tree from that of some other n edge, $(n + 1)$-node graph. One property which necessarily holds if the matrix represents a tree, easily distinguishes many nontrees.

LEMMA 3. Each column (and row) of the adjacency matrix of a tree must contain one or more "1" entries.

Proof: The connected nature of a tree requires it. □

Unfortunately any disconnected graph with no isolated nodes has an adjacency matrix with no all-zero columns. A necessary and sufficient discrimination between trees and nontrees is somewhat more unwieldy to apply. Let A be the adjacency matrix and let D be the $n \times n$ matrix with d_{ii} = degree of v_i and $d_{ij} = 0$, $i \neq j$.

LEMMA 4. If A is a canonical adjacency matrix of an n-edge, $(n + 1)$-node graph G with no column containing all zero entries and if $\det(A - D) = 1$, then G is a tree.

Proof: This follows immediately from Kirchhoff's theorem stating that in any graph $\det(A - D)$ = the number of spanning trees. For any disconnected graph, $\det(A - D) = 0$, and an n-edge, $(n + 1)$-node graph must be either disconnected or a tree. □

LEMMA 4 and THEOREM 1 prove a characterization of the canonical adjacency matrices of gracefully numbered trees.

THEOREM 2. A canonical adjacency matrix of a gracefully numbered tree has "1"'s in each column (moreover, it satisfies Kirchhoff's condition for one spanning tree), and contains exactly one "1" on each diagonal.

If this characterization of trees' graceful adjacency matrices could be used to count the trees, it would be possible to determine if all trees had graceful numberings, since the number of trees is known. Unfortunately the number of graceful numberings varies for different trees and at present no practical general census methods have been devised. Nevertheless, some classes of trees can be more completely described. For example, we have the following theorem:

THEOREM 3. A canonical adjacency matrix of order $n + 1$ (with nodes numbered from 0 to n) specifies a graceful numbering for an $n + 1$ node caterpillar if (1) $a_{0n} = 1$ and (2) $a_{ij} = 1$ (for $j > i + 1$) only if *either* $a_{i-1,j} = 1$ *or* $a_{i,j-1} = 1$.

Any matrix of this form will be termed a *chained, graceful caterpillar matrix* (CGCM).

Before proving this theorem we shall briefly examine an example of its meaning and application. Consider the adjacency matrix of FIGURE 25a which represents the numbered caterpillar illustrated in FIGURE 16. FIGURE 25b is a redrawing of FIGURE 16c for convenience.

The circled entries in the matrix identify the path which serves as the backbone of the caterpillar. The nodes in this snake can be read directly from the matrix by starting with the n'th, i.e., 17'th, column and alternatively listing the columns and rows of the circled "corner" entries proceeding always left and down toward the main diagonal. The entries in the matrix between the "corner" entries are the caterpillar's "legs." This intuitive observation can be formalized into a proof of the theorem.

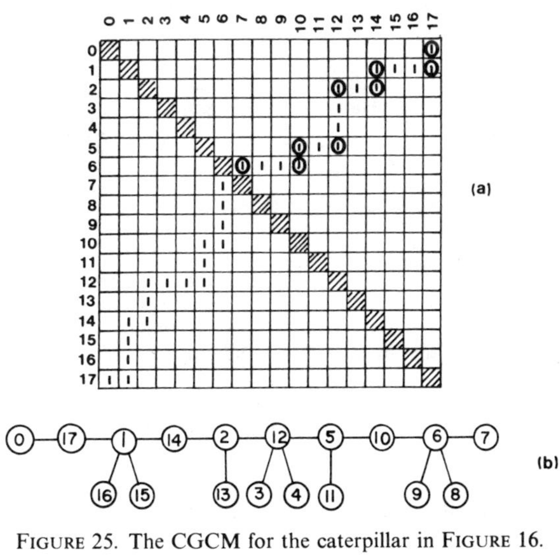

FIGURE 25. The CGCM for the caterpillar in FIGURE 16.

Proof of Theorem 3: A CGCM certainly contains exactly one "1" on each diagonal making it a graceful numbering of a graceful graph.

The caterpillar nature of a graph represented by a CGCM is established by "growing" the chain of "1"'s from the n'th diagonal entry in the upper right corner down to the first diagonal. The $(0, n)$ entry joins the 0 and n nodes by an edge. This 1-edge caterpillar can have nodes attached to one end or the other (but not both). A "1" in the $(0, n-1)$ entry indicates that the rest of column n will be empty, and thus node n will be adjacent to no other nodes. If $a_{1,n-1} = 1$, then $a_{0,j} = 0$, for $j \neq n$, and the 0 node is one "end" of the caterpillar. Because of the symmetry of these two cases we need consider only one of them, say the former, where node n adjoins only node 0, and 0 must adjoin $n - 1$. The caterpillar so grown is shown in FIGURE 26a. FIGURE 26b and c shows the graphs that occur if $a_{1,n-1} = 1$ or $a_{0,n-2} = 1$, respectively. In each case the new graph grown is a caterpillar and *one node of the newest edge* added is the new *end* of the caterpillar. More specifically, we saw that the second "end" of the 3-node caterpillar could be either $n - 1$ (when node 1 was adjoined) or 0 (when node $n - 2$ was adjoined).

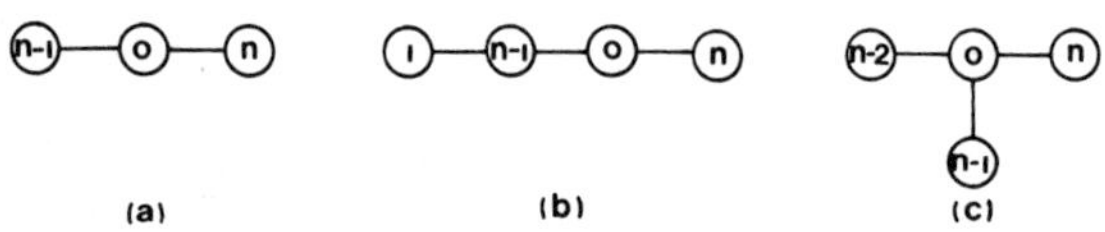

FIGURE 26. The earliest stages in a gracefully numbered caterpillar's growth.

Inductively, we continue adjoining to an existing caterpillar one of the pair of qualifying nodes on the $(k - 1)$'th diagonal after the node from the k'th diagonal is adjoined. Suppose that $a_{ij} = 1$ (where $j - i = k$) is the k'th diagonal entry, then either $a_{i-1,j} = 1$ in which case edge $(i - 1, j)$ will adjoin to node j, or $a_{i,j-1} = 1$ in which case the edge $(i, j - 1)$ will adjoin i. These four possibilities are shown in FIGURE 27. In none of these cases can a node be added farther than one edge away from the backbone of the previously grown caterpillar. It is this criterion that defines the caterpillar and finally establishes that a CGCM always represents one. □

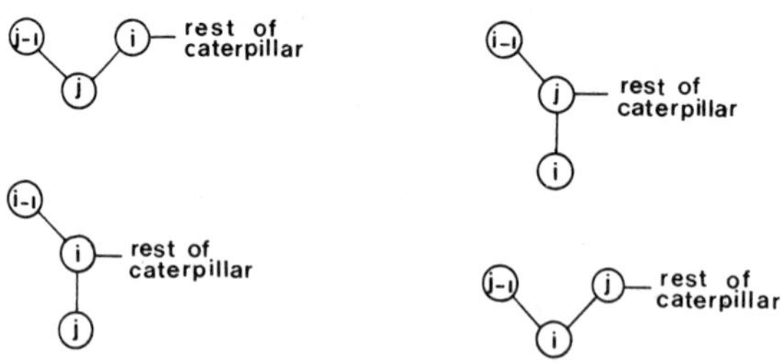

FIGURE 27. All ways permitted by a CGCM to adjoin new nodes to existing caterpillars.

Not only does every CGCM represent a gracefully numbered caterpillar, but indeed the following is true.

THEOREM 4. Every caterpillar can be gracefully numbered by a CGCM.

Proof: The caterpillar numbering shown in the third section (*Early Results*) which is valid for all caterpillars, leads directly to a matrix of this form. □

Counting Caterpillars

A spin-off of the adjacency matrix approach is a direct calculation of the number of n-edge, $(n + 1)$-node caterpillars by a count of appropriate CGCM's.

THEOREM 5. Let $C(n)$ represent the number of n-edge caterpillars, then for $n \geq 4$

$$C(n) = 2^{n-3} + 2^{\lceil (n-4)/2 \rceil}$$

where $\lceil\ \rceil$ represents the ceiling or greatest integer function.

Proof: We shall count the CGCM's that are associated with distinct caterpillars to calculate $C(n)$. First, we count the number of CGCM's, ignoring the fact that several CGCM's may be associated with one tree. Since the nth diagonal must contain a "1", $a_{nn} = 1$, and there are $n - 1$ "1"'s to be located in the adjacency matrix. Since the "1" on the k'th diagonal must be either to the left or below the entry on the $(k + 1)$'th diagonal, $1 \leq k \leq n - 1$, there are a total of 2^{n-1} CGCM's.

We now observe that there remain in the 2^{n-1} at least twice as many CGCM as caterpillars. Every CGCM may be reflected about the axis between $(0, n)$ and $(n, 0)$ to give a different CGCM representing the same caterpillar.‡ Consequently, the 2^{n-2} CGCM's with $a_{1n} = 1$ must represent every caterpillar at least once.

The nature of the chaining of "1"'s in the building up of a CGCM gives the needed correspondence with the set of caterpillars. Each edge added must attach to one end of the previous edge. By the previous induction, one gets a caterpillar. The only other way that this same caterpillar could be generated is by starting to generate it at the other end of its backbone. (Starting the edge-adjacent-to-the-previous-edge construction elsewhere, e.g., on a leg or in the middle of the backbone certainly cannot generate the caterpillar.)

Consequently, most caterpillars will have two CGCM's, among the 2^{n-2} being considered, one representing a numbering from each end. Since a caterpillar symmetric about the middle of its backbone is identical starting at either end, such symmetric caterpillars are represented by just one CGCM in this set. Since

$$C(n) = C_s(n) + C_u(n)$$

where $C_s(n)$ = symmetric caterpillars and $C_u(n)$ = unsymmetric caterpillars, we have, using the set of 2^{n-2} CGCM's being considered, the following relationships:

$$\begin{aligned} C(n) &= A_s(n) + \tfrac{1}{2}A_u(n) \\ &= A_s(n) + \tfrac{1}{2}(A(n) - A_s(n)) \\ &= \tfrac{1}{2}(A(n) + A_s(n)) \end{aligned}$$

‡ Such a reflection gives a "complementary" renumbering of the nodes in which a node labelled i is relabelled $n - i$, and in which the edge numbers are unchanged. The reflection can also be seen as a relabelling of the bicolored caterpillar in which the "low-number colored nodes" and the "high-number colored nodes" exchange places.

where $A(n) =$ the total number of CGCM's being considered, $A_s(n) =$ the number of CGCM's for symmetric caterpillars, and $A_u(n) =$ the number of CGCM's for unsymmetric caterpillars.

We know that $A(n) = 2^{n-2}$. The CGCM's for symmetric caterpillars are also easy to count. Once $\lceil n/2 \rceil + 1 = \lceil (n+2)/2 \rceil$ of the n edges have been connected from one end of the caterpillar, the entire structure is determined. Since $a_{0n} = a_{1n} = 1$, the first two edges have been set, and there are $\lceil (n+2)/2 \rceil - 2 = \lceil (n-2)/2 \rceil$ edges to be chosen, and so

$$A_s(n) = 2^{\lceil (n-2)/2 \rceil}$$

Consequently,

$$C(n) = \tfrac{1}{2}(2^{n-2} + 2^{\lceil (n-2)/2 \rceil}$$

$$= 2^{n-3} + 2^{\lceil (n-4)/2 \rceil} \quad \square §$$

§ $C(n)$ may also be directly evaluated from the caterpillars themselves without the CGCM's using the relation

$$C(n) = \tfrac{1}{2}(A(n) + A_s(n))$$

where $A(n) =$ the total number of ways in which $n - k$ legs may be affixed to the (labelled) interior nodes of a length k backbone, summed over k $(2 \leq k \leq n)$.

$A_s(n) =$ the number of symmetric caterpillars. (Just as with the CGCM's, $A(n) - A_s(n) =$ twice the number of unsymmetric caterpillars.) $A(n)$ is determined by first counting the number of ways that $n - k$ legs can be affixed to the $k - 1$ potential sites (interior nodes) of the backbone. This is,

$$\binom{(n-k) + (k-1) - 1}{n-k} = \binom{n-2}{n-k}$$

Summing this over all backbone lengths k, $2 \leq k \leq n$ gives $A(n)$:

$$A(n) = \sum_{k=2}^{n} \binom{n-2}{n-k} = \sum_{l=0}^{n-2} \binom{n-2}{(n-2)-l} = 2^{n-2}$$

The unsymmetric caterpillars appear twice each in $A(n)$ because the head-to-tail structure of one is the tail-to-head structure of another. Symmetric caterpillars are the same coming and going and appear only once in $A(n)$. These latter may be counted in a way similar to the evaluation of $A(n)$. Roughly speaking, half of the legs in each case are located on the (say, right) half of the caterpillar. The left half will be a mirror reflection of the right. The approximate nature of these last comments comes from the need to consider parity (i.e., an odd number of legs cannot symmetrically be placed at an even number of sites). Working through these details leads one (once again) to

$$A_s(n) = 2^{\lceil (n-2)/2 \rceil}$$

An exposition of the independently derived details of this calculation as well as one from the more general machinery of Polya's theorem can be found in Harary and Schwenk [20].

New Graceful Trees from Old

A graceful adjacency matrix of any tree (not just caterpillars) may be used to generate new gracefully numbered trees. For example, the gracefully numbered, 6-node tree *T* in FIGURE 28a has adjacency matrix *A* shown in FIGURE 28b. When *A* is made the upper right-hand corner of adjacency matrix *B* (FIGURE 28c), matrix *B* represents two copies of *T* numbered as shown in FIGURE 28d, in which the dotted lines represent edges that correspond to entries on the (unfilled) sixth diagonal of *B* (which corresponds to the main diagonal of *A*). A new graceful tree will be created when one of the dotted lines is made the "6" labelled edge of the graph.

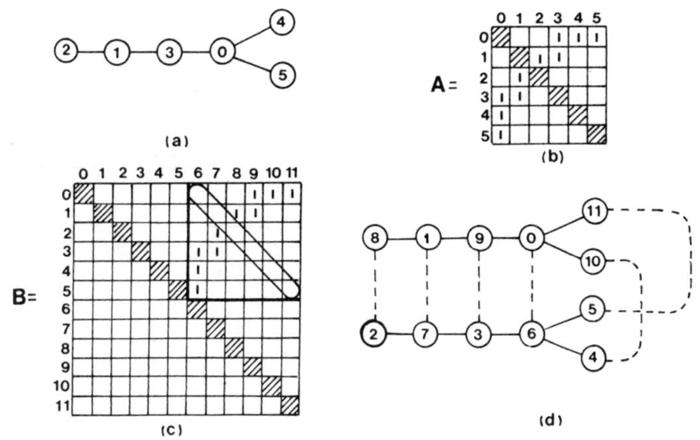

FIGURE 28. A doubled graceful tree construction.

The example in FIGURE 28 was previously seen in FIGURES 17–20 as an illustration of Stanton and Zarnke's technique of graphting together isomorphs of a given gracefully numbered tree. Indeed, one can show that the Stanton and Zarnke techniques may be viewed as special cases of assembling the adjacency matrices of gracefully numbered trees into larger matrices of larger trees with an associated graceful numbering.

As the final example, FIGURE 29, shows, these techniques are not limited simply to graphting isomorphs of one tree, but can, in fact, grapht distinct and duplicated trees into one gracefully numbered tree. A compilation of techniques and results based on the adjacency matrix approach is currently being assembled and will constitute a separate paper by Bloom *et al.* [21].

ACKNOWLEDGMENTS

Helpful suggestions for the presentation of this material were given to me by various colleagues. I wish specifically to thank R. Guy, G. Haggard, F. Harary, and A. Rosa for their comments and encouragement.

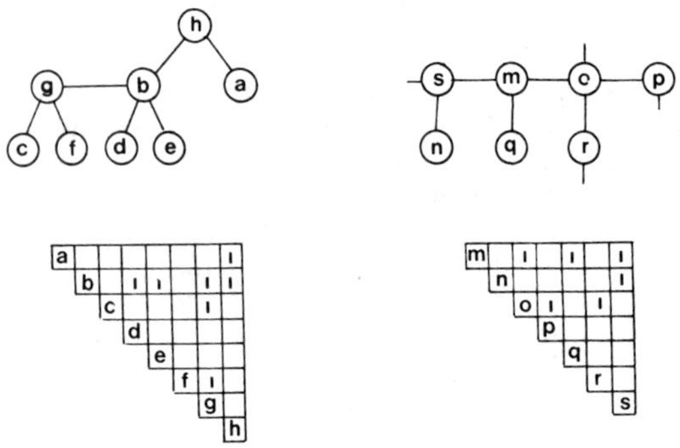

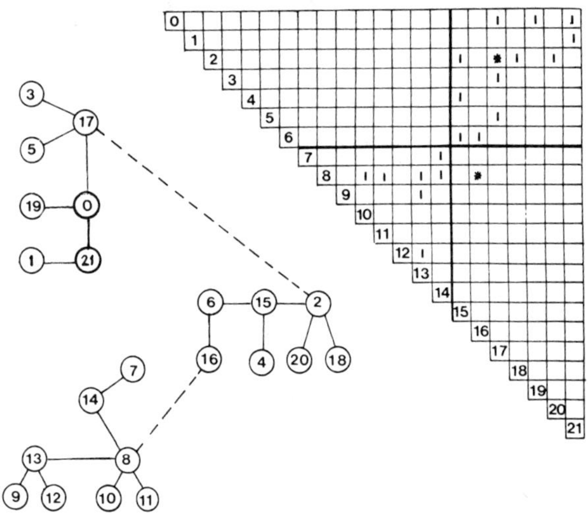

FIGURE 29. The graceful graphting of a tree with two isomorphs of a second tree.

REFERENCES

1. RINGEL, G. 1964. Problem 25. Theory of Graphs and Its Applications. Proc. Int. Symp. Smolenice (June 1963). Czech. Acad. Sci. Prague, Czech. p. 162.
2. ROSA, A. 1967. On certain valuations of the vertices of a graph. Theory of Graphs. Gordon and Breach, New York, N.Y. pp. 349–355.
3. SHEPPARD, D. A. 1976. The factorial representation of balanced labeled graphs. Discrete Math. **15:** 379–388.
4. ROSA, A. 1977. Labeling snakes. Ars Combinatoria. **3:** 67–74.
5. CHUNG, F. R. K. & F. K. Hwang. Rotatable graceful graphs. To appear.
6. GOLOMB, S. W. 1972. How to number a graph. *In* Graph Theory and Computing. R. C. Read, Ed. Academic Press, New York, N.Y. pp. 23–37.
7. DUKE, R. A. 1969. Can the complete graph with $2n + 1$ vertices be packed with copies of an arbitrary tree having n edges? Am. Math. Mon. **76:** 1128–1130.
8. KOTZIG, A. 1973. On certain vertex-valuations of finite graphs. Utilitas Math. **4:** 261–290.

9. STANTON, R. G. & C. R. ZARNKE. 1973. Labelling of balanced trees. Proc. Fourth S. E. Conf. Combinatorics Graph Theory Comput. Utilitas Math., Winnipeg, Manitoba, Canada. pp. 479–495.
10. GUY, R. K. & V. KLEE. 1971. Monthly research problems, 1969–1971. Am. Math. Mon. **78:** 1113–1122.
11. GUY, R. K. 1973. Monthly research problems, 1969–1973. Am. Math. Mon. **80:** 1120–1128.
12. GUY, R. K. 1975. Monthly research problems, 1969–1975. Am. Math. Mon. **82:** 995–1004.
13. GUY, R. K. 1977. Monthly research problems, 1969–1977. Am. Math. Mon. **84:** 807–815.
14. CAHIT, I. & R. CAHIT. 1975. On the graceful numbering of spanning trees. Info. Proc. Lett. **3:** 115–118.
15. GABOW, H. 1975. How to gracefully number certain symmetric trees. SIGACT News. **7:** 33–36.
16. CAHIT, I. 1976. Are all complete binary trees graceful? Am. Math. Mon. **83:** 259–260.
17. BRUALDI, R. A. 1976. Comments and complements. Am. Math. Mon. **83:** 798–801.
18. HAGGARD, G. & P. MCWHA. 1975. Decomposition of complete graphs into trees. Czech. Math. J. **25:** 31–36.
19. BLOOM, G. S. 1975. Numbered undirected graphs and their uses: A survey of a unifying scientific and engineering concept and its use in developing a theory of non-redundant, homometric sets relating to some ambiguities in X-ray diffraction analysis. University of Southern California (1975). Doctoral dissertation.
20. HARARY, F. & A. J. SCHWENK. 1973. The number of caterpillars. Discrete Math. **6:** 359–365.
21. BLOOM, G. S., G. HAGGARD & H. TAYLOR. The adjacency matrices of graceful trees. In preparation.

GRAPH FUNCTIONS WHICH ARE RAMSEY FUNCTIONS

Fred Buckley*

Department of Mathematics
Graduate Center
City University of New York
New York, New York 10036

Introduction

Ramsey's Theorem, stated informally, says that if $|V(G)|$ is large enough then G contains a large K_n or a large K_n as an induced subgraph. This prompts the pursuit of a very new kind of Ramsey problem which can be stated informally as follows: for a given nonnegative invariant f can we find a finite number of classes of graphs ($f \to \infty$ on each class) so that any graph with a very large f value has a large member of one of the classes as an induced subgraph.

The notation we use is that of [2], where definitions can be found for invariants not specifically defined here. In order to present results for a large number of invariants, only a few proofs will be given here. Theorem 1 displays the strong relationship of our results to Ramsey's Theorem. The notion of a Ramsey function was first presented in [6].

Ramsey Functions

As it is not unlikely that Ramsey functions will be of interest in other investigations, we define them in terms of partially ordered sets, although we concern ourselves only with the set of all graphs (including the null graph) partially ordered by inclusion as an induced subgraph. If H and G are graphs, we use $H \subset G$ to mean H is an induced subgraph of G.

Let P be a partially ordered set with $0 \in P$, $0 \leq a$ for all $a \in P$. Let f be a nonnegative monotonic function defined on P.

Definition 1. A chain $0 = a_0 < a_1 < a_2 < \cdots$ for which $f(a_i) \to \infty$ is called an *f-chain*.

Let $F = \{(a_{ij}), i = 1, 2, \ldots, m; j = 0, 1, 2, \ldots\}$ be a finite set of f-chains.

Definition 2. For a given f and $F \neq \phi$, $\underline{t_F(a)} = \max_{a_{ij}, F}\{j \mid a_{ij} \leq a\}$.

Note: This is defined for all $a \in P$ since the chains contain 0. If we list the chains of F in matrix form, then for a graph G, $t_F(G)$ is the column index j of an induced subgraph of G farthest out in any of the rows of the matrix.

Definition 3. A nonnegative monotonic function f defined on P for which there is a sequence $(a_i) \subset P$, where $f(a_i) \to \infty$, is a *Ramsey function* if there exists a

* Current address: Mathematics and Science Division, St. John's University, 300 Howard Avenue, Staten Island, New York 10301.

0077-8923/79/0328-0052$1.75/1 © 1979, NYAS

finite collection of f-chains F such that for any sequence $S = s_1, s_2, s_3, \ldots$ in P $\sup_S f(s_i) = \infty$ implies $\sup_S t_F(s_i) = \infty$.

EXAMPLE. If $f = |V(G)|$ then f is a Ramsey function and the corresponding $F = \{(0 \subset K_1 \subset K_2 \subset \cdots), (0 \subset \bar{K}_1 \subset \bar{K}_2 \subset \cdots)\}$. If $G = K_2 + K_{1,2}$ then $t_F(G) = 4$.

Requiring the existence of some sequence on which f is unbounded in DEFINITION 3 eliminates trivial cases. We shall need:

LEMMA 1 (Ramsey's Theorem). There exists a smallest positive integer $R(n; m)$ such that any coloring of the edges of the complete graph on $t \geq R(n; m)$ vertices with m colors has a monochromatic complete subgraph of size n.

We write $R(n)$ for $R(n; 2)$.

LEMMA 2 [9, p. 45]. There exists a function $Q(n)$ such that any square (0,1)-matrix of order at least $Q(n)$ with all 1's on the diagonal contains a principal submatrix of order n of form I, J, T, or T^t, where I is the identity matrix, J is the matrix of all 1's, and $T = (t_{ij})$ has $t_{ij} = 1$ iff $i \leq j$.

Clearly $R(n) \geq n$ and $Q(n) \geq n$.

FUNCTIONS WHICH ARE RAMSEY FUNCTIONS

At the end of this section we give a table of functions which we know to be Ramsey functions, along with their f-chains. To give the flavor of the type of proof involved we prove one of these results in THEOREM 1.

We first describe some graphs needed as f-chains below. An *n-legged spider* consists of n distinct P_3's incident at a vertex. nG means n disjoint copies of G. The graph H_n has $2n + 1$ vertices and consists of a K_{2n} and a K_{n+1} with exactly n of their vertices in common. The graph $(K_n \cup \bar{K}_n) + nK_2$ has $2n$ vertices and consists of a K_n and a $\bar{K}_n$ with additional adjacencies: each vertex of K_n is adjacent to exactly one vertex of $\bar{K}_n$. T_n is a bipartite graph with $2n$ vertices which we can label $1, 2, \ldots, n, 1', 2', \ldots, n'$ so that $E(T_n) = \{(i, j') \mid i \leq j\}$. W_n is a bipartite graph with $2n$ vertices which we can label $1, 2, \ldots, n, 1', 2', \ldots, n'$ so that $E(W_n) = \{(i, j') \mid i \neq j\}$.

We describe several of the invariants for which results are given. $\lambda^1(G)$ is the least eigenvalue of $A(G) - xI$. $Match^*(G) = \min\{\beta_1(G), \beta_1(\bar{G})\}$, where $\beta_1(G)$ is the line independence number.

DEFINITION 4. When f is not monotonic we consider $f'(G) = \max_{H \subset G} f(H)$.

DEFINITION 5. $v \in V(G)$ is a *cutpoint* of G if the removal of v increases the number of components.

The following are easy consequences of DEFINITION 5:

(a) For any cutpoint v_i, there exist points u_i and l_i adjacent to v_i such that v_i is on every $u_i - l_i$ path.

(b) Given n mutually adjacent cutpoints $v_i \in V(G)$, there exists an x_i adjacent to v_i for each i such that $(x_i, v_j) \notin E(G)$, $i \neq j$, and the x_i's are mutually nonadjacent.

(c) Given n mutually nonadjacent cutpoints $v_i \in V(G)$, all adjacent to the same point w, then there exist points x_i distinct from w such that $(x_i, v_i) \in E(G)$ for each i and (x_i, w), (x_i, x_j), $(x_i, v_j) \notin E(G)$.

Let $\text{Cut}(G)$ = the number of cutpoints in G. When we refer to a *chain* G_n we mean the chain $0 \subset G_1 \subset G_2 \subset \cdots$, where we specify the nth element in the chain. We use (a), (b), and (c) in:

THEOREM 1. $\text{Cut}'(G)$ is a Ramsey function with F consisting of the three chains: nP_3, $(K_n \cup \bar{K}_n) \dotplus nK_2$, and an n-legged spider.

Proof: Each chain described is a Cut′-chain since $\text{Cut}'(G_i) \to \infty$ for each chain. We establish that $\text{Cut}'(G)$ is a Ramsey function by showing that $\text{Cut}'(G) \geq R(nQ(R(Q(R(Q(n))))))$ implies $t_F(G) \geq n$. If $\text{Cut}'(G)$ is as stated then there exists $H \subset G$ with $\text{Cut}(H) \geq R(nQ(R(Q(R(Q(n))))))$. Apply LEMMA 1 to the cutpoints of H and find $nQ(R(Q(R(Q(n)))))$ of them are mutually adjacent or that same number are mutually nonadjacent. In the former case (b) implies $(K_n \cup \bar{K}_n) \dotplus nK_2 \subset H \subset G$. In the latter case, consider points u_i and l_i adjacent to each of the mutually nonadjacent cutpoints v_i so that v_i is on every $u_i - l_i$ path. We thus have $nQ(R(Q(R(Q(n)))))$ paths of length 2, with a cutpoint in the center of each path. Either n of the paths are mutually incident (via endpoints) or $Q(R(Q(R(Q(n)))))$ of the paths are mutually disjoint. In the former case (c) implies an n-legged spider $\subset H \subset G$. In the latter case, consider the $Q(R(Q(R(Q(n)))))$ paths lined up in a row with u_i vertices as upper vertices, v_i in the middle and l_i as lower vertices (FIGURE 1). We check for addi-

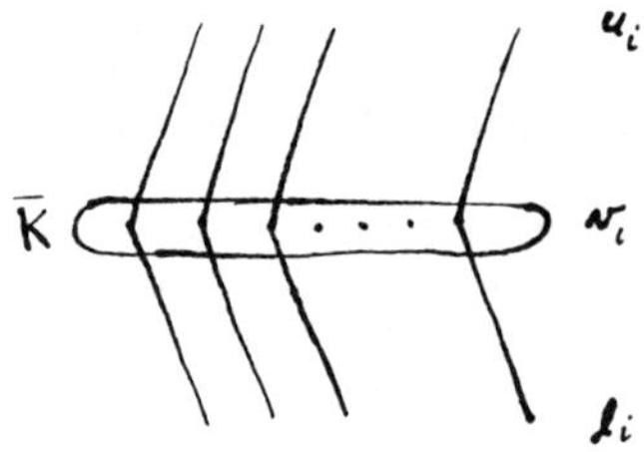

FIGURE 1.

tional adjacencies. Recall that the v_i's are mutually nonadjacent. We may assume that the v_i's have indices $i = 1, 2, \ldots, Q(R(Q(R(Q(n)))))$. Consider the (0, 1)-incidence matrix M, where row i and column i correspond to u_i and v_i, respectively. Matrix M has all 1's on the diagonal since $(u_i, v_i) \in E(G)$. LEMMA 2 applied to M implies M has a principal submatrix M' of order $R(Q(R(Q(n))))$ of form I, J, T, or T^t. If $M' = J$, T, or T^t, there is a vertex u_i adjacent to n of the mutually nonadjacent v_j, thus (c) implies an n-legged spider $\subset H \subset G$. Suppose $M' = I$ and, without loss of generality, assume the u_i and v_i involved have lowest indices. Applying LEMMA 1 to the u_i's we find either $Q(R(Q(n)))$ of them are mutually adjacent or that same number are mutually nonadjacent. In the former case, $(K_n \cup \bar{K}_n) \dotplus nK_2 \subset H \subset G$. In the latter case, again assume the u_i's have lowest possible indices. We have an induced matching between u_i and v_i vertices (FIGURE 2). Consider the (0, 1)-incidence matrix N, where row i and column i correspond to v_i and l_i, respectively. Matrix N has all 1's on the diagonal and is of order $Q(R(Q(n)))$. LEMMA 2 implies N has a principal submatrix N' of order $R(Q(n))$ having form I, J, T, or T^t. If $N' = J$, T, or T^t, (c) implies an n-legged spider $\subset H \subset G$. Thus suppose $N' = I$. LEMMA 1 applied to the l_i corresponding to columns of N' implies $Q(n)$ of the l_i are mutually adjacent or $Q(n)$ of the l_i are mutually nonadjacent. In the former case, $(K_n \cup \bar{K}_n) \dotplus$

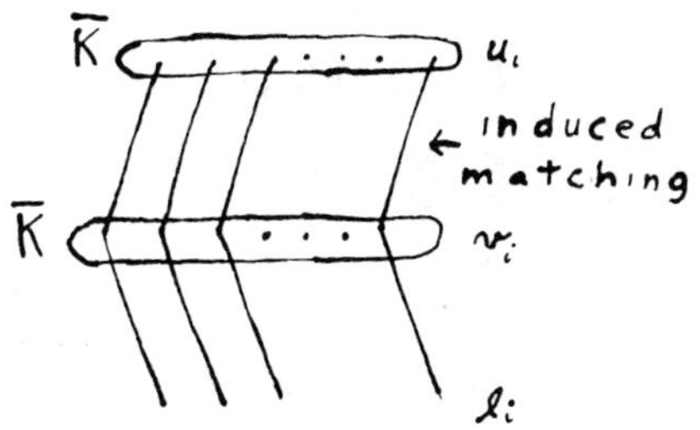

FIGURE 2.

$nK_2 \subset H \subset G$. In the latter case, we may assume the l_i have lowest indices. We are thus left to consider the paths (u_i, v_i, l_i), $i = 1, 2, \ldots, Q(n)$ (FIGURE 3). The only possible additional adjacencies are between vertices u_i and l_j. $(u_i, l_i) \notin E(H)$ for this would contradict the fact that v_i is on every $u_i - l_i$ path. Consider the "antimatching" edges (u_i, l_i) in $\bar{H}$. $(u_i, l_i) \in E(\bar{H})$. Let $\bar{M}$ be the (0, 1)-incidence matrix for $\bar{H}$. Matrix $\bar{H}$ has all 1's on the diagonal and is of order $Q(n)$. LEMMA 2 implies $\bar{M}$ has a principal submatrix, $\bar{M}'$, of order n with form I, J, T, or T^t (for $\bar{H}$). These give a complementary incidence matrix M' (interchange roles of 0 and 1), respectively, of form $J - I$, 0, $T^t - I$, and $T - I$ (for H). If $M' = J - I$, or if $M' = T^t - I$ or $T - I$ and $n = 1$ or 2, n-legged spider $\subset H \subset G$. Otherwise, these are impossible by the $u_i - l_i$ path condition. If $M' = 0$ $nP_3 \subset H \subset G$. Thus for all possible cases, $t_F(G) \geq n$ and Cut$'(G)$ is a Ramsey function. □

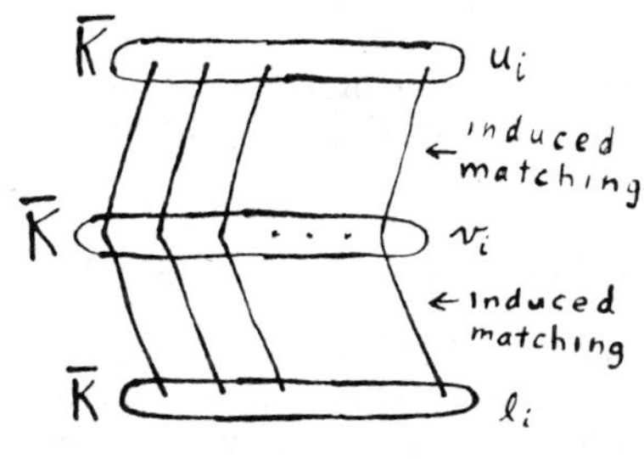

FIGURE 3.

NOTE. We note the strong dependence of the proof on Ramsey's Theorem. This dependence is strengthened by the fact that $Q(n) \leq R(n; 4)$ (see [9]).

Below we list functions we know to be Ramsey functions. LEMMA 1, LEMMA 2, and other Ramsey style theorems (see [4]) are used in proving these results.

Ramsey function f	*f-chains in F*
$\lvert\lambda^1(G)\rvert$, in [5]	$K_{1,n}$, H_n.
diameter$'$ or radius$'$	P_n.
intersection number	K_n, $\bar{K}_n$.
line chromatic number	K_n, $K_{1,n}$.
line covering number	K_n, $K_{1,n}$, nK_2.
line independence number	K_n, $K_{n,n}$, nK_2.
match*(G), in [7]	nK_2, $\overline{nK_2}$, $K_{n,n}$, $2K_n$, $K_n \cup \bar{K}_n$, $K_n + \bar{K}_n$.
maximum degree of a vertex	K_n, $K_{1,n}$.
number of blocks$'$	$K_{1,n}$, nK_2, $(K_n \cup \bar{K}_n) \dotplus nK_2$.
number of cutpoints$'$	nP_3, $(K_n \cup \bar{K}_n) \dotplus nK_2$, n-legged spider.

number of edges in the line graph of G	K_n, $K_{1,n}$, nK_3, nP_3.
number of vertices of odd degree′	K_n, $K_{1,n}$, nK_2.
point covering number	K_n, $K_{n,n}$, nK_2.
achromatic number, in [7]	K_n, nK_2, W_n, T_n.

Functions Which Are Not Ramsey Functions

The following gives a method of determining when a function is not a Ramsey function.

THEOREM 2. If for any n a_i's in P with $f(a_i)$ large, there exists an $\bar{a} \subset P$, incomparable to each a_i with $f(\bar{a})$ as large as we wish, then f is not a Ramsey function.

In [1] a theorem much stronger than the following, dealing with Hamiltonian circuits is proved:

LEMMA 3. Given $r \geq 3$, $g \geq 3$, there exists a graph G which is regular of degree r and has girth g.

LEMMA 4. If G and H are graphs, each having finite girth, and f is monotonic (nondecreasing), then $f(H) \geq f(G)$ and $g(H) > g(G)$ together imply that H is incomparable to G.

Proof: If H were comparable to G, f monotonic and $f(H) \geq f(G)$ would imply $G \subset H$. Since G is an induced subgraph of H, $g(H) \leq g(G)$, a contradiction. So H is incomparable to G. □

THEOREM 3. The arboricity, $\gamma(G)$, is not a Ramsey function.

Proof: We show that, for any n graphs G_i with $\gamma(G_i)$ large, we can find a graph B incomparable to each of the G_i's and having $\gamma(B)$ as large as we wish. If B is regular of degree r with girth $g(B)$, then using a formula of Nash-Williams we find

$$\gamma(B) = \max_{C \subset B} \left[\frac{|E(C)|}{|V(C)| - 1}\right] \geq \left[\frac{|E(B)|}{|V(B)| - 1}\right] \geq \left[\frac{nr}{2(n-1)}\right] \geq \left[\frac{r}{2}\right]$$

Now LEMMA 3 implies that we can choose B with $\gamma(B)$ as large as we wish and $g(B) > g(G_i)$, $i = 1, 2, \ldots, n$. Note that $\gamma(G_i)$ large implies $g(G_i)$ is finite. LEMMA 4 shows B is incomparable to each G_i and THEOREM 2 thus implies that $\gamma(G)$ is not a Ramsey function. □

It should be noted that the only method we have at this time for showing that a graph function is not a Ramsey function depends on the properties of girth.

The following are not Ramsey functions: chromatic number, girth, genus, minimum degree of a vertex′, number of spanning trees, and the length of the longest trail. The proofs for these rely on THEOREM 2 and the lemmas of this section. The result for chromatic number uses a well-known theorem of Erdös involving girth. The genus result uses an inequality in [8].

The results of this paper are part of the work done in the author's doctoral dissertation, where various other invariants are examined. Several other very interesting invariants are defined and examined in [7] where the concept of *equivalence of Ramsey functions* is explored, as well as whether or not certain Ramsey functions are *spectral functions*. The author would like to thank the referee for helpful comments on the presentation of this material.

References

1. Sachs, H. 1963. Regular graphs with given girth and restricted circuits. J. London Math. Soc. **38:** 423–429.
2. Harary, F. 1969. Graph Theory. Addison-Wesley, Reading, Mass.
3. Hoffman, A. J. 1972. Eigenvalues and partitionings of the edges of a graph. Lin. Alg. Appl. **5:** 137–146.
4. Hoffman, A. J. 1974. Applications of Ramsey style theorems to eigenvalues of graphs. Proc. Adv. Stud. Inst. Combinatorics, Breukelen, the Netherlands (Part 2). pp. 43–57.
5. Hoffman, A. J. 1973. On spectrally bounded graphs. Surv. Combinatorial Theory Symp. Colo. State Univ. (1971). pp. 277–283.
6. Hoffman, A. J. & Peter Joffe. Nearest $\mathscr{S}$-matrices of given rank and the Ramsey problem for eigenvalues of bipartite $\mathscr{S}$-graphs. To appear.
7. Hoffman, A. J. & Fred Buckley. On the mixed achromatic number and other functions of graphs. To appear.
8. Nordhaus, E. A. 1972. On the girth and genus of a graph. Conf. Graph Theory, W. Mich. Univ. (Kalamazoo). pp. 207–214.
9. Ryser, H. J. 1963. Combinatorial Mathematics. The Mathematical Association of America. Distributed by John Wiley, New York, N.Y.

A SURVEY OF NONCOMPLETE RAMSEY THEORY FOR GRAPHS

Stefan A. Burr

Department of Mathematics
City College
City University of New York
New York, New York 10031

1. Introduction

In recent years, the study of generalized Ramsey theory for graphs has grown to become a sizable and sprawling field. In the case of generalized Ramsey numbers and related problems, surveys have been made [4, 47, 48]. It seems timely, then, to survey another part of the field. In this paper we shall study what may be called "noncomplete Ramsey theory" in that the graphs to be colored are not, in general, complete. To clarify this, we must make some definitions. Generally, notation not defined here follows that of Harary [46], except that our graphs will have no isolates. In particular, we distinguish between subgraphs and induced subgraphs.

Definition. If F, $G_1, \ldots, G_k$ are graphs, write $F \to (G_1, \ldots, G_k)$ to mean that however the edges of F are colored with k colors, it must happen for some i, $1 \le i \le k$, that the subgraph of F with color i contains a copy of G_i.

The meaningfulness of this definition is assured by the finite version of the famous theorem of Ramsey [79]. For definiteness, we state this fact formally; the result is of course a trivial consequence of Ramsey's theorem.

Theorem 1.1. If $G_1, \ldots, G_k$ are any graphs, then there exists a graph F for which $F \to (G_1, \ldots, G_k)$.

The above relationship is often called the weak Ramsey arrow. Another kind of Ramsey arrow, often called the strong arrow, has received considerable attention lately. It is defined as follows.

Definition. If F, $G_1, \ldots, G_k$ are graphs, write $F \rightarrowtail (G_1, \ldots, G_k)$ to mean that however the edges of F are colored with k colors, it must happen for some i, $1 \le i \le k$, that the subgraph of F with color i contains a copy of G_i that is an induced subgraph of F.

To see that the strong arrow is meaningful in general is much more difficult than for the weak; but it has been done, settling a conjecture of Henson [51].

Theorem 1.2 [21, 29, 80, 81]. If $G_1, \ldots, G_k$ are any graphs, then there exists a graph F for which $F \rightarrowtail (G_1, \ldots, G_k)$.

Nešetřil and Rödl [69] give a simpler proof of Theorem 1.2. This theorem was actually first stated for $k = 2$; however, the extension to $k > 2$ follows immediately from the case $k = 2$. Indeed, many of the results of this survey were stated for the 2-color case only, when extension to the multicolor case is trivial. Here, these results will be stated in their general forms without comment. (We also warn the reader that we have freely changed many of the original notations in the works cited, also without comment.) However, when such a general statement cannot be made,

0077-8923/79/0328-0050$01.75/1

two points of notation are in order. The case $k = 2$ and $G_1 = G_2$ is very common. For this reason we write $F \to G$ for $F \to (G, G)$ and similarly for the strong arrow, calling this the *diagonal* case. In addition, the argument list $(G, \ldots, G)$, where k G's appear, will be written $(G; k)$. Thus $F \to G$ is equivalent to $F \to (G; 2)$.

The question of when $K_n \to (G_1, \ldots, G_k)$ is the same as the question of determining the (generalized) Ramsey number $r(G_1, \ldots, G_k)$; for this sort of question one can consult the surveys mentioned in the first paragraph. When one asks questions about F in general for which $F \to (G_1, \ldots, G_k)$ or $F \rightarrowtail (G_1, \ldots, G_k)$, it is appropriate to call the subject *noncomplete Ramsey theory*. It is with this subject that we will be concerned. Note that since we take graphs to be finite by definition, only the finite versions of certain results will be considered. Indeed, THEOREMS 1.1 and 1.2 have infinite counterparts.

Certain graphical parameters that have been studied in other contexts can be cast in terms of an arrow relation. Most prominent among these is the edge chromatic number $\chi_1(F)$, which may be defined as the smallest k such that $F \not\to (K_{1,2}; k)$. Fiorini and Wilson [36] have written an entire book about this parameter, and Fiorini [35] has given an extensive bibliography.

Another graphical parameter which more or less fits the mold is the (edge-) arboricity $\Upsilon(F)$. Let $\mathscr{C}$ denote the class of cycles. Then $\Upsilon(F)$ is the smallest k such that $F \not\to (\mathscr{C}; k)$. (Here we have extended the definition of the weak arrow in an obvious way, namely, that color i should induce some member of $\mathscr{C}$, for some i; we will use this sort of extended Ramsey arrow or number from time to time in what follows.) The famous formula of Nash-Williams [59] can therefore be thought of as a result in Ramsey theory. A further such parameter is the biparticity of F, which comes from taking $\mathscr{C}$ to be the class of odd cycles in the above and which was shown by Harary *et al.* [49] to be $\{\log_2 \chi(F)\}$, the least integer not less than $\log_2 \chi(F)$.

We shall not discuss parameters of this general type in this survey, except to point out the connection to arrow relations and to mention that Lesniak and Roberts [57] develop some of the theory of such parameters and apply the theory to questions about certain Ramsey numbers involving them. Chvátal and Harary [17] also relate Ramsey theory to other well-known parameters.

Section 2 discusses relatively direct analogs of the ordinary Ramsey numbers. Section 3 considers questions in which the minimality of the arrowing graphs comes into play. The final section discusses results, puts forth some open questions, and gives a few theorems that do not fit conveniently into the other sections. Generally, no proofs will be given, with the exception of a few simple new results which illuminate the exposition.

An attempt has been made to cover the subject at hand more or less as completely as possible within the limits set. Nevertheless, the spread of the field is so great that some significant results have probably been omitted inadvertently. An advance apology for any such omissions and any errors is hereby made; the author will be very glad to have any omissions or errors called to his attention. It should be noted that the strong arrow will be somewhat deemphasized. Nešetřil and Rödl [66] give a survey of problems and results involving the strong arrow relation and generalizations of it.

2. ANALOGS OF RAMSEY NUMBERS

In this section we consider results that involve more or less direct analogs of the ordinary (generalized) Ramsey numbers. In particular, many of the results involve studying

$$r_\phi(G_1, \ldots, G_k) = \min \{\phi(F)\colon F \to (G_1, \ldots, G_k)\}$$

or

$$r^*_\phi(G_1, \ldots, G_k) = \min \{\phi(F)\colon F \rightarrowtail (G_1, \ldots, G_k)\}$$

where ϕ is a graphical parameter satisfying $\phi(F_1) \le \phi(F_2)$ whenever $F_1 \subset F_2$. Thus $r_p(G_1, \ldots, G_k) = r(G_1, \ldots, G_k)$, where $p = p(F) = |V(F)|$. (We shall, of course, not consider r_p.) The condition on ϕ assures that r_ϕ is well behaved. Of course, r^*_ϕ need not be so well behaved, but at least $r^*_\phi \ge r_\phi$ in all cases. Benedict [2] has defined a similar, but more general concept.

One of the first results in this area shows that cl(F), the order of the largest clique of F, can be made as small as possible in the 2-color case for complete graphs. This result answers a question of Erdös and Hajnal [27].

THEOREM 2.1 [37]. $r_{\mathrm{cl}}(K_m, K_n) = \max(m, n)$.

Thus, there is an F for which $F \to K_3$ and $K_4 \not\subset F$. It would be of interest to have a reasonable upper bound on how small $p(F)$ can be made for such a graph. The upper bound implied by Folkman's proof is huge. If we relax the condition, reasonable bounds are known. We make a simplification to notation similar to that in Section 1 by writing $r_\phi(G)$ for $r_\phi(G, G)$, and similarly for r^*_ϕ.

THEOREM 2.2 [58, 54], $10 \le \min \{p(F)\colon F \to K_3, K_5 \not\subset F\} \le 18$.

The lower bound is due to Lin, the upper bound to Irving. Earlier Graham and Spencer [42] had found an upper bound of 23 and, even earlier, Pósa (unpublished) had shown the minimum to be finite. Relaxing the condition further, we have an exact result.

THEOREM 2.3 [39]. $\min \{p(F)\colon F \to K_3, K_6 \not\subset F\} = 8$.

The unique such F is $K_8 - C_5$. Note that this minimum equals $r(K_3) + 2$. Lin [58] proved similar results for $F \to (K_3, K_5)$, (K_4, K_4), and (K_3, K_3, K_3).

THEOREM 2.1 has been greatly generalized as follows; Henson [51] had earlier obtained partial results.

THEOREM 2.4 [63]. $r^*_{\mathrm{cl}}(G_1, \ldots, G_k) = \max_{1 \le i \le k} \mathrm{cl}(G_i)$.

Note that from this, $r_{\mathrm{cl}} = r^*_{\mathrm{cl}}$. Earlier, Nešetřil and Rödl [60] had proved THEOREM 2.4 for the special case $\mathrm{cl}(G_1) = 2$, and Erdös and Hajnal [28] had solved the case $k = 2$, $G_1 = G_2 = C_5$, answering a question of F. Galvin.

The above result is an example of what is often called the "Galvin-Ramsey property" of a class of graphs. Such a class $\mathcal{G}$ has this property if for any $G \in \mathcal{G}$, there is an $F \in \mathcal{G}$ for which $F \to \mathcal{G}$. Similarly, one may speak of the strong Galvin-Ramsey property. THEOREM 2.4 asserts the strong Galvin-Ramsey property for the class of graphs G with $\mathrm{cl(G)} \le t$, or for that matter, with $\mathrm{cl(G)} = t$.

Of course, THEOREMS 1.1 and 1.2 show that the class of all graphs has the Galvin-Ramsey property in the weak and strong senses, respectively. Erdös and Hajnal first discussed the Galvin-Ramsey property in [28], where they gave a certain class $\mathcal{G}$ with this property; however, this class is rather tedious to describe. F. Galvin had first posed the idea in a letter. Henson [51] showed that the class of (finite) comparability graphs has the strong, and hence the weak, Galvin-Ramsey property. (A comparability graph is one whose edges can be oriented to form the digraph of a partial ordering.) It is easily seen that if $\mathscr{G}$ has the Galvin-Ramsey property, it necessarily has the analogous property for any finite number of colors.

Nešetřil and Rödl have given another instance of the Galvin-Ramsey property. Write oc(G) for the length of the shortest odd cycle of G.

THEOREM 2.5 [68]. $r^*_{\text{oc}}(G_1, \ldots, G_k) = \min_{1 \le i \le k} \text{oc}(G_i)$.

Again, it immediately follows that $r_{\text{oc}} = r^*_{\text{oc}}$.

Another graphical parameter of interest is the chromatic number χ. Here it is not possible to give as final a result as THEOREM 2.4; nevertheless, a rather definitive theorem can be given. To state the result, we must make a definition. Define a homomorphism of G to be the result of any sequence of identifications of non-adjacent points of G, and define hom(G) to be the set of all homomorphisms of G.

THEOREM 2.6 [11]. $r_\chi(G_1, \ldots, G_k) = r(\text{hom}(G_1), \ldots, \text{hom}(G_k))$.

We note that Benedict [2] independently discovered some special cases of THEOREM 2.6.

One can raise the question of r^*_χ and of minimizing $\chi(F)$ and cl(F) simultaneously. The following result answers both questions.

THEOREM 2.7 [38, 72]. For any $G_1, \ldots, G_k$ there is an F for which $F \rightarrowtail (G_1, \ldots, G_k)$, $\text{cl}(F) = \max_{1 \le i \le k} \text{cl}(G_i)$, and $\chi(F) = r(\text{hom}(G_1), \ldots, \text{hom}(G_k))$.

Thus in particular $r^*_\chi(G_1, \ldots, G_k) = r_\chi(G_1, \ldots, G_k)$, since $r^*_\phi \ge r_\phi$ for any ϕ. Nešetřil and Rödl [72] have pointed out that THEOREMS 2.5 and 2.6 can also be combined in a similar way.

THEOREMS 2.6 and 2.7 have a number of corollaries. (We shall not always call attention to corollaries of THEOREM 2.7.) The following was actually used in the proof of THEOREM 2.6.

THEOREM 2.8. If $G_1, \ldots, G_k$ are bipartite, then $r_\chi(G_1, \ldots, G_k) = 2$; that is, there is a bipartite F for which $F \to (G_1, \ldots, G_k)$.

It is not clear who first stated the above as a Ramsey-theoretic result. It follows directly from an extremal result due to Erdös and Kövari *et al.* [56] connected to the famous problem of Zarankiewicz [85]. THEOREM 2.8 can also be proved along the lines of Ramsey's theorem for graphs. It is interesting that extremal results are insufficient to establish Ramsey's theorem, but that they are effective in the bipartite case. Note that, by THEOREM 2.7, THEOREM 2.8 also holds with r^* and $\rightarrowtail$ replaces the corresponding weak Ramsey symbols. Of course, THEOREM 2.8 shows that the class of bipartite graphs has the (weak or strong) Galvin-Ramsey property.

Another corollary of THEOREM 2.6 follows from the fact that $\text{hom}(K_n) = \{K_n\}$; the first proof used a different approach.

THEOREM 2.9 [58]. $r_\chi(K_{n_1}, \ldots, K_{n_k}) = r(K_{n_1}, \ldots, K_{n_k})$.

Benedict [2] discovered this result independently. Note that this result can be strengthened, in that each K_{n_i} can be replaced by any G_i satisfying $\text{cl}(G_i) = \chi(G_i) = n_i$. Also, THEOREM 2.7 can be used to get a stronger result; namely, that the clique number can be held to $\max_i n_i$. (Of course, there is no distinction between the weak and strong cases when the sought-after graphs are all complete.)

THEOREM 2.10 [11]. If $\chi(G) = 3$, then

$$r_\chi(G) = \begin{cases} 5, & \text{if } G \text{ is homomorphic to } C_5, \\ 6, & \text{otherwise.} \end{cases}$$

THEOREMS 2.8 and 2.10 are special cases of a general conjecture of [11]; namely, that, for any n, there is a graph G with $\chi(G) = n$ for which $r_\chi(G) = (n-1)^2 + 1$. In [11] it is shown that this conjecture also holds for $n = 4$. In any case, if $\chi(G) = n$, it is necessarily true that $r_\chi(G) \geq (n-1)^2 + 1$. Hence, if $n \geq 3$, and $\mathscr{G}_n$ is the class of graphs with chromatic number n, then $F \not\rightarrow G$ for any F, $G \in \mathscr{G}_n$, an anti-Galvin-Ramsey property. Of course, if $n = 2$, $\mathscr{G}_n$ has the Galvin-Ramsey property.

The next corollary involves Δ, the maximum degree of a graph; the original statement was less general.

THEOREM 2.11 [11]. $r_\Delta(K_{n_1}, \ldots, K_{n_k}) = r(K_{n_1}, \ldots, K_{n_k}) - 1$.

Setting $r = r_\chi(G_1, \ldots, G_k)$, one can ask how small n can be and have $K(n; r) \rightarrow (G_1, \ldots, G_k)$, where $K(n; r)$ is a complete r-partite graph with n points in each part. Benedict [2] has considered this question and the related one in which $K(n; r)$ is replaced by $K(n; s)$, $s > r$. Powers [74] has greatly extended Benedict's work. The results of [74] are very interesting but are too recent to be discussed here, having arrived when this survey was in proof. The following is also a natural problem, and should be easier: How small can n be and have $K(N, \ldots, N, n) \rightarrow (G_1, \ldots, G_k)$, where $K(N, \ldots, N, n)$ is a complete r-partite graph with $r - 1$ parts of size N, with N arbitrarily large? This problem has applications to generalized Ramsey numbers; in a paper by Erdös *et al.* [24] the connection and some relevant results appear implicitly without actually being developed. When the G_i are bipartite the answer is straightforward (see THEOREM 2.28).

Another function that has received considerable attention is r_q, where $q(G)$ is the number of edges of G. Erdös *et al.* [25] call this the "size Ramsey number," writing it as $\hat{r}$.

THEOREM 2.12 [25]. If $r = r(K_{n_1}, \ldots, K_{n_k})$, then $r_q(K_{n_1}, \ldots, K_{n_k}) = \binom{r}{2}$.

This result is, of course, a direct corollary of THEOREM 2.9, and hence is yet another corollary of THEOREM 2.6. This theorem was proved for $k = 2$ by Chvátal using another method (see [25]).

THEOREM 2.13 [8]. $r_q(mK_{1,k}, nK_{1,l}) = (m + n - 1)(k + l - 1)$.

The case $m = n = 1$ was treated in [25]. In [8], it is also shown that if $G \rightarrow (mK_{1,k}, nK_{1,l})$ and G has $r_q(mK_{1,k}, nK_{1,l})$ edges, then $G = (m + n - 1)K_{1,k+l-1}$, except when $k = l = 2$, in which case, any of the copies of $K_{1,3}$ may be replaced by copies of K_3.

THEOREM 2.14 [82]. If $n \geq 5$, then $r_q(K_{2,2}, K_{1,n}) = 4n$; if $m \geq 9$ and n is sufficiently large, then $r_q(K_{2,m}, K_{1,n}) = 4n + 2m - 4$.

In [82] it is also shown that $r_q(C_4) = 15$; the proof is rather tedious. This appears to be the only known case beyond THEOREM 2.12 for which $r_q = \binom{r}{2}$. Rousseau and Sheehan point out that $r(K_{2,m}, K_{1,n})$ is not known in most cases; this situation is a reversal of the general rule that r_q is more difficult to deal with than r.

Erdös *et al.* [25] consider several asymptotic questions, expressed here as THEOREMS 2.15–2.18. In particular, they call a sequence $\{G_n\}$ of graphs an o-sequence if

$$\lim_{n\to\infty} \frac{r_q(G_n)}{\binom{r(G_n)}{2}} = 0$$

THEOREM 2.15 [25]. Let G be a graph with v a point of G. Let G_n be formed by adjoining a $K_{1,n}$ centered at v. Then $\{G_n\}$ is an o-sequence if and only if G is bipartite.

Recall that $G + \bar{K}_n$ is formed from G by adjoining n points and joining each of the n points to each point of G by an edge.

THEOREM 2.16 [25]. The sequence $\{G + \bar{K}_n\}$ is an o-sequence if and only if G consists of isolated points.

THEOREM 2.17 [25]. Let G be a graph and let G_n be formed from G by adjoining a copy of $K_{1,n}$ centered at every point of G so that each star is disjoint from the others. (Note the difference from these G_n's and the G_n of THEOREM 2.15.) Then $\{G_n\}$ is an o-sequence if and only if G consists of isolated points.

Several upper and lower bounds are derived in [25] for r_q and r, relevant to the above three theorems. These bounds are rather technical and will not be given here. However, the following compendium of special cases may be of interest.

THEOREM 2.18 [25]. There exist absolute constants $a_1, \ldots, a_{10}$ such that the following hold for m fixed and n sufficiently large.

(1) Let $G = K_m$ and form G_n, as in THEOREM 2.17. Then $a_1 m^2 n^2 \le r_q(G_n) \le a_2 m^2 n^2$.
(2) $a_3 m^2 n^2 \le r_q(K_m + \bar{K}_n) \le a_4 2^{4m} n^2$.
(3) Let $G = K_m$ and form G_n as in THEOREM 2.16. Then $a_5 m^3 n^2 \le r_q(G_n) \le a_6 m^4 n^2$.
(4) $a_7 m 2^m n \le r_q(K_{m,n}) \le a_8 m^2 2^m n$.
(5) $a_9 n^2 2^{n/2} \le r_q(K_{n,n}) \le a_{10} n^3 2^n$.

Note that, except for (1), there is a significant gap in the bounds in each of the above. It would be of interest to close, or at least narrow, these gaps, especially those in (2) and (5).

Another question of interest is the following: If $F \to (mG, nH)$, how many disjoint copies of G (or H) must F contain? THEOREMS 2.19–2.23, taken from [8], give partial answers to this question. Recall that $\beta_0(H)$ is the point independence number of H.

THEOREM 2.19 [8]. If $F \to (mG, H)$, then $sG \subset F$, where $s = [(mp(G) + p(H) - \beta_0(H) - 1)/p(G)]$.

As a corollary, we have the following theorem.

THEOREM 2.20 [8]. Let $p(G) = k$, $p(H) = l$, $\beta_0(G) = i$, and $\beta_0(H) = j$ If $F \to (mG, nH)$, then

(1) $sG \subset F$, where $s = [(mk + nl - nj - 1)/k]$;
(2) $tH \subset F$, where $t = [(mk + nl - mi - 1)/l]$.

If G and H are fixed, then either (1) or (2) is within a bounded amount of being sharp for all m and n, depending on whether $mi \geq nj$ or $mi < nj$, respectively. This fact follows from a result of Burr *et al.* [12]:

$$r(mG, nH) \leq mk + nl - \min(mi, nj) + c,$$

where c depends only on G and H. For the same reason, the following is within a bounded amount of being sharp for fixed G and all m and n.

THEOREM 2.21 [8]. If $p(G) = k$, $\beta_0(G) = i$, and $m \geq n$, then $F \to (mG, nG)$ implies that F contains at least $[(mk + nk - ni - 1)/k]$ disjoint copies of G.

It is plausible to guess that if $F \to (mG, nH)$, then F contains at least $[r(mG, nH)/p(G)]$ disjoint copies of G. This is perhaps too much to hope for, but it does happen in the following two theorems. We note that, by results of [12], if $m \geq n$, then $r(mK_2, nK_2) = 2m + n - 1$, and $r(mK_3, nK_3) = 3m + 2n$, except $r(K_3) = 6$. Recall that $\beta_1(F)$ is the line-independence number.

THEOREM 2.22 [8]. If $m \geq n$, then $r_{\beta_1}(mK_2, nK_2) = [(2m + n - 1)/2]$.

This, of course, follows immediately from THEOREM 2.21, but the next result requires some additional argument. Define $t(G)$ to be the maximum number of disjoint triangles in G.

THEOREM 2.23 [8]. If $m \geq n$, then $r_t(mK_3, nK_3) = [(3m + 2n)/3]$, except that $r_t(K_3) = 2$.

We now turn to a topic suggested by THEOREM 2.8. If $G_1, \ldots, G_k$ are bipartite, for which m and n does $K_{m,n} \to (G_1, \ldots, G_k)$? Write $\mathrm{Br}(G_1, \ldots, G_k)$ for the set of all pairs (m, n) satisfying $m \leq n$, $K_{m,n} \to (G_1, \ldots, G_k)$, and neither $K_{m-1,n} \to (G_1, \ldots, G_k)$ nor $K_{m,n-1} \to (G_1, \ldots, G_k)$. Not too much is known about Br; but the two-color problem has been completely solved for paths. Note that P_n denotes a path on n points.

THEOREM 2.24 [44, 32].

$\mathrm{Br}(P_{2m}, P_{2n}) = \{(m + n - 1, m + n - 1)\}$;
$\mathrm{Br}(P_{2m+1}, P_{2n}) = \{(m + n - 1, m + n)\}$, for $m \geq n - 1$;
$\mathrm{Br}(P_{2m+1}, P_{2n}) = \{(m + n - 1, m + n - 1)\}$, for $m < n - 1$;
$\mathrm{Br}(P_{2m+1}, P_{2n+1}) = \{(m + n - 1, m + n)\}$, for $m \neq n$;
$\mathrm{Br}(P_{2m+1}, P_{2m+1}) = \{(2m - 1, 2m + 1)\}$.

Gyárfás and Lehel only gave a proof of an equivalent of the first of the above, stating the others without proof. Faudree and Schelp [33] give a sketch of their proof, which is given in full in [32].

A case that can be treated completely is that of stars.

THEOREM 2.25. $\mathrm{Br}(K_{1,n_1}, \ldots, K_{1,n_k}) = \{(1, n)\}$, where $n = \sum_{i=1}^{k} (n_i - 1) + 1$.

Proof: It is clear that $K_{1,n} \to (K_{1,n_1}, \ldots, K_{1,n_k})$. On the other hand, $K_{n-1,n-1} \not\to (K_{1,n_1}, \ldots, K_{1,n_k})$, since $K_{n-1,n-1}$ can obviously be factored into $(n - 1)$ 1-factors, and coloring $(n_i - 1)$ 1-factors with color i gives the desired coloring. □

In view of the above, it might seem possible that Br is always a singleton; but the next result lays this possibility to rest.

THEOREM 2.26 (S. Niven, see [43]). $\mathrm{Br}(C_4) = \{(3, 7), (5, 5)\}$.

Determining Br is likely to be tedious even in most simple cases, so it is attractive to study some related parameters. Rousseau and Sheehan [82] have studied $\min\{mn: (m, n) \in \mathrm{Br}(G, H)\}$, which they designate $\hat{r}_B(G, H)$. They prove that $\hat{r}_B$ takes on the same values as r_q in THEOREM 2.13; on the other hand, they prove that $\hat{r}_B(C_4) = 21 \neq 15 = r_q(C_4)$. Note that the left-hand equality is equivalent to part of THEOREM 2.26. In addition, they prove the following.

THEOREM 2.27 [82]. If l and m are fixed, with $l \leq m$, then if n is sufficiently large,

$$\hat{r}_B(K_{l,m}, K_{1,n}) = 2(l-1)(2n-1) + 2(m-1)\binom{2l-2}{l}.$$

Perhaps surprisingly, it does not appear that the quantity $\min\{m + n: (m, n) \in \mathrm{Br}(G, H)\}$ has been studied. On the other hand, $r_m(G_1, \ldots, G_k) = \min\{m: (m, n) \in \mathrm{Br}(G_1, \ldots, G_k)\}$ is easy to deal with. The following theorem is related to results of [24].

THEOREM 2.28. Let $G_1, \ldots, G_k$ be bipartite graphs with the minimum values of their smaller parts being $m_1, \ldots, m_k$, respectively. (That is, if G_i is disconnected, m_i is the sum of the smaller parts of its components.) Then

$$r_m(G_1, \ldots, G_k) = \sum_{i=1}^{k} (m_i - 1) + 1.$$

Proof: Denote the right-hand side of the above by m_0. To see that $K_{m-1,n} \nrightarrow (G_1, \ldots, G_k)$, for any n, choose $(m_i - 1)$ points of the $m_0 - 1$, for each $1 \leq i \leq k$, and color all points incident with them in color i. It is clear that with this coloring, no G_i occurs in color i.

To prove that $K_{m_0,n_0} \to (G_1, \ldots, G_k)$ for some n_0, it suffices to assume that, for each i, $G_i = K_{m_i,n}$, for some fixed n. We claim that the choice $n_0 = n \cdot k^{m_0}$ is large enough for the desired result. Denote the points of the smaller part of K_{m_0,n_0} by $v_1, \ldots, v_{m_0}$. A straightforward induction shows that if K_{m_0,n_0} is k-colored, then in it there must occur a $K_{m_0,n}$ with the property that, for each v_i, all edges incident with v_i are the same color. The value of m_0 now guarantees that, for some $1 \leq i \leq k$, $K_{m_i,n}$ is found in color i. □

In contrast to the above simple result, the value of $r_n = \min\{n: (m, n) \in \mathrm{Br}(G_1, \ldots, G_k)\}$ appears difficult to deal with. Beineke and Schwenk [1], and Irving [55] have (in effect) considered this problem in the two-color, diagonal case. In addition to the case of stars, which follows from THEOREM 2.25, THEOREMS 2.29 and 2.30 summarize their results.

THEOREM 2.29 [1].

(1) $r_n(K_{2,n}) \leq 4n - 3$, for all n.
(2) $r_n(K_{2,n}) \geq 4n - 4$, if there exists a Hadamard matrix of order $4n - 4$.
(3) $r_n(K_{2,n}) = 4n - 3$, if $n = 2$ or if n is odd and there exists a Hadamard matrix of order $4n - 4$.

THEOREM 2.30 [1, 55].

(1) $r_n(K_{3,3}) = 17$.

(2) $r_n(K_{3,n}) \leq 8n - 7$.

(3) $r_n(K_{3,n}) = 8n - 7$, if there exists a Hadamard matrix of order $4n - 4$, so that $r_n(K_{3,4}) = 25$, $r_n(K_{3,5}) = 33$.

The upper bound (2) is due to Irving, who also shows that $r_n(K_{4,4}) \leq 48$, disproving a conjecture of [1].

Irving [55] considers in addition, as have Erdös and Moon [30], Guy [43], and Chvátal [15], what are called "polarized partition relations." In a polarized relation, one requires not only that $K_{m,n} \rightarrow K_{k,l}$, but that in the monochromatic $K_{k,l}$, the k points of the "left-hand" part be a subset of the m points in the left-hand part of the $K_{m,n}$. Although this is perhaps a more natural relation, it is outside the scope of this survey. When $k = l$, their relation coincides with our arrow; the above authors derive certain inequalities and asymptotic results which could be specialized to the case $k = l$. We shall omit these results, referring the reader to the above papers.

The reader may have noticed that we have nowhere mentioned the function r_p^*, in spite of the importance of $r_p = r$. The reason for this is that nothing of substance seems to be known about it. Erdös [23] has asked the following interesting question: What is max $r_p^*(G)$, where G runs over all graphs with n points? Erdös suggests that the answer may be greater than $r(K_n)$. A good case to try would be $K_{n-1} \cdot K_2$, that is, a copy of K_{n-1} to which a free edge has been added. In particular, is $r_p^*(K_3 \cdot K_2) > 18$?

Similarly, one may ask for min $r_p^*(G)$, say for connected G or for arbitrary G, but in either case with n points. For connected G a very plausible candidate for the minimizing graph is $K_{1,n-1}$. It is easy to see from THEOREM 3.16 that $r_p^*(K_{1,n-1}) = 2n - 2$. The above extremal questions for r_p^* are analogous to those for $r_p = r$ considered in [7].

One may also ask, when does it happen that $r_p^*(G) = r(G)$? This is the case when $G = K_n$ or $G = K_{1,2n+1}$. On the other hand, $r_p^*(K_{1,2n}) = r(K_{1,2n}) + 1 = 4n$. It is also clear that $r_p^*(K_n \cdot K_2) > r(K_n \cdot K_2)$, for instance by THEOREM 2.9.

3. RAMSEY-MINIMAL GRAPHS

The results of this section all involve the concept of minimality. Say that F is $(G_1, \ldots, G_k)$-*minimal* if $F \rightarrow (G_1, \ldots, G_k)$, but $F' \nrightarrow (G_1, \ldots, G_k)$, for any proper subgraph F' of F. Say that F is $(G_1, \ldots, G_k)^*$-minimal if the above holds with strong arrows. Write $\mathscr{R}(G_1, \ldots, G_k)$ for the set of $(G_1, \ldots, G_k)$-minimal graphs, and $\mathscr{R}^*(G_1, \ldots, G_k)$ for the set of $(G_1, \ldots, G_k)^*$-minimal graphs. Most of the work on $\mathscr{R}$ and $\mathscr{R}^*$ has involved two colors. Of the work on two colors, some of it generalizes easily to multiple colors and some of it definitely does not.

A central question in Ramsey theory for noncomplete graphs, due to Nešetřil and Rödl (see [23]), at least in the strong case, is: For which $(G_1, \ldots, G_k)$ is $\mathscr{R}(G_1, \ldots, G_k)$ or $\mathscr{R}^*(G_1, \ldots, G_k)$ infinite? In the 2-color, weak case, we make the following conjecture.

CONJECTURE. The set $\mathscr{R}(G, H)$ is finite if and only if: (1) either G or H is a matching, or (2) G and H are both forests of the form $S \cup M$, where S is a star of odd degree and M is a (possibly empty) matching.

In the multiple-color, or strong, case it is not so clear what to guess, but something similar is likely to be true. (Nešetřil and Rödl [62, 66] have conjectured that every graph with at least two edges has infinitely many *vertex*-minimal strong Ramsey graphs.) THEOREMS 3.1–3.11 summarize what is known about the above conjecture.

THEOREM 3.1 [9]. If either G or H is a matching, then $\mathscr{R}(G, H)$ is finite.

The proof of the above theorem yields a (large) upper bound on the order of graphs in such $\mathscr{R}(G, H)$. THEOREM 3.1 generalizes a result of Burr *et al.* [11], in which G and H were both matchings.

THEOREM 3.2 If S_1 and S_2 are stars of odd degree, and M_1 and M_2 are (possibly empty) matchings, then $\mathscr{R}(S_1 \cup M_1, S_2 \cup M_2)$ is finite.

The remaining results concern cases in which $\mathscr{R}$ is finite.

THEOREM 3.3 [10]. If G and H are forests of stars containing no star of degree 1, and if besides they are not both single stars of odd degree, then $\mathscr{R}(G, H)$ is infinite.

THEOREM 3.4 [10]. If G is a forest of stars and H is a forest not consisting entirely of stars, then $\mathscr{R}(G, H)$ is infinite.

THEOREM 3.5 [70]. If G and H are forests, neither consisting entirely of stars, then $\mathscr{R}(G, H)$ is infinite.

The following definition will be useful: Write Γ_3 for the class of all 3-connected graphs, together with K_3.

THEOREM 3.6 [71]. If $G \in \Gamma_3$ and $H \in \Gamma_3$ then $\mathscr{R}(G, H)$ is infinite.

The above result was proved first by Burr *et al.* [11], when G and H were both complete; the proof in [11] has a gap in the off-diagonal case, an omission which is corrected in [13].

THEOREM 3.7 [71]. If $\chi(G) \geq 3$ and $\chi(H) \geq 3$, then $\mathscr{R}(G, H)$ is infinite.

THEOREM 3.8 [10]. Let $\mathscr{G}$ be the class of all graphs all of whose blocks are single edges or are in Γ_3, and which have at least one block in Γ_3. Then if $G \in \mathscr{G}$ and $H \in \mathscr{G}$, $\mathscr{R}(G, H)$ is infinite.

THEOREM 3.9 [10]. Let G be a graph formed by joining two or more graphs in Γ_3 by identifying an edge in each and let H be formed similarly. Then $\mathscr{R}(G, H)$ is infinite.

THEOREM 3.10 [10]. If $n \geq 2$ and G is 2-connected, then $\mathscr{R}(K_{1,n}, G)$ is infinite.

In addition to this, it has very recently been shown [34] that $\mathscr{R}(K_{1,2}, G)$ is infinite whenever G is not a matching.

THEOREM 3.11 [72]. If $n \geq 3$, then $\mathscr{R}(C_n)$ is infinite.

Quite a number of other results could be added to the above, but they would tend to be tedious to state, or rather narrow. Also, a number of the above results can be extended rather easily to the multicolor, or strong, case. For instance, THEOREMS 3.2 and 3.5 extend to the multicolor case, and THEOREMS 3.5 and 3.7 hold with $\mathscr{R}$ replaced by $\mathscr{R}^*$. THEOREMS 3.6, 3.8, and 3.9 can be pushed even further: Write Γ_2' for all graphs G which are 2-connected and have the further property that the removal of no two adjacent points disconnects the graph (so that any cycle is in Γ_2'). Then THEOREMS 3.6, 3.8, and 3.9 hold with $\mathscr{R}$ replaced by $\mathscr{R}^*$ and Γ_3 replaced by Γ_2'. Note that the condition is thus weaker in the strong Ramsey case than in the weak case. This apparent anomaly comes from the fact that $\mathscr{R}$ and $\mathscr{R}^*$ contain only minimal graphs.

Another problem of interest is whether $\mathscr{R}(G_1, \ldots, G_k)$ contains infinitely many graphs with a given property. This is the case, for instance, in some results of Section 2. Thus Nešetřil and Rödl [70] have shown that in THEOREM 2.4 there are infinitely many arrowing graphs achieving the minimum clique number. The same sort of result is shown in [72] for THEOREM 2.5. THEOREM 2.7 can also be extended to yield infinitely many arrowing graphs.

It is now convenient to extend the definitions r_ϕ and r_ϕ^* in Section 2. Set

$$r_\phi(G_1, \ldots, G_k) = \min_{F \in \mathscr{R}(G_1, \ldots, G_k)} \phi(F)$$

$$r_\phi^*(G_1, \ldots, G_k) = \min_{F \in \mathscr{R}^*(G_1, \ldots, G_k)} \phi(F)$$

It is clear that these definitions are consistent with the previous ones. Two parameters of graphs, considered by Burr *et al.* [11], are the minimum degree δ and the (point-) connectivity κ. Note that these parameters are uninteresting in the context of Section 2; it requires minimality to make these Ramsey questions meaningful. The following two theorems, as stated, did not consider the question of infinitude. The proofs of these theorems are incomplete in [11] when $m \neq n$; Burr *et al.* [13] fill in the gap.

THEOREM 3.12 [11]. If $m, n \geq 3$, then $r_\delta(K_m, K_n) = (m-1)(n-1)$; this value is achieved for infinitely many F.

THEOREM 3.13 [11]. If $m, n \geq 3$, then $r_\kappa(K_m, K_n) = 2$, if $m \neq n$, and $r_\kappa(K_m, K_n) = 3$, if $m = n$; in either case this value is achieved for infinitely many F.

Of course, in these two theorems r and r^* are equivalent. In THEOREM 3.13, K_m and K_n can be replaced by arbitrary 4-connected graphs G and H with the condition $m \neq n$ replaced by $G \neq H$ (see [14]). No such convenient generalization applies to THEOREM 3.12, although something could be said in that direction. However, at present no way is known to extend these two theorems to multiple colors.

It is clear that if $G_1, \ldots, G_k$ are connected, then so are all members of $\mathscr{R}(G_1, \ldots, G_k)$ and $\mathscr{R}^*(G_1, \ldots, G_k)$. Suppose, on the contrary, that $G_1, \ldots, G_k$ are all disconnected. Now are all members of $\mathscr{R}$ or $\mathscr{R}^*$ disconnected? The following theorem shows that this need not be the case.

THEOREM 3.14 [13]. If $m, n \geq 1$ and $s, t \geq 3$, then there exist infinitely many graphs in $\mathscr{R}(mK_s, nK_t)$ that are connected.

(The reader is warned that the meanings of $\mathscr{R}$ and $\mathscr{R}^*$ in [13] are different from their meanings here.) The above theorem can be extended to the strong case, as well as that in which K_s and K_t are replaced by arbitrary 3-connected graphs. The same is true of the following theorem, which partially settles a conjecture of Nešetřil and Rödl [71].

THEOREM 3.15 [13]. Let $s, t \geq 3$ and suppose $F \not\rightarrow (K_s, K_t)$. Then F is an induced subgraph of infinitely many members of $\mathscr{R}(K_s, K_t)$.

In this result, it is possible to replace K_s and K_t by arbitrary 3-connected graphs [14].

The reader will have noticed by now that we have seen many facts about $\mathscr{R}(G_1, \ldots, G_k)$, but no complete characterizations, except for THEOREM 3.2. Indeed, not many have been given, nor are many likely, as will be seen. A little, however is known, as shown by THEOREMS 3.16–3.20. The next theorem partly overlaps THEOREM 3.2.

THEOREM 3.16 (U. S. R. Murty, see [11]). When n is odd, $\mathscr{R}(K_{1,n}) = \{K_{1,2n-1}\}$; when n is even, $\mathscr{R}(K_{1,n})$ consists of $K_{1,2n-1}$, together with all connected graphs which are regular of degree $2n-2$ and have an odd number of points.

It is even possible to deal with odd stars in the multicolor case, generalizing THEOREM 3.2.

THEOREM 3.17. If n_i, $i = 1, \ldots, k$, are all odd, then

$$\mathscr{R}(K_{1,n_1}, \ldots, K_{1,n_k}) = \{K_{1,m}\},$$

where

$$m = \sum_{i=1}^{k} (n_i - 1) + 1.$$

Proof: Clearly $K_{1,m} \rightarrow (K_{1,n_1}, \ldots, K_{1,n_k})$. On the other hand, suppose $G \in \mathscr{R}(K_{1,n_1}, \ldots, K_{1,n_k})$, $G \neq K_{1,m}$. Then $\Delta(G) \leq m - 1$, and hence G is contained in some $(m-1)$-regular graph G'. Clearly $G' \rightarrow (K_{1,n_1}, \ldots, K_{1,n_k})$; but, by a result of Petersen [73]. G' is factorable into $(m-1)/2$ 2-factors, since $m - 1$ is even. This factorization easily yields a coloring in which color i has maximum degree $n_i - 1$. Therefore G', and hence G, does not arrow $(K_{1,n_1}, \ldots, K_{1,n_k})$. □

Whenever $\mathscr{R}(G, H)$ has been proved constructively to be finite, as in THEOREM 3.1, $\mathscr{R}(G, H)$ can of course be determined in principle. In practice, to do so is difficult, and only the following two nontrivial results are known involving matchings.

THEOREM 3.18 [11]. $\mathscr{R}(2K_2) = \{C_5, 3K_2\}$.

THEOREM 3.19 [9]. $\mathscr{R}(2K_2, K_3)$ consists of three graphs; namely, those in FIGURE 1.

The complexity of graph (3) in the figure suggests that any systematic treatment of $\mathscr{R}(nK_2, G)$ is unlikely. (In [9] it is also shown that $\mathscr{R}(2K_2, K_n)$ and $\mathscr{R}(2K_2, K_{1,n})$ contain quite a number of graphs.) One might hope at least for a polynomial-bounded algorithm to determine, given F, n, and G, whether $F \rightarrow (nK_2, G)$, or at least whether $F \in \mathscr{R}(nK_2, G)$. It seems likely that none exist, although clearly THEOREM 3.1 shows that such algorithms exist when n and G are fixed. However, for stars the situation is different.

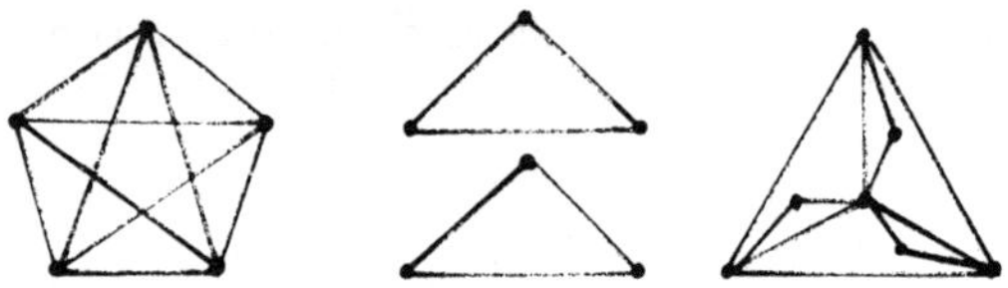

(1) (2) (3)

FIGURE 1. $\mathscr{R}(2K_2, K_3)$.

THEOREM 3.20 [5]. There is an algorithm which determines, for any F, m, and n, whether $F \to (K_{1,m}, K_{1,n})$ in time bounded by a polynomial in $p(F)$, m, and n.

The relationship of this theorem to the theory of c-matchings is clear; it was J. A. Edmonds who pointed out how the problem could actually be reduced to one of c-matching. However, in general NP-completeness seems to be the rule, as the next result suggests.

THEOREM 3.21 [5]. Let G be a fixed graph in Γ_3. Then the problem of determining, for arbitrary F, whether $F \in \mathscr{R}(G, H)$, is NP-complete.

Note that, consequently, under the same conditions the problem of determining whether $F \to (G, H)$ is also NP-complete. Similar results apply to the strong Ramsey case. NP-completeness is likely to be even more prevalent in the multicolor case, possibly even for as simple a Ramsey-type parameter as the edge-chromatic number. (At present, however, the edge-chromatic number is not known to be an NP-complete property.)

A result of [5] which is related to THEOREM 3.21 is that with G and H as above, the problem for infinite graphs F is undecidable, even for F which are essentially periodic.

A question of considerable interest is that of when $\mathscr{R}(G_1, \ldots, G_k)$ contains a complete graph. In other words, when is it true that $K_r - e \not\to (G_1, \ldots, G_k)$, for $r = r(G_1, \ldots, G_k)$? This is the case when the G_i's are all complete, by THEOREM 2.9, and when $k = 2$, $G_1 = G_2 = K_{1,2n}$, by THEOREM 3.16. On the other hand, it is very much not the case when $k = 2$, $G_1 = G_2 = K_{1,2n+1}$, also by THEOREM 3.16. Probably it is usually true that $\mathscr{R}$ does not contain a complete graph.

The above question is of significance for the problem of Ramsey multiplicities. Harary and Prins [50] have shown that when $G = K_2$ or $G = K_{1,2n}$ and $r = r(G)$, there exists a 2-coloring of K_r with just one monochromatic copy of G. This can be expressed by saying that these G's have Ramsey multiplicity 1. They conjecture that these graphs are the only such. It is obvious that any G with Ramsey multiplicity 1 has the property that $K_r - e \not\to G$. Therefore, information about one question could help the study of the other.

4. CONCLUSION

In this final section we shall discuss some promising areas for further study. Various open problems have been mentioned in previous sections, and many others are apparent from the incompleteness of certain theorems; we do not discuss these here.

One area that merits attention is the following, which is in some sense "dual" to the types of problems of earlier sections. Let ϕ be a graphical parameter satisfying $\phi(F_1) \leq \phi(F_2)$, whenever $F_1 \subset F_2$, and let $(G_1, \ldots, G_k)$ be graphs, or perhaps classes of graphs. Then does there exist an n_0 (and what is the smallest such) with the property that $\phi(F) \geq n_0$ implies that $F \to (G_1, \ldots, G_k)$?

One very simple result in this area is the following.

THEOREM 4.1. Let T_i, $1 \leq i \leq k$, denote some tree on n_i points, and set $n = \prod_{i=1}^{k} (n_i - 1) + 1$. Then if $\chi(F) \geq n$, $F \to (T_1, \ldots, T_k)$.

Proof: Let $\mathscr{K}_m$ denote the class of graphs with chromatic number m. Then it is straightforward that $F \to (\mathscr{K}_{n_1}, \ldots, \mathscr{K}_{n_k})$, if $\chi(F) \geq n$. (Actually the converse is true as well [6].) From this it is immediate that $F \to (T_1, \ldots, T_k)$. □

THEOREM 4.1 certainly does not give the best possible result, for instance when the T_i's are stars. It seems probable that in fact n can be taken to be linear in the n_i in all cases. Another theorem of somewhat similar character is the following.

THEOREM 4.2 [3]. Let $p(F) \geq k(s-1) + s + 1$ and $\delta(F) \geq k(s-1) + 1$. Then $F \to (sK_2; k)$.

A result of Cockayne and Lorimer [18] shows that $r(sK_2, \ldots, sK_2) = k(s-1) + s + 1$, so all F, as above, satisfying $p(F) = k(s-1) + s + 1$ are vertex-minimal, but not generally minimal, for the arrowing property.

The following result is somewhat analogous to the two preceding results, but in this case Ramsey properties are made to depend on each other. The work cited was presented without reference to Ramsey theory.

THEOREM 4.3 [52, 53]. If m and k are both odd then $G \to (K_{1,m}; k-1)$ if and only if $G \to (K_{1,k}; m-1)$. If m is odd and k is even or if m and k are both even and $m > k$, then $G \to (K_{1,m}; k-1)$ implies that $G \to (K_{1,k}; m-1)$; the reverse implication is false.

Note that the first part of the above follows from THEOREM 3.17.

Another area of interest is that of considering other kinds of arrow relations. A "semistrong" arrow may be defined as follows: Write $F \hookrightarrow G$ if for any 2-coloring of F, there must occur a monochromatic G which is induced in its color class. (For convenience we consider the 2-color, diagonal case for the rest of the section.) This sort of arrow was considered by Erdös *et al.* [29] in the infinite case, but no work on the finite case appears to have been done. (Of course, many results follow immediately from what is known about the weak and strong cases.)

One may also consider an "extra strong" arrow. Write $F \twoheadrightarrow G$ if $F \to G$ and every occurrence of G as a subgraph of F is induced. An important question is: For which G does there exist an F for which $F \twoheadrightarrow G$? When G is complete, all four arrows coincide, and there is nothing to prove. The answer is also in the affirmative when G is an odd cycle (by THEOREM 2.5) or when G is a complete bipartite graph (by THEOREM 2.8); some other cases, such as matchings, come to mind. On the other hand, if G is a triangle with one extra point and one extra edge adjoined, it is trivial that there is no F for which $F \twoheadrightarrow G$, since any F for which $F \to G$ must have two triangles sharing an edge. For this problem it would be very valuable even to have a plausible conjecture.

Note that it is not strictly fair to consider $\twoheadrightarrow$ to be another arrow relation, since it involves in part a local condition on F that is independent of any questions concerning colorings of F. However, the fact that $\twoheadrightarrow$ implies all the other arrows seems to justify such terminology. Many other local questions can be asked.

A very general question of this sort is: If G is fixed, what is the maximal class of graphs $\mathscr{F}$ with the property that if $F \to G$, then $F' \subset F$ for every $F' \in \mathscr{F}$? Note that $\mathscr{F}$ is finite, since $F' \in \mathscr{F}$ implies $p(F') \leq r(G)$. Even though $\mathscr{F}$ is finite, determining it may nevertheless be very difficult. The following result bears on this question, although its significance goes well beyond that one problem. The theorem settles a conjecture of Erdös, Nešetřil, and Rödl (see [23]).

THEOREM 4.4 [66]. For any $n \geq 3$, $k \geq 2$, there exists an F such that $F \to (K_n; k)$, but such that no two copies of K_n in F share more than two points.

We observe that this is a strengthening of THEOREM 2.1. When $k \geq 3$, Nešetřil and Rödl have an even stronger result. Note that when $n = 3$, THEOREM 4.4 is trivial; nevertheless, the $\mathscr{F}$ corresponding to K_3 is unknown. It seems that to determine it may not be too hard, in view of THEOREM 2.1. We note that results in this survey bear on many other local questions.

We close with one final type of question. Does there exist a labelled graph F such that $F \not\to K_3$, but such that there exists only one 2-coloring of the edges of F (up to exchange of colors) without a monochromatic K_3? If F is to be unlabelled, K_5 works, but are there any other F's which work? One can ask this sort of question for any graph G in place of K_3; for instance, if $G = K_4$, nothing is known. Oddly enough, if $G = K_3$ or G is 3-connected, it is possible to construct an infinite labelled graph F which does have the desired property.

For work related to, but beyond the scope of this paper, the reader may consult a forthcoming book by Graham *et al.* [41], as well as [16, 19—22, 26, 31, 40, 43, 45, 61, 62, 64—67, 75—78, 83, 84].

SUMMARY

If F, $G_1, \ldots, G_k$ are graphs, write $F \to (G_1, \ldots, G_k)$ to mean that however the edges of F are colored with k colors, it must happen, for some i, that color i contains a copy of G_i as a subgraph. We also consider a strong arrow in which G_i must be an induced subgraph of F. Such arrow relations have been much studied in recent years. In this survey we consider the case in which F is in general not complete. This case divides roughly into two types of problems: the ones more or less analogous to those of Ramsey numbers and those involving minimality.

ACKNOWLEDGMENTS

A number of people made valuable contributions to this survey by reading and commenting on early drafts of the manuscript, and in other ways. At the risk of leaving someone out, I would like to thank J. M. Benedict, P. Erdös, R. J. Faudree, F. Galvin, R. L. Graham, F. Harary, L. Lesniak-Foster, S. Lin, J. Nešetřil, V. Rödl, C. C. Rousseau, R. H. Schelp, J. Sheehan, and J. H. Spencer for their help. (Naturally, any errors are my responsibility.)

References

1. Beineke, L. W. & A. J. Schwenk. 1975. On a bipartite form of the Ramsey problem. Proc. Fifth British Combinatorial Conf., Aberdeen. Utilitas Math. Winnipeg, Manitoba. pp. 17–22.
2. Benedict, J. M. 1976. On Ramsey numbers defined by factorizations of regular complete multi-partite graphs. Western Michigan University. Doctoral dissertation.
3. Bondy, J. A. & V. Chvátal. 1976. A method in graph theory. Discrete Math. **15:** 111–135.
4. Burr, S. A. 1974. Generalized Ramsey theory for graphs—a survey. *In* Graphs and Combinatorics. Springer, Berlin, Federal Republic of Germany. pp. 52–75.
5. Burr, S. A. On the computational complexity of Ramsey-type problems. To appear.
6. Burr, S. A. A Ramsey-theoretic result involving chromatic numbers. Graph Theory, to appear.
7. Burr, S. A. & P. Erdös. 1976. Extremal Ramsey theory for graphs. Utilitas Math. **9:** 247–258.
8. Burr, S. A., P. Erdös, R. J. Faudree, C. C. Rousseau & R. H. Schelp. 1978. Ramsey-minimal graphs for multiple copies. Nederl. Akad. Wetensch. Proc. Ser. A **81:** 187–195.
9. Burr, S. A., P. Erdös, R. J. Faudree & R. H. Schelp. 1978. A class of Ramsey-finite graphs. Proc. Ninth Southeastern Conf. Combinatorics, Graph Theory, Comput. Utilitas Math., Winnipeg, Manitoba. pp. 171–180.
10. Burr, S. A., P. Erdös, R. J. Faudree & R. H. Schelp. To appear.
11. Burr, S. A., P. Erdös & L. Lovász. 1976. On graphs of Ramsey type. Ars Combinatoria **1:** 167–190.
12. Burr, S. A., P. Erdös & J. H. Spencer. 1975. Ramsey theorems for multiple copies of graphs. Trans. Am. Math. Soc. **209:** 87–99.
13. Burr, S. A., R. J. Faudree & R. H. Schelp. 1977. On Ramsey-minimal graphs. Proc. Eighth Southeastern Conf. Combinatorics, Graph Theory, Comput. Utilitas Math., Winnipeg, Manitoba, Canada. pp. 115–124.
14. Burr, S. A., J. Nešetřil & V. Rödl. To appear.
15. Chvátal, V. 1969. On finite polarized partition relations. Can. Math. Bull. **12:** 321–326.
16. Chvátal, V. 1972. Monochromatic paths in edge-colored graphs. J. Comb. Theory (B). **13:** 69–70.
17. Chvátal, V. & F. Harary. 1973. Generalized Ramsey theory for graphs I. diagonal numbers. Period. Math. Hung. **3:** 113–122.
18. Cockayne, E. J. & P. J. Lorimer. 1975. The Ramsey graph number for stripes. J. Aust. Math. Soc. (A) **19:** 252–256.
19. Deuber, W. 1973. Partitionen und lineare Gleichungssysteme. Math. Z. **133:** 109–123.
20. Deuber, W. 1975. A generalization of Ramsey's theorem for regular trees. J. Comb. Theory (B). **18:** 18–23.
21. Deuber, W. 1975. Generalizations of Ramsey's theorem. *In* Infinite and Finite Sets. Colloq. Math. Soc. János Bolyai 10. North-Holland, Amsterdam, the Netherlands. pp. 323–332.
22. Deuber, W. 1975. Partitionstheoreme für Graphen. Comment. Math. Helvetici. **50:** 311–320.
23. Erdös, P. 1975. Problems and results on finite and infinite graphs. Recent Adv. Graph Theory. Academia, Prague, Czechoslovakia. pp. 183–192.
24. Erdös, P., R. J. Faudree, C. C. Rousseau & R. H. Schelp. 1976. Generalized Ramsey theory for multiple colors. J. Comb. Theory (B). **20:** 250–264.
25. Erdös, P., R. J. Faudree, C. C. Rousseau & R. H. Schelp. 1978. The size Ramsey number. Period. Math. Hung. **9:** 145–161.
26. Erdös, P. & A. Hajnal. 1967. On decomposition of graphs. Acta Math. Acad. Sci. Hung. **18:** 359–377.
27. Erdös, P. & A. Hajnal. 1967. Research Problem 2-5. J. Comb. Theory **2:** 104.
28. Erdös, P. & A. Hajnal. 1970. Problems and results in finite and infinite combinatorial analysis. Ann. N.Y. Acad. Sci. **175(1):** 115–124.
29. Erdös, P., A. Hajnal & L. Pósa. 1975. Strong embeddings of graphs into colored graphs. *In* Infinite and Finite Sets. Colloq. Math. Soc. János Bolyai 10. North-Holland, Amsterdam, the Netherlands. pp. 585–595.

30. ERDÖS, P. & J. W. MOON. 1964. On subgraphs of the complete bipartite graph. Can. Math. Bull. **7:** 35–39.
31. ERDÖS, P. & R. RADO. 1956. A partition calculus in set theory. Bull. Am. Math. Soc. **62:** 427–489.
32. FAUDREE, R. J. & R. H. SCHELP. 1975. Path-path Ramsey-type numbers for the complete bipartite graph. J. Comb. Theory (B). **19:** 161–173.
33. FAUDREE, R. J. & R. H. SCHELP. 1975. Ramsey type results. *In* Infinite and Finite Sets. Colloq. Math. Soc. János Bolyai 10. North-Holland, Amsterdam, the Netherlands. pp. 657–665.
34. FAUDREE, R. J. & R. H. SCHELP. Personal communication.
35. FIORINI, S. 1978. A bibliographic survey of edge-colorings. J. Graph Theory. **2:** 93–106.
36. FIORINI, S. & R. J. WILSON. 1977. Edge-Colorings of Graphs. Pitman, London.
37. FOLKMAN, J. 1970. Graphs with monochromatic complete subgraphs in every edge coloring. SIAM J. Appl. Math. **18:** 19–24.
38. GALVIN, F. Personal communication.
39. GRAHAM, R. L. 1968. On edgewise 2-colored graphs with monochromatic triangles and containing no complete hexagon. J. Comb. Theory. **4:** 300.
40. GRAHAM, R. L. & B. L. ROTHSCHILD. 1974. Some recent developments in Ramsey theory. Math. Centre Tracts 56. Combinatorics. Mathematisch Centrum. Amsterdam, the Netherlands. pp. 61–76.
41. GRAHAM, R. L., B. L. ROTHSCHILD & J. H. SPENCER. Finite Ramsey theorems. To appear.
42. GRAHAM, R. L. & J. H. SPENCER. 1971. On small graphs with forced monochromatic triangles. *In* Recent Trends in Graph Theory. Springer, Berlin, Federal Republic of Germany. pp. 137–141.
43. GUY, R. K. 1969. A many-facetted problem of Zarankiewicz. *In* The Many Facets of Graph Theory. Proc. Conf. Western Michigan University, Kalamazoo, Michigan, 1968. Springer, New York, N.Y. pp. 129–148.
44. GYÁRFÁS, A. & J. LEHEL. 1973. A Ramsey-type problem in directed and bipartite graphs. Period. Math. Hung. **3:** 299–304.
45. HALES, A. W. & R. I. JEWETT. 1963. Regularity and positional games. Trans. Am. Math. Soc. **106:** 222–229.
46. HARARY, F. 1969. Graph Theory. Addison-Wesley, Reading, Mass.
47. HARARY, F. 1974. A survey of generalized Ramsey theory. *In* Graphs and Combinatorics. Springer, Berlin, Federal Republic of Germany. pp. 10–17.
48. HARARY, F. 1975. The foremost open problems in generalized Ramsey theory. Proc. Fifth British Combinatorial Conf., Aberdeen. Utilitas Math., Winnipeg, Manitoba, Canada. pp. 269–282.
49. HARARY, F., D. HSU & Z. MILLER. 1977. The biparticity of a graph. J. Graph Theory. **1:** 131–133.
50. HARARY, F. & G. PRINS. 1974. Generalized Ramsey theory for graphs IV. the Ramsey multiplicity of a graph. Networks. **4:** 163–173.
51. HENSON, C. W. 1973. Edge partition properties of graphs. Can. J. Math. **25:** 603–610.
52. HILTON, A. J. W. & R. PRICE JONES. 1978. Further remarks on partitioning the edges of a graph. J. Comb. Theory (B). **24:** 249–250.
53. HILTON, A. J. W. & R. PRICE JONES. 1978. Partitioning the edges of a graph. J. Comb. Theory (B). **25:** 290–294.
54. IRVING, R. W. 1973. On a bound of Graham and Spencer for a graph-coloring constant, J. Comb. Theory (B). **15:** 200–203.
55. IRVING, R. W. 1978. A bipartite Ramsey problem and the Zarankiewicz numbers. Glasgow Math. J. **19:** 13–26.
56. KÖVÁRI, T., V. SÓS & P. TURÁN. 1954. On a problem of K. Zarankiewicz. Colloq. Math. **3:** 50–57.
57. LESNIAK, L. & J. A. ROBERTS. 1977. On Ramsey theory and graphical parameters. Pac. J. Math. **68:** 105–114.
58. LIN, S. 1972. On Ramsey numbers and K_r-coloring of graphs. J. Comb. Theory (B). **12:** 82–92.
59. NASH-WILLIAMS, C. ST. J. A. 1961. Edge-disjoint spanning trees of finite graphs. J. London Math. Soc. **36:** 445–450.

60. NEŠETŘIL, J. & V. RÖDL. 1975. A Ramsey graph without triangles exists for any graph without triangles. *In* Infinite and Finite Sets. Colloq. Math. Soc. János Bolyai 10. North-Holland, Amsterdam, the Netherlands. pp. 1127–1132.
61. NEŠETŘIL, J. & V. RÖDL. 1975. Type theory of partition properties of graphs. Recent Adv. Graph Theory. Academia, Prague, Czechoslovakia. pp. 405–412.
62. NEŠETŘIL, J. & V. RÖDL. 1976. Partitions of vertices. Comment. Math. Charles University **17:** 85–95.
63. NEŠETŘIL, J. & V. RÖDL. 1976. The Ramsey property for graphs with forbidden complete subgraphs. J. Comb. Theory (B). **20:** 243–249.
64. NEŠETŘIL, J. & V. RÖDL. 1977. Partitions of relational and set systems. J. Comb. Theory (A). **22:** 289–312.
65. NEŠETŘIL, J. & V. RÖDL. 1977. A structural generalization of the Ramsey theorem. Bull. Am. Math. Soc. **83:** 127–128.
66. NEŠETŘIL, J. & V. RÖDL. 1978. Partition (Ramsey) theory—A survey. *In* Combinatorics. Colloq. Math. Soc. János Bolyai 18. North-Holland, Amsterdam, the Netherlands. pp. 759–792.
67. NEŠETŘIL, J. & V. RÖDL. On sets of integers with the Schur property. Acta Arith. To appear.
68. NEŠETŘIL, J. & V. RÖDL. 1978. Ramsey property of graphs without short odd cycles. Math. Slovaca. To appear.
69. NEŠETŘIL, J. & V. RÖDL. 1978. A simple proof of the Galvin-Ramsey property of the class of all finite graphs and a dimension of a graph. Discrete Math. **23:** 49–55.
70. NEŠETŘIL, J. & V. RÖDL. 1978. On Ramsey minimal graphs. Colloq. Int. C.N.R.S. 260, Problèmes Combinatoires et Theorîe des Graphes. pp. 307–308.
71. NEŠETŘIL, J. & V. RÖDL. The structure of critical Ramsey graphs. Acta Math. Acad. Sci. Hung. To appear.
72. NEŠETŘIL, J. & V. RÖDL. Personal communication.
73. PETERSEN, J. 1891. Die Theorie der regulären Graphen. Acta Math. **15:** 193–220.
74. POWERS, G. Personal communication.
75. RADO, R. 1933. Studien zur Kombinatorik. Math Z. **36:** 424–480.
76. RADO, R. 1943. Note on combinatorial analysis. Proc. London Math. Soc. **48:** 122–160.
77. RADO, R. 1970. Some partition theorems. *In* Combinatorial Theory and Its Applications. Colloq. Math. Soc. János Bolyai 4. North-Holland, Amsterdam, the Netherlands. pp. 929–936.
78. RADO, R. 1970. Theorems on the colouring of the edges of a graph. Proc. Second Chapel Hill Conf. Combinatorial Math. Appl. University of North Carolina. pp. 385–390.
79. RAMSEY, F. P. 1930. On a problem of formal logic. Proc. London Math. Soc. **30:** 264–286.
80. RÖDL, V. 1973. Dimension of a graph and a generalization of Ramsey's theorem (Czech). Charles University. Doctoral dissertation.
81. RÖDL, V. 1976. A generalization of Ramsey's theorem. *In* Graphs, Hypergraphs and Block Systems. Zielona Gora. pp. 211–220.
82. ROUSSEAU, C. C. & J. SHEEHAN. Size Ramsey numbers for certain bipartite graphs. To appear.
83. SPENCER, J. H. 1975. Restricted Ramsey configurations. J. Comb. Theory (A). **19:** 278–286.
84. TOMOVÁ, E. 1975. On the decomposition of complete bipartite graphs into factors with given diameters. Recent Adv. Graph Theory. Academia, Prague, Czechoslovakia. pp. 507–509.
85. ZARANKIEWICZ, K. 1951. Problem P101. Colloq. Math. **2:** 301.

ON EXTREMAL GRAPHS WITH GIVEN DIAMETER AND CONNECTIVITY

Louis Caccetta

Department of Combinatorics and Optimization
University of Waterloo
Waterloo, Ontario, Canada N2L 3G1

INTRODUCTION

A communication network is any complex of centers receiving and transmitting information, together with a set of links which convey information. In any network, the method of construction is usually dictated by cost, efficiency, and reliability considerations. For efficiency and reliability, a natural condition to impose is that information can be sent from any center to any other center by using at most k links, and, even if s centers (links) are out of order or engaged, information can be sent from any center to any other center by using at most λ links. It is natural to require that $\lambda \geq k$, since one cannot expect the efficiency of the network to increase when there is a breakdown. We would like to build up the links in such a way that the cost would be minimal, subject to the requirements.

Before formulating this problem as a graph theoretical problem, we introduce some terminology. All graphs considered in this paper are finite, undirected, and without loops and multiple edges. Notions not defined here can be found in Bondy and Murty [10]. Let G be a graph with vertex set $V(G)$ and edge set $E(G)$. The *distance* $d(x, y)$ between two vertices x and y is defined as the length of the shortest (x, y)-path in G; if there is no path connecting x and y we define $d(x, y)$ to be infinite. The *diameter* $D(G)$ of the graph G is defined as

$$D(G) = \operatorname*{Max}_{x,y \in V(G)} d(x, y)$$

Thus a connected graph has a finite diameter. The *degree* $d(x)$ of a vertex x of G is defined as the number of edges incident to that vertex.

If, in the communication network problem, the cost of connecting any two centers is constant, then the problem of constructing the minimal-cost network reduces to the following graph theoretical problem: *determine the least number of edges for a graph on n vertices and diameter at most k having the property that, after deleting any s or fewer vertices (edges), s < n, the resulting subgraph (partial graph) has diameter at most λ, λ ≥ k, and find those graphs with the minimum number of edges.* It is this formulation which is due to Vijayan and Murty [25] that has received attention in the literature.

The first case to be studied was the case $k = \lambda = 2$, that is, when information can be sent from any center to any other center using at most two links, even if any s centers are engaged or out of order. This case was studied by Vijayan and Murty [25]. We now discuss this particular case fully as it is the simplest one to describe, and also it illustrates some important aspects of the problem.

Let G be a graph with minimum number of edges satisfying the following property: G has diameter 2 and remains of diameter 2 after the deletion of any s, or fewer vertices, $s < n$. Clearly

(i) any two nonadjacent vertices of G must be connected to at least $s + 1$ common vertices, and

0077-8923/79/0328-0076$1.75/1

(ii) every vertex of G has degree at least $s + 1$, and if a vertex x, say, of G has degree $s + 1$, then every vertex of $G - x$ that is not adjacent to x must be adjacent to each of the $s + 1$ vertices of G that are adjacent to x.

The complete bipartite graph $K(s + 1, n - s - 1)$ with $s + 1$ vertices in one set and $n - s - 1$ vertices in the other, satisfies the diameter constraints provided $n \geq 2s + 2$. In fact, if G has a vertex of degree $s + 1$ then, by virtue of (ii), above, $G = K(s + 1, n - s - 1)$. It turns out that G has a vertex of degree $s + 1$, a fact which we now prove for $n > (2 + \sqrt{2})(s + 1)$.

Let m denote the minimum degree in G and suppose $m > s + 1$. Then the number of edges $e(G)$ in G satisfies

$$e(G) \geq \frac{mn}{2} \tag{1}$$

and

$$e(G) \geq m + (n - m - 1)(s + 1) + \frac{(n - m - 1)(m - s - 1)}{2} \tag{2}$$

By construction,

$$e(G) \leq (s + 1)(n - s - 1)$$

Hence,

$$n - 2s - 2 \leq m \leq \frac{2(s + 1)(n - s - 1)}{n}$$

that is,

$$n^2 - 4n(s + 1) + 2(s + 1)^2 \leq 0$$

From this quadratic inequality it follows that $n \leq (2 + \sqrt{2})(s + 1)$. Consequently for $n > (2 + \sqrt{2})(s + 1)$ G must have a vertex of degree $s + 1$, and hence $G = K(s + 1, n - s - 1)$. Bollobás and Eldridge [6] showed, using a more complicated argument, that this result is in fact true for every $n \geq 2s + 2$.

To date, most of the known results deal with $k = 2$, $s = 1$, or $\lambda = n - 1$. Both exact values and order of magnitude (i.e., asymptotic results) are included among known results. Before we proceed to discuss other results we introduce some notation. This we do in the next section. Following that our treatment proceeds as follows. In the third section we discuss the results for $k = 2$. For $k \geq 3$, we distinguish two cases according to whether or not $s = 1$; the case $s = 1$ is considered in the fourth section and that for $s > 1$ is considered in the fifth section. We conclude the survey with a discussion of some other related problems.

Preliminaries

Throughout this paper $[x]$ denotes the integer part of x. Moreover, if a class G of graphs has only one member, it is convenient to refer to this member as G, since no ambiguity results.

Let n, k, λ, and s be positive integers such that $n > \lambda \geq k$ and $n > s$. Denote by $G_V(n, k, \lambda, s)$ the class of graphs with n vertices and diameter at most k having the property that the subgraph obtained by deleting any s, or fewer, vertices has diameter at most λ, $\lambda \geq k$. A graph with the minimum number of edges within the class $G_V(n, k, \lambda, s)$ is denoted by Min $G_V(n, k, \lambda, s)$ and the minimum possible number of edges is denoted by $M_V(n, k, \lambda, s)$. Similarly the class $G_E(n, k, \lambda, s)$ consists of those graphs with n vertices and diameter at most k having the property that the partial graph obtained by deleting any s or fewer edges has diameter at most λ, $\lambda \geq k$. A graph with the minimum possible number of edges within the class $G_E(n, k, \lambda, s)$ is denoted by Min $G_E(n, k, \lambda, s)$, and the minimum possible number of edges is denoted by $M_E(n, k, \lambda, s)$. The graphs with the minimal number of edges within these classes will be called *extremal graphs*.

The most general formulation of our problem is the following which is due to Murty [21].

PROBLEM. Given n centers of communication $x_1, x_2, \ldots, x_n$ and an $n \times n$ matrix $C = \{c_{ij}\}$, which describes the cost of connecting center x_i with center x_j in the network, what is the optimum set of links which minimizes the total establishment cost when it is required that the resulting network be either (a) a $G_V(n, k, \lambda, s)$ or (b) a $G_E(n, k, \lambda, s)$?

We observed in the previous section that if $c_{ij} = c$, a constant for every $i \neq j$ ($c_{ii} = 0$), the above problem reduces to finding a Min $G_V(n, k, \lambda, s)$ or a Min $G_E(n, k, \lambda, s)$. Except for the case $k = \lambda = 2$, the only known results are for $c_{ij} = c$, for every $i \neq j$. For the case $k = \lambda = 2$, Murty [21] showed that the problem is equivalent to minimizing a pseudo-Boolean function. Unfortunately, for diameters larger than 2 it is difficult, if not impossible, to find pseudo-Boolean expressions.

We conclude this section by making some rather simple but useful observations.

(1) If $s = 0$, then the only extremal graphs are trees of diameter at most k.

(2) In a $G_E(n, k, \lambda, s)(G_V(n, k, \lambda, s))$ there will be at least $s + 1$ edge (vertex) disjoint paths of length at most λ between any two vertices (any two nonadjacent vertices), at least one of which has length at most k.

(3) The functions $M_E(n, k, \lambda, s)$ and $M_V(n, k, \lambda, s)$ are monotonic nondecreasing in s and monotonic nonincreasing in k and λ.

(4) Let G be an extremal graph with minimum degree $\delta(G)$. The degree of every vertex of G is at least $s + 1$, i.e., $\delta(G) \geq s + 1$. Furthermore, if a vertex α has degree $s + 1$, then every vertex of G which is not adjacent to α must be connected to each of the $s + 1$ vertices adjacent to α by a path of length $\leq \lambda - 1$, by (2).

(5) Let $G \in G_E(n, k, \lambda, 1)$. G may or may not have a cut vertex. If it has, then $G \in G_V(n, k, \lambda', 1)$ for some λ'.

To make any progress towards a solution of our problem, one has to make careful use of the above properties. Indeed, they supply about the only tools one has to work with. Recall that in the introduction we observed properties (2) and (4) for the particular case $k = \lambda = 2$. In fact, these two properties proved extremely useful in finding the structures of Min $G_V(n, 2, 2, s)$, for $n > (2 + \sqrt{2})(s + 1)$. Making use of the above properties Caccetta [11–17] was able by simple, but somewhat lengthy counting arguments, to obtain the structures of Min $G_V(n, k, \lambda, s)$ and Min $G_E(n, k, \lambda, 1)$ for $k = 3$ and 4.

EXTREMAL GRAPHS OF DIAMETER 2

In the introduction we discussed the following result.

THEOREM 1. If $n \geq 2s + 2$, then $M_V(n, 2, 2, s) = (s + 1)(n - s - 1)$ and the only member of Min $G_V(n, 2, 2, s)$ is the complete bipartite graph $K(s + 1, n - s - 1)$ with $s + 1$ vertices in one set and $n - s - 1$ vertices in the other.

The corresponding result for edge deletion is due to Murty [22]. Let $G_0 = K_{s+1} + \bar{K}_{n-s-1}$, so that G_0 is obtained from $K(s + 1, n - s - 1)$ by adding all edges within the first class. Clearly $G_0 \in G_E(n, 2, 2, s)$, since each pair of vertices are connected by $s + 1$ edge disjoint paths of length 2. Using an argument analogous to that used in proving THEOREM 1 (see first section) Murty [22] proved

THEOREM 2. If $n > \frac{1}{2}(3 + \sqrt{5})(s + 1)$, then $M_E(n, 2, 2, s) = (s + 1)(n - \frac{1}{2}s - 1)$ and G_0 is the only member of Min $G_E(n, 2, 2, s)$.

At first it was suspected that the bound on n in THEOREM 2 above could be lowered substantially. However, Bollobás and Erdös [7] recently showed that the bound on n is almost necessary for G_0 to be extremal. In particular, they proved:

THEOREM 3. Let c be fixed in the range $1 < c < \frac{1}{2}(3 + \sqrt{5})$ and let $n = [c(s + 1)]$. Then

$$M_E(n, 2, 2, s) = \tfrac{1}{2}c^{3/2}(s + 1)^2 + 0((s + 1)^2)$$

We now turn our attention to the case $\lambda \geq 3$. Murty [23, 24] solved the problem completely for the case $s = 1$. The next two theorems report his results.

Two vertices are *neighborly* if and only if they have the same neighbor set. Consider the graphs displayed in FIGURE 1. The class C_1^n, $n \geq 6$, (C_2^n, $n \geq 8$) is obtained from the class $C_1^{n-1}(C_2^{n-1})$ by adding a vertex that is neighborly to a vertex of $C_1^{n-1}(C_2^{n-1})$ having degree 2. The graph C_3^{10} is of course the Petersen graph on ten vertices. Clearly the elements of C_i^n, $i = 1, 2, 3$, belong to the class $G_V(n, 2, \lambda, 1)$, $\lambda \geq 3$, and they have $2n - 5$ edges. By first proving that every Min $G_V(n, 2, \lambda, 1)$ has at least one vertex of degree 2, except possibly for $n = 10$ (for this case the only other possibility is the Petersen graph), Murty [23, 24] proved

THEOREM 4. If $n \geq 5$ and $\lambda \geq 3$, then $M_V(n, 2, \lambda, 1) = 2n - 5$ and the graphs of C_i^n, $i = 1, 2, 3$ are the elements of Min $G_V(n, 2, \lambda, 1)$.

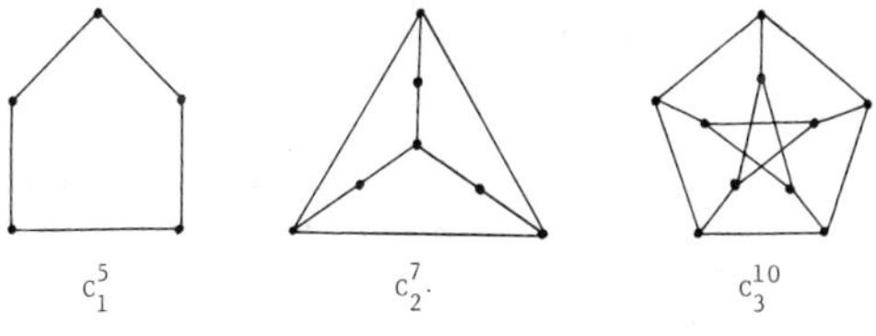

FIGURE 1.

Consider the graphs of FIGURE 2. The class $S^{2m+1}(S^{2m})$, $m \geq 5$, is obtained from the class $S^{2m-1}(S^{2m-2})$ by adding two adjacent vertices and connecting both of them to c. Clearly the graphs S^{2m} and S^{2m+1} belong to the class $G_E(n, 2, \lambda, 1)$, $\lambda \geq 3$, and they have $[(3n - 2)/2]$ edges. We now prove that these graphs are the only extremal ones. Our proof illustrates the use of observation (5) made in the previous section.

THEOREM 5. If $n > 8$ and $\lambda \geq 3$, then $M_E(n, 2, \lambda, 1) = [(3n - 2)/2]$ and the graphs of S^{2m} and S^{2m+1} are the elements of Min $G_E(n, 2, \lambda, 1)$ for $n = 2m$ and $n = 2m + 1$, respectively.

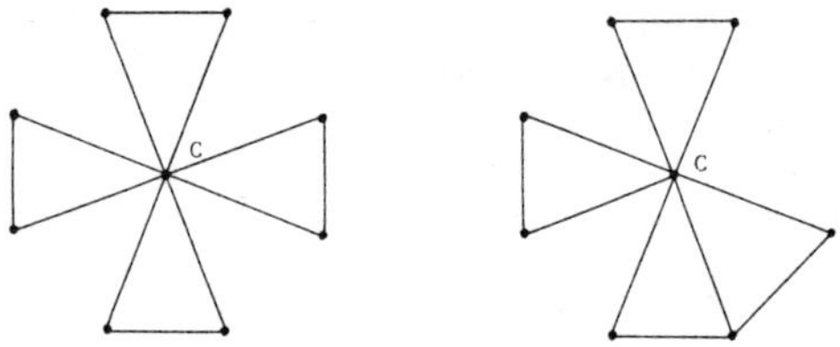

S^9 S^8

FIGURE 2.

Proof: Let $G \in \text{Min } G_E(n, 2, \lambda, 1)$. If G has a cut vertex, c say, then the theorem is immediate, since every vertex of $G - c$ must be adjacent to c as $D(G) \leq 2$. Hence we can suppose that G has no cut vertex. Then $G \in G_V(n, 2, \lambda', 1)$, for some λ', and hence, by THEOREMS 1 and 4, G must have at least $2n - 5$ edges. But $2n - 5 > [(3n - 2)/2]$, for every $n > 8$, hence the theorem. □

For $s > 1$ and $\lambda \geq 3$ our first result is on $(s + 1)$-connected graphs of diameter 2 and is due to Bollobás and Harary [8].

THEOREM 6. Let G be an $s + 1$ (≥ 3) connected graph of diameter 2 with n vertices and q edges.

(i) If N is a prime power and $s = N$, then there exists a graph G with $n = N^2 + N + 1$ and $q = \frac{1}{2}n(s + 1)$.
(ii) Every G has $q \geq \frac{1}{2}\{(n - 1)(s + 2) - (s + 1)^2 + (s + 1)\}$.
(iii) If $n > 3(s + 1)^2 + 5$, then G has $q \geq [((n - 1)(s + 2) + 1)/2]$.

Equality can be attained for every n and s. Equality holds for a graph G if and only if G is a join of the form $G = K_1 + H$, where H is an s-connected graph of order $n - 1$ with minimum number of edges, that is, with $[(s(n - 1) + 1)/2]$ edges.

Part (iii) of THEOREM 6 was proved by Bondy and Murty [9] for the case $n > (s + 1)^3 + \alpha(n) \cdot \alpha(s + 1) \cdot (s + 1) + 1$, where the function $\alpha(t)$ is defined by the rule that $\alpha(t) = 0$ or 1 according as t is odd or even.

THEOREM 6 gives us information about $M_V(n, 2, n - 1, s)$ and $M_E(n, 2, n - 1, s)$. In fact, it is much stronger; for consider the graph $S = K_1 + H$, where H is an s-connected graph of order $n - 1$ with minimum number of edges. Clearly $S \in G_E(n, 2, \lambda, s)$, for $\lambda \geq 4$ and $n \geq s + 1$, since any two vertices of S are connected by $(s + 1)$-edge disjoint paths of length at most 4, at least one of which has length at most 2. Hence the following result follows from THEOREM 6, (iii), and Observation 3.

THEOREM 7. If $n > 3(s + 1)^2 + 5$ and $\lambda \geq 4$, then

$$M_E(n, 2, \lambda, s) = \left[\frac{(n - 1)(s + 2) + 1}{2}\right]$$

and the graph S is the only extremal graph.

It is not unreasonable to conjecture that the bound on n in the above theorem could be lowered to $n \geq s + 1$, but we have not been able to prove this. However, it follows from THEOREM 6 that

$$\lim_{n\to\infty} \frac{M_E(n, 2, \lambda, s)}{n} = \frac{1}{2}(s+2)$$

for $n > \frac{1}{2}(s+1)^2 + 1$ and $\lambda \geq 4$.

THEOREMS 1–7 above sum up most of the known results for the case $k = 2$. We conclude this section by mentioning two of the conjectures, both concerning the vertex version of the problem made by Caccetta [14].

Let t, r, and b be natural numbers such that $n = t(r-1) + b$, where $r = n - s$ and $0 \leq b \leq r - 1$. Let $K_t' = K_t(r-1) + \bar{K}_b$, where $K_t(r-1)$ denotes the complete t-partite graph with $r - 1$ vertices in each set. Clearly the graph $K_t' \in G_V(n, 2, 2, s)$, $s + 1 \leq n \leq 2s + 1$, has $\frac{1}{2}n(s+1) + \frac{1}{2}b(r-b-1)$ edges (see FIGURE 3). Bollobás and Eldridge [6] proved that if $b = 0$, then K_t' is the only element of Min $G_V(n, 2, 2, s)$. Caccetta [14] proved that K_t' is the only element of Min $G_V(n, 2, 2, s)$ for $s + 1 < n < s + 3 + 2\sqrt{s+2}$ and also for $b = 1$ and 2. If we could prove that every Min $G_V(n, 2, 2, s)$, $s + 1 < n \leq 2s + 1$, has a vertex of degree $s + 1$, then we could easily settle the following conjecture.

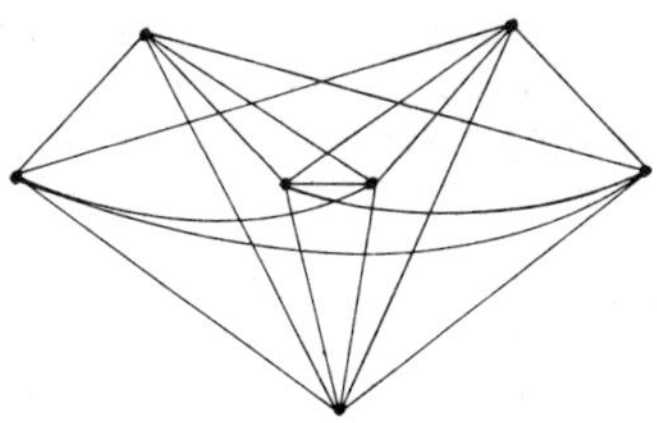

FIGURE 3. A $G_V(7, 2, 2, 4)$ graph.

CONJECTURE 1. If $s + 1 < n \leq 2s + 1$, then

$$M_V(n, 2, 2, s) = \tfrac{1}{2}n(s+1) + \tfrac{1}{2}b(r-b-1)$$

and K_t' is the only element of Min $G_V(n, 2, 2, s)$.

Next we consider the case $\lambda \geq 4$. Let $l = [\frac{1}{2}\lambda]$. If λ is even, then denote by S_1^n the class of graphs G defined as follows: We let $V(G) = \{x_1, x_2, \ldots, x_n\}$, and take $\{x_1x_i, i = 2, 3, \ldots, n\} \subset E(G)$. Furthermore, vertices $x_2, x_3, \ldots, x_{n-2}$ are partitioned into sets containing l or fewer vertices such that the vertices in each set span a tree with two end vertices. One end vertex of each tree is connected to x_{n-1} and the other to x_n. If λ is odd, i.e., $\lambda = 2l + 1$, we denote by S_2^n the class of graphs on n vertices obtained in exactly the same way as the class S_1^n except that here, in partitioning vertices $x_2, x_3, \ldots, x_{n-2}$, we let one set have $l + 1$ vertices. It is easily seen that these graphs belong to the class $G_V(n, 2, \lambda, 2)$, $\lambda \geq 5$, and they have at least $2n - 3 + [[n - \alpha(\lambda - 1) - 4]/[\frac{1}{2}\lambda]]$ edges (see FIGURE 4), where $\alpha(t)$ is defined by the rule that $\alpha(t) = 0$ or 1 according as t is odd or even. Caccetta [11] proved that these are the best possible constructions for $n \geq 43$ and $\lambda \geq 2[(n-2)/3]$.

CONJECTURE 2. If $n \geq 5$ and $\lambda \geq 5$, then

$$M_V(n, 2, \lambda, 2) = 2n - 3 + \left[\frac{n - \alpha(\lambda - 1) - 4}{[\frac{1}{2}\lambda]}\right]$$

and the graphs of S_1^n and S_2^n defined above are the elements of Min $G_V(n, 2, \lambda, 2)$.

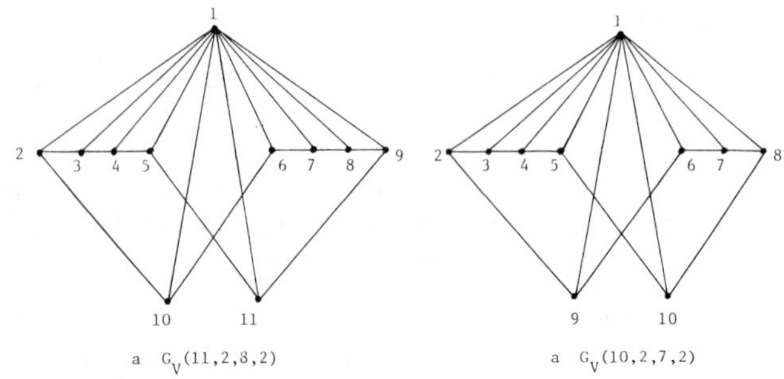

FIGURE 4.

EXTREMAL GRAPHS OF DIAMETER ≥ 3—CASE $s = 1$

As we mentioned in the introduction, most of the known results for $s = 1$ deal with $k \leq 4$. First we present the results for diameters 3 and 4, and then we discuss some asymptotic results as well as some conjectures for the larger diameters.

Consider the graphs displayed in FIGURE 5. For $n \geq 9$ we obtain the class J_1^n $(i = 1, 2, 3, 4)$ from the class J_i^{n-1} by adding a vertex that is neighborly (recall that two vertices are neighborly if they have the same neighbor set) to a vertex of J_i^{n-1} having degree 2. It is easily seen that the elements of J_i^n $(i = 1, 2, 3, 4)$; $n \geq 8$) belong to the class $G_V(n, 3, 3, 1)$ and they have $2n - 6$ edges. It turns out that these are the only extremal graphs except for $n = 12$ for which the graph of FIGURE 6 is also extremal. By first proving that every Min $G_V(n, 3, 3, 1)$ has at least one

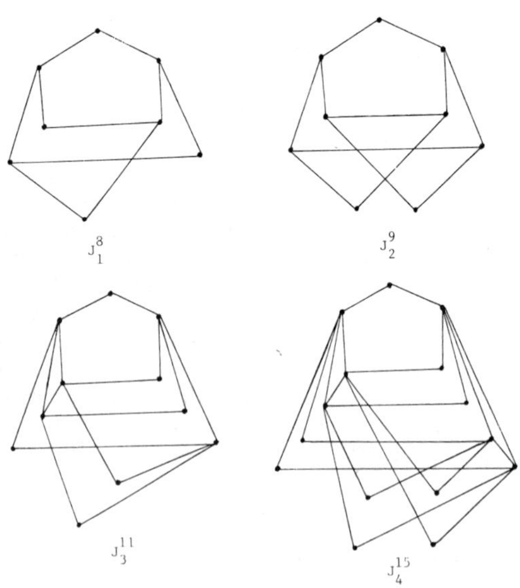

FIGURE 5.

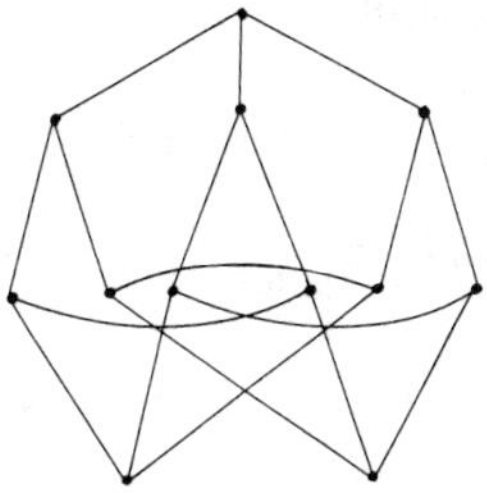

FIGURE 6. Graph H^{12}.

vertex of degree 2, except possibly for $n = 12$ (for this case the only other possibility is the graph of FIGURE 6) Caccetta [13], established

THEOREM 8. If $n \geq 8$, then $M_V(n, 3, 3, 1) = 2n - 6$, and the graphs of H^{12} and J_i^n $(i = 1, 2, 3, 4)$ are the elements of Min $G_V(n, 3, 3, 1)$.

Naturally, observations (2) and (4) in the second section (PRELIMINARIES) prove very useful in establishing THEOREM 8.

Consider the graphs S^{2m} and S^{2m+1} of THEOREM 5 and the graphs displayed in FIGURE 7. Clearly these graphs belong to the class $G_E(n, 3, 3, 1)$, and they have $[(3n - 2)/2]$ edges. Bollobás [2] proved that these were the best possible constructions. It turns out that these are the only extremal graphs. Caccetta [13] proved that if a Min $G_E(n, 3, 3, 1)$ has no cut vertex, then $n \leq 6$, and the only possibility is one of the graphs of FIGURE 7. Once this is established, it is only a simple exercise to prove

THEOREM 9. If $n \geq 5$, then $M_E(n, 3, 3, 1) = [(3n - 2)/2]$ and the graphs of S^{2m}, S^{2m+1}, H^4, H_j^5 $(j = 1, 2)$, and H_j^6 $(j = 1, 2, 3, 4)$ are the elements of Min $G_E(n, 3, 3, 1)$, for $n = 2m$, $2m + 1$, 4, 5, and 6, respectively.

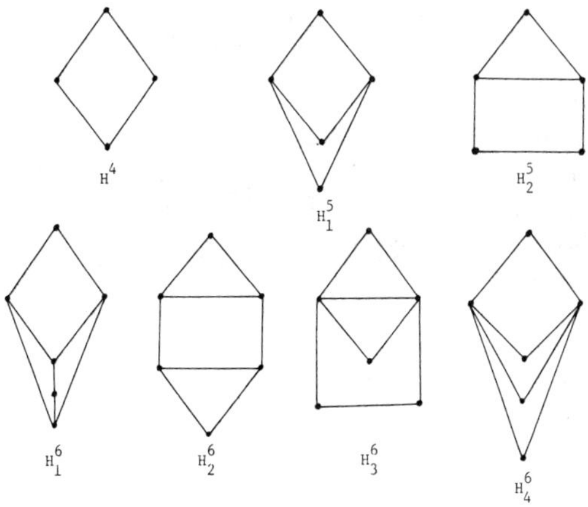

FIGURE 7.

For $\lambda \geq 4$ and $s = 1$, Bollobás [1] proved that

$$M_V(n, 3, \lambda, 1) = \begin{cases} \left[\dfrac{3n-5}{2}\right], & \text{for } \lambda = 4 \\ \left[\dfrac{3n-6}{2}\right], & \text{for } \lambda \geq 5 \end{cases}$$

Caccetta [15] found all the elements of Min $G_V(n, 3, \lambda, 1)$, $\lambda \geq 4$. These results are described in the next two theorems.

Denote by K^{2m} the class of graphs on $2m$ vertices that are obtained by connecting two vertices x_1 and x_2 by m disjoint paths of length 3. The class K^{2m+1} is obtained from the class K^{2m} by adding one vertex and connecting it to any two nonadjacent vertices of K^{2m}. Clearly the graphs of K^{2m} and K^{2m+1} belong to the class $G_V(n, 3, 4, 1)$ and they have $[(3n-5)/2]$ edges (see FIGURE 8; we have labelled the vertex x_i as i). Note that there are three variations K_j^{2m+1}, $j = 1, 2, 3$, of K^{2m+1}. Bollobás [1] conjectured that these are the only extremal graphs. By first proving that every Min $G_V(n, 3, 4, 1)$ graph has two adjacent vertices having degree 2 for every $n \geq 6$, except possibly $n = 9$, Caccetta [15] proved that these are indeed the only extremal graphs of $G_V(n, 3, 4, 1)$. For $n = 9$ the only other possibility turns out to be the graph displayed in FIGURE 9.

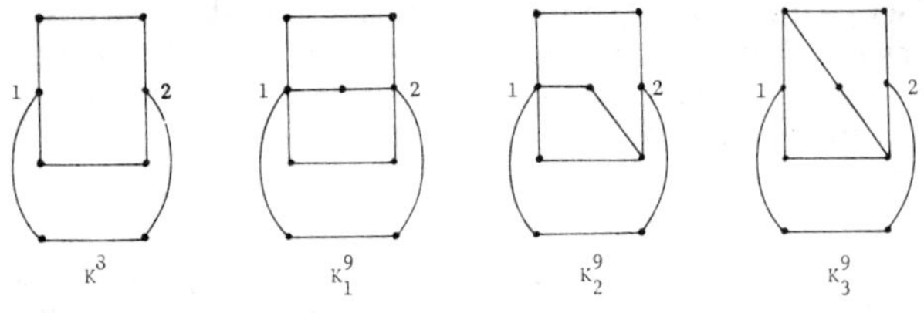

FIGURE 8.

THEOREM 10. *If $n \geq 6$, then $M_V(n, 3, 4, 1) = [(3n-5)/2]$ and the graphs of H^9, K^{2m} and K_j^{2m+1} ($j = 1, 2, 3$) are the elements of Min $G_V(n, 3, 4, 1)$ for $n = 9$, $2m$, and $2m+1$, respectively.*

Now consider the graphs displayed in FIGURE 10. The class L_1^{2m+1} (L_j^{2m}, $j = 1, 2, 3, 4$), $m \geq 5$, is obtained from the class L_1^{2m-1} (L_j^{2m-2}) by adding an (x_1, x_2)-path of length 3. The class L_2^{2m+1} is obtained from the class $L_1^{2m'+1}$, $m' < m$, by adding $m - m'$ vertex disjoint (x_1, x_3)-paths of length 3. We could similarly obtain graphs from $L_1^{2m'}$, $L_2^{2m'}$, and $L_3^{2m'}$. For $m \geq 4$, put $L_5^{2m} = K^{2m}$. It

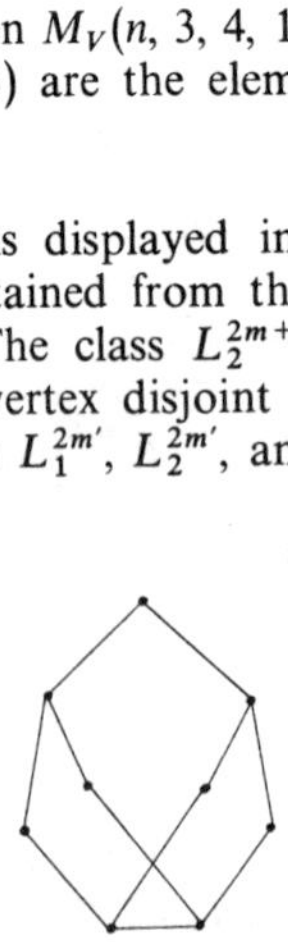

FIGURE 9.

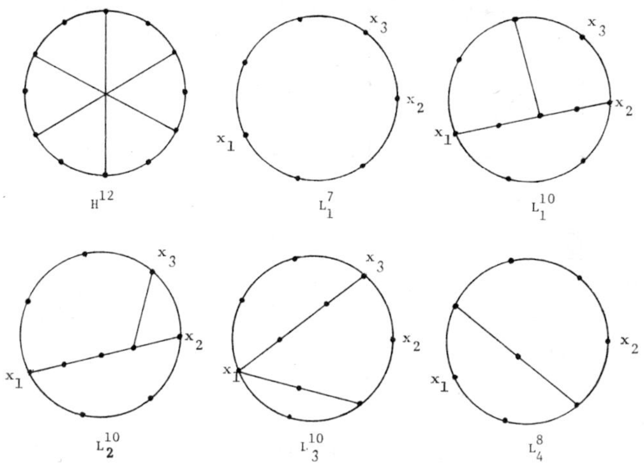

FIGURE 10.

is easily observed that the graphs of H^{12}, L_j^{2m+1} $(j = 1, 2)$, and L_j^{2m} $(j = 1, 2, 3, 4, 5)$ belong to the class $G_V(n, 3, \lambda, 1)$, $\lambda \geq 5$, for $n = 12$, $n = 2m + 1$, and $n = 2m$, respectively. In fact, these graphs with possibly some slight variations (see Caccetta [15]) turn out to be the only extremal graphs of $G_V(n, 3, \lambda, 1)$, $\lambda \geq 5$. This was proved in [15].* As in the previous theorem, the main exercise was to prove the existence of adjacent vertices having degree 2.

THEOREM 11. If $n \geq 7$ and $\lambda \geq 5$, then $M_V(n, 3, \lambda, 1) = [(3n - 6)/2]$, and the graphs of H^{12}, L_j^{2m+1} $(j = 1, 2)$, and L_j^{2m} $(j = 1, 2, 3, 4, 5)$ are the elements of Min $G_V(n, 3, \lambda, 1)$, for $n = 12$, $2m + 1$, and $2m$, respectively.

The corresponding results for the edge case were also obtained by Caccetta [15]. These results are described in the next three theorems. As in the vertex case, the main exercise in establishing these results is to prove the existence of adjacent vertices of degree 2.

Consider the graphs displayed in FIGURE 11. For $m > 4$, we obtain the class M_j^{2m} $(j = 1, 2, \ldots, 6)$ from the class M_j^{2m-2} by adding an (x_1, x_2)-path of length 3. The class M_7^{2m} is obtained from the class $M_1^{2m'}$, $m' < m$, by adding $m - m'$ triangles with x_1 being a common vertex in all triangles and the other two being new vertices. It is easily seen that these graphs belong to the class $G_E(n, 3, 4, 1)$ and have $[(3n - 4)/2]$ edges. It turns out that these graphs will exhaust all the extremal graphs in the class $G_E(n, 3, 4, 1)$, except for $n = 8$, 10, and 12 for which the graphs of FIGURE 12 also belong to the class $G_E(n, 3, 4, 1)$ and have $[(3n - 4)/2]$ edges.

THEOREM 12. If $n \geq 4$, then $M_E(n, 3, 4, 1) = [(3n - 4)/2]$ and the graphs of H_1^8, H_j^{10} $(j = 1, 2, 3, 4)$, H_j^{12} $(j = 1, 2, 3)$, M_j^{2m} $(j = 1, 2, \ldots, 7)$, and K_1^{2m+1} are the elements of Min $G_E(n, 3, 4, 1)$ for $n = 8$, 10, 12, $2m$, and $2m + 1$, respectively.

* Drs. R. Tindell and F. Boesch inform me that they have obtained results along the lines of THEOREM 11. The graph H^{12} in FIGURE 10 was pointed out to me by Dr. Tindell.

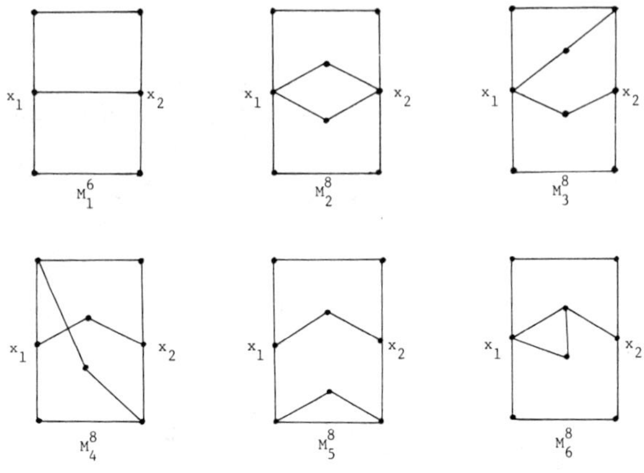

FIGURE 11.

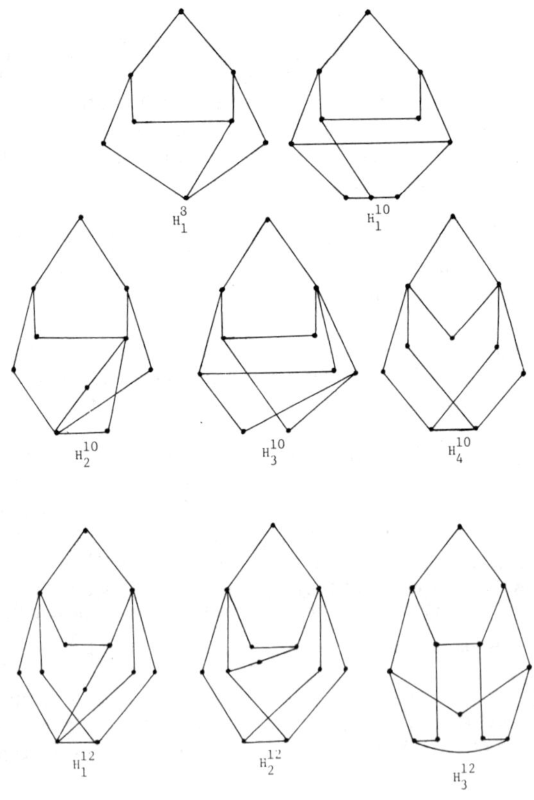

FIGURE 12.

Denote by P_j^{2m} ($j = 1, 2$) the class of graphs obtained from the class L_j^{2m-1} by adding a vertex x_{2m} and connecting it to x_1 and one of x_2 or x_3 (see FIGURE 10). The class K_4^{2m+1} (K_5^{2m+1}) is obtained from the class P_1^{2m} (P_2^{2m}) by adding a vertex x_{2m+1} that is neighborly to x_{2m}. The class K_6^{2m+1} is obtained from the class $K_1^{2m'+1}$, $m' < m$, by adding $m - m'$ triangles with x_1 being a common vertex in all triangles and the other two being new vertices. Put $P_3^{2m} = K^{2m}$. Clearly the graphs of P_j^{2m} ($j = 1, 2, 3$) and K_j^{2m+1} ($j = 1, 2, \ldots, 6$) belong to the class $G_E(n, 3, 5, 1)$, for $n = 2m$ and $n = 2m + 1$, respectively, and they have $[(3n - 5)/2]$ edges. In [15] it was proved that these are the only extremal graphs of $G_E(n, 3, 5, 1)$ other than the graph H^9 of FIGURE 9.

THEOREM 13. If $n \geq 5$, then $M_E(n, 3, 5, 1) = [(3n - 5)/2]$ and the graphs of H^9, P_j^{2m} ($j = 1, 2, 3$), and K_j^{2m+1} ($j = 1, 2, \ldots, 6$) are the elements of Min $G_E(n, 3, 5, 1)$ for $n = 9$, $2m$, and $2m + 1$, respectively.

Bollobás [2] proved that if $\lambda \geq 6$ and $n \geq 6$, then $M_E(n, 3, \lambda, 1) = [(3n - 6)/2]$. Caccetta [15] proved

THEOREM 14. If $\lambda \geq 6$ and $n \geq 6$, then $M_E(n, 3, \lambda, 1) = [(3n - 6)/2]$ and the structures of Min $G_E(n, 3, \lambda, 1)$ and Min $G_V(n, 3, \lambda - 1, 1)$ coincide.

We now discuss the results for diameter-4 graphs. Caccetta [12, 16, 17] completely solved the problem for all $\lambda \geq 4$ except $\lambda = 5$.

First we discuss the results for the vertex version of the problem. Our first result concerns the case $\lambda \geq 6$ which was proved by Caccetta [12].

THEOREM 15. If $n \geq 6$ and $\lambda \geq 6$, then

$$M_V(n, 4, \lambda, 1) = \left[\frac{4n - \text{Min}(\lambda, 6)}{3}\right]$$

The extremal graphs for $G_V(n, 4, \lambda, 1)$, $\lambda \geq 6$, can be obtained by joining two vertices with disjoint paths of length 4 and altering slightly the graphs obtained in this way. As one would expect from the results for $k = 3$, the number of variations in the extremal structures is quite large. Caccetta [16] found all the extremal graphs for $G_V(n, 4, \lambda, 1)$, $\lambda \geq 6$. We do not give the details of these graphs here, because of the number of variations.

For $\lambda = 4$ and 5, Caccetta [16] proved the following two theorems.

THEOREM 16. If $n \geq 14$, then $M_V(n, 4, 4, 1) = [(3n - 5)/2]$ and the structures of Min $G_V(n, 4, 4, 1)$ coincide with the structures of Min $G_V(n, 3, 4, 1)$.

THEOREM 17. If $n \geq 12$, then $M_V(n, 4, 5, 1) = [(3n - 5)/2]$.

The corresponding results for the edge version of the problem were given by Caccetta [17]. These are described in the next three theorems.

Consider the graphs displayed in FIGURE 13. Clearly these graphs belong to the class $G_E(n, 4, \lambda, 1)$, for $\lambda \geq 6$, and they have $[(5n - 2)/4]$ edges. Now making use of observation (4) in the second section and THEOREMS 15–17, it is easily seen that Min $G_E(n, 4, \lambda, 1)$, $\lambda \geq 6$, has a cut vertex for every $n \geq 25$, except possibly $n = 26$, 27, and 30. Let $G \in$ Min $G_E(n, 4, \lambda, 1)$, $\lambda \geq 6$. By first proving that if G has

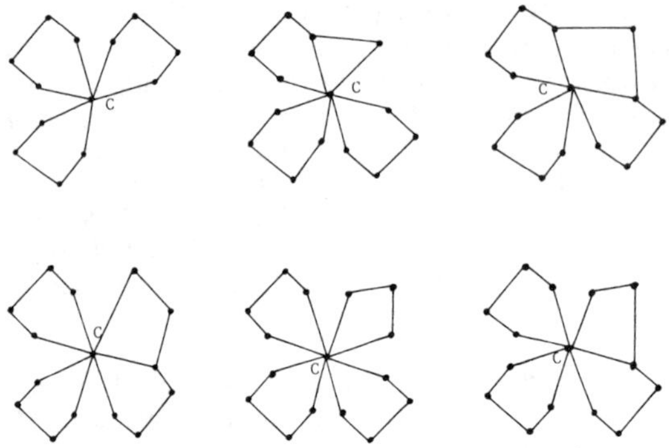

FIGURE 13.

a cut vertex, c say, then every vertex of $G - c$ is at distance at most 2 from c. Caccetta [17] proved

THEOREM 18. If $\lambda \geq 6$ and $n > 19$, $M_E(n, 4, \lambda, 1) = [(5n - 2)/4]$, and for $n = 25$, 28, 29, and $n > 30$ the extremal graphs can be obtained from the graphs of FIGURE 13 (with possibly some slight variations) by adding distinct pentagons, any two of which have exactly one common vertex, namely c.

For $n < 19$ it is not difficult to establish

THEOREM 19.
(a) $M_E(n, 4, \lambda, 1) = [(4n - 7)/3]$, for $\lambda \geq 8$ and $7 \leq n \leq 18$.
(b) $M_E(n, 4, 7, 1) = [(4n - 6)/3]$, for $6 \leq n \leq 18$.
(c) $M_E(n, 4, 6, 1) = \text{Min}\ \{[(5n - 2)/4], [(4n - 5)/3]\}$, for $5 \leq n \leq 18$.

For $\lambda = 4$ and 5 we have the following two theorems.

THEOREM 20. If $n \geq 13$, $M_E(n, 4, 5, 1) = [(4n - 3)/3]$, and when Min $G_E(n, 4, 5, 1)$ has a cut vertex, the extremal graphs are obtained from the graphs of FIGURE 14 by adding 4-gons, any two of which have exactly one common vertex, namely c.

THEOREM 21. For every $n \geq 4$, $M_E(n, 4, 4, 1) = [(3n - 4)/2]$, and for $n > 10$, the structures on Min $G_E(n, 4, 4, 1)$ coincide with the structures of Min $G_E(n, 3, 4, 1)$.

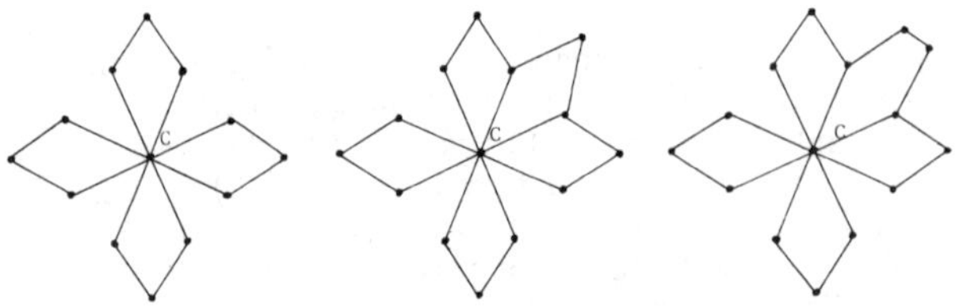

FIGURE 14.

As one would expect there are few exact results for $k \geq 5$. However, there are some asymptotic results which we now discuss.

Bollobás [2] made some conjectures concerning the case $k \geq 3$ and $s = 1$. We now restate his conjecture in a more explicit form.

CONJECTURE 3.

(i) If $\lambda \geq 2k - 2$ and $n \geq \lambda + 1$, then

$$M_V(n, k, \lambda, 1) = \begin{cases} n + \left[\dfrac{n-k-2}{k-1}\right], & \text{if } \lambda = 2k-2 \\ n + \left[\dfrac{n-k-3}{k-1}\right], & \text{otherwise} \end{cases}$$

(ii) If $n \geq 4(k-1)$ and $\lambda = 2k - 3$, then

$$M_V(n, k, \lambda, 1) = n + \left[\frac{n-k-3}{k-2}\right]$$

(iii) If $n \geq 2k$ and $\lambda < 2k - 3$, then

$$M_V(n, k, \lambda, 1) = M_V(n, k-1, \lambda, 1)$$

(iv) If k is odd, $\lambda \geq 2k - 1$, and $n \geq 2k$, then

$$M_E(n, k, \lambda, 1) = M_E(n, k, \lambda - 1, 1)$$

(v) If k is odd and $\lambda = 2k - 2$, then

$$M_E(n, k, \lambda, 1) = n + \left[\frac{n-k-1}{k-1}\right]$$

(vi) If k is even, $\lambda \geq \frac{3}{2}k - 1$, and $n \geq 2k^2 - k + 2$, then

$$M_E(n, k, \lambda, 1) = \begin{cases} n + \left[\dfrac{n-3}{k-1}\right], & \text{if } \lambda = \frac{3}{2}k - 1 \\ n + \left[\dfrac{n-2}{k}\right], & \text{otherwise} \end{cases}$$

(vii) If k is even, $\lambda \geq 2k - 2$, and $n \geq 2k$, then

$$M_E(n, k, \lambda, 1) = \text{Min}\left\{ M_V(n, k, \lambda - 1, 1), n + \left[\frac{n-2}{k}\right]\right\}$$

(viii) For other values of λ,

$$M_E(n, k, \lambda, 1) = M_E(n, k-1, \lambda, 1)$$

Naturally, the results for $k = 3$ and 4, described above, support this conjecture. In fact, it is not difficult to see how the constructions for $k = 3$ and 4 can be extended to construct graphs satisfying the diameter constraints and having the number of edges indicated by the above conjecture. Bollobás [4, 5] added substantial support to this conjecture by proving

THEOREM 22.

(a) If $\lambda \geq 2k - 2$, then

$$\lim_{n \to \infty} \frac{M_V(n, k, \lambda, 1)}{n} = \frac{k}{k-1}$$

(b) If $\lambda \geq \frac{3}{2}(k-1)$, then

$$\lim_{n\to\infty} \frac{M_E(n, k, \lambda, 1)}{n} = 1 + \frac{1}{2[\frac{1}{2}k]}$$

Caccetta [11] verified parts (i), (v), and (vi) of the CONJECTURE for the case $k = 5$. He also proved the following lemma which may prove useful in any attempt to verify the conjecture.

LEMMA. If $\lambda \geq 2k - 1$, then any Min $G_E(n, k, \lambda, 1)$ and Min $G_V(n, k, \lambda - 1, 1)$ possesses at least $[(k-3)/(k-1)]n + 2$ vertices of degree 2.

For the particular case of $k = 6$, the above lemma tells us that any Min $G_V(n, k, \lambda, 1)$, $\lambda \geq 11$, must have at least two adjacent vertices of degree 2. Making use of this fact Caccetta [11] was able, by extending the arguments for the case $k = 4$ (see [12, 16]), to verify part (i) of the CONJECTURE 3.

Perhaps the most likely way to establish at least part of CONJECTURE 3 would be to first try and prove CONJECTURE 2 (see the third section). For suppose CONJECTURE 2 were true, then there exists among the extremal graphs of $G_V(n, 2, \lambda, 2)$ a graph, G say, with a vertex x of degree $n - 1$. But then $G - x \in G_V(n-1, k', \lambda, 1)$, $k' \leq \lambda$, and has

$$(n-1) + \left[\frac{(n-1) - [\frac{1}{2}\lambda] - \alpha(\lambda - 1) - 3}{[\frac{1}{2}\lambda]}\right]$$

edges. Moreover, from the construction of the graphs S_1^n and S_2^n (see FIGURE 4) it follows that $k' = [(\lambda + 2)/2]$. Therefore, if CONJECTURE 2 is true, then part (i) of CONJECTURE 3 would also be true. Incidentally, once part (i) of CONJECTURE 3 is established it is not difficult to establish some of the other parts of the conjecture.

EXTREMAL GRAPHS OF DIAMETER ≥ 3—CASE $s > 1$

In this section we describe some results for $s > 1$. For simplicity we let $m = [\frac{1}{2}k]$, and put

$$S(s, r) = \sum_{i=0}^{r-1} (s-1)^i$$

Bollobás [5] proved the following two theorems.

THEOREM 23. Let G be a graph of order n with minimal degree at least $s + 1$ (≥ 3) and diameter at most $2m \geq 2$. Then G has at least

$$\frac{n}{2}(s + 1 + S(s+1, m)^{-1}) - c(s+1, m)$$

edges, where

$$c(s+1, m) = \begin{cases} 4^{m+1}, & \text{if } s = 2 \\ (s+1)^{m+1}, & \text{otherwise} \end{cases}$$

THEOREM 24. Let G be a graph of order n with minimal degree at least $s + 1$ (≥ 2) and diameter at most $2m + 1 \geq 3$. Then G has at least

$$\frac{n}{2}(s + 1 + S(s+1, m)^{-1}) - (s+2)n^{(m+1)/(m+2)}$$

edges.

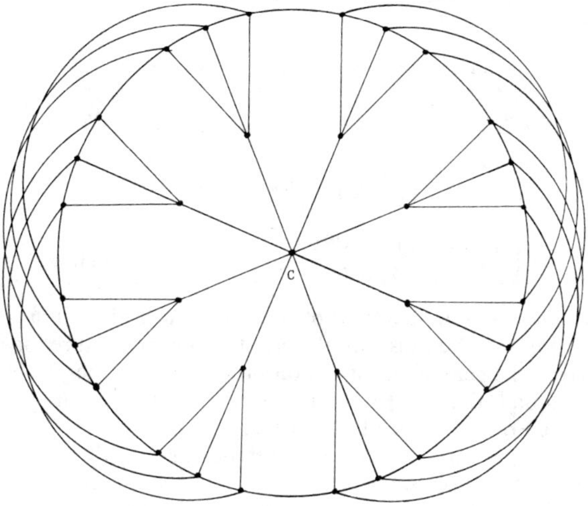

FIGURE 15. A $G_E(33, 4, 8, 3)$ graph.

From the constructions in Bollobás and Harary [8], it is easily seen that for $n > (s+1)^m$ there are $G_E(n, k, \lambda, s)$, $\lambda \geq 4m$, graphs with $(n/2)(s + 1 + S(s+1, m)^{-1}) + A(n, s)$ edges, where $\lim_{n\to\infty} [A(n, s)]/n = 0$. A typical member of $G_E(n, k, \lambda, s)$ is given in FIGURE 15. Using this fact together with THEOREMS 23 and 24 we have

THEOREM 25. Let $n > (s + \alpha(k))^{m+\alpha(k)}$, where $\alpha(k) = 1$ or 2 according as k is even or odd. Then for $k \geq 3$ and $\lambda \geq 2k$

$$\lim_{n\to\infty} \frac{M_E(n, k, \lambda, s)}{n} = \frac{1}{2}(s + 1 + S(s+1, m)^{-1})$$

The above result was noted in [14]. For the vertex case, Bollobás [5] proved that

THEOREM 26.

$$\lim_{n\to\infty} \frac{M_V(n, k, n-1, s)}{n} = \frac{1}{2}(s + 1 + S(s+1, m)^{-1})$$

We conclude this section with two conjectures which are due to Caccetta [14].

CONJECTURE 4. For $k \geq 3$ and $n > (s+1)^2$, there exists a function $\phi(s)$ of s such that, for $2k - 2 \leq \lambda \leq n - \phi(s)$,

$$\lim_{n\to\infty} \frac{M_V(n, k, \lambda, s)}{n} = \begin{cases} \dfrac{(s+1)}{2}\left(\dfrac{k}{k-1}\right), & \text{if } s \text{ is odd} \\[2ex] \dfrac{k(s+2)^2 - 4(2s+3)}{2\{k(s+2) - (s+6)\}}, & \text{otherwise} \end{cases}$$

CONJECTURE 5. If $3 \leq k \leq \lambda \leq 2k - 2$ and $n > (s+1)^2$, then

(i) $$\lim_{n\to\infty} \frac{M_V(n, 3, 3, s)}{n} = (s + 1)$$

(ii) For $\lambda \geq 4$,

$$\lim_{n\to\infty} \frac{M_V(n, k, \lambda, s)}{n} = \lim_{n\to\infty} \frac{M_V(n, k-1, \lambda, s)}{n}$$

In the case of $\lambda \geq 2k - 2$, we can construct $G_V(n, k, \lambda, s)$'s with

$$\begin{cases} \dfrac{n(s+1)}{2}\left(\dfrac{k}{k-1}\right) - A(k, s), \text{ if } s \text{ is odd} \\ \dfrac{n}{2}\left\{\dfrac{k(s+2)^2 - 4(2s+3)}{k(s+2) - (s+6)}\right\} - A(k, s), \text{ otherwise} \end{cases}$$

edges, where $A(k, s)$ is a function of k and s such that if $n > (s+1)^2$, then $\lim_{n\to\infty} [A(k, s)]/n = 0$. We illustrate the construction procedure for the case $k = 3$; once this is given the general construction procedure becomes evident.

Suppose s is odd and $n = \frac{1}{2}(s+1)r + b$, where r is a natural number ≥ 6 and $0 \leq b \leq \frac{1}{2}(s-1)$. For the moment, we further suppose $b = 0$. Partition the vertices into r distinct sets $S_1, S_2, \ldots, S_r$, each set having exactly $\frac{1}{2}(s+1)$ vertices. We connect the vertices as follows. For $j = 1, 2, \ldots, 5$, connect every vertex of S_j to every vertex of S_{j+1}. Connect every vertex of S_6 to every vertex of S_1. For $j = 4, 5, \ldots, [\frac{1}{2}r]$, connect every vertex of S_{2j} (S_{2j+1}) to every vertex of S_1 (S_4) and S_{2j+1}. If r is odd or $b > 0$, then connect the remaining $\frac{1}{2}(s+1) + b$ vertices to every vertex of S_1 and S_4. It is easily seen that the graphs so constructed belong to the class $G_V(n, 3, \lambda, s)$, $\lambda \geq 4$, and they have $\frac{3}{4}n(s+1) - c$ edges, where $s + 1 < c < (s+1)^2$. Notice that all but $s + 1$ vertices have degree $s + 1$.

If s is even, then in the above construction we take $|S_i| = \frac{1}{2}(s+2)$, for $i = 1, 2, \ldots, 6$, and $|S_i| = \frac{1}{2}s$, for $i \geq 7$. Naturally, we are now taking $n = 3(s+2) + (ts/2) + b$, where $0 \leq b < \frac{1}{2}s$ and $r = t + 6$. It is easily seen that the resulting graph is a $G_V(n, 3, \lambda, s)$, $\lambda \geq 4$, and has $\frac{1}{4}(3s+4) - c$ edges, where $s + 1 \leq c < (s+1)^2$.

For the case of CONJECTURE 5(i) it is not difficult to construct $G_V(n, 3, 3, s)$'s with $n(s+1) - c$ edges, where $s + 1 \leq c < (s+1)^2$.

SOME RELATED PROBLEMS

One constraint we have not considered is that of the capacity of a center which we may define as *the maximum number of links incident to it.* In graph theoretic terms this is just an upper bound on the degrees of the vertices. Murty [21] considered this problem and obtained some results for the case $k = \lambda = 2$. Erdös *et al.* [18] and Bollobás [3] considered the graph theoretical problem of determining the smallest number of edges for a graph which is of diameter k and in which every vertex has a prescribed maximum degree.

Throughout this survey we have considered only undirected graphs. The corresponding problem for directed graphs received some attention in Murty [21]. He obtained some results for the case of diameter 2.

We conclude this survey by posing the following problem: *given n centers of communication and q links, construct a network with the highest possible reliability and satisfying the constraint that each center can be connected to every other center using at most k links.* In graph theoretic terms, we ask for the maximum connectivity $K(n, q, k)$ of a graph with n vertices, q edges, and diameter at most k. From a result of Harary [19], we conclude that

$$K(n, q, k) \leq \left[\frac{2q}{n}\right]$$

For the particular case $k = 2$, we can construct a graph with connectivity $[2q/(n-1)] - 1$ as follows. First construct a graph S with $n - 1$ vertices, $q - n + 1$ edges, and maximum connectivity $[2q/(n-1)] - 2$, using the construction procedure of Harary [19]. Then the graph $G = K_1 + S$ has the desired properties. Now since $K(n, q, k)$ is clearly a monotonic, nondecreasing function of k we have

$$\left[\frac{2q}{n-1}\right] - 1 \leq K(n, q, k) \leq \left[\frac{2q}{n}\right]$$

Notice that for connected graphs $[2q/n] = [2q/(n-1)]$ or $[2q/(n-1)] - 1$. So the question is when can we construct $[2q/n]$-connected graphs with n vertices, q edges, and diameter k. We suspect that for small values of k, at least, this is rare when $[2q/n] = [2q/(n-1)]$.

Summary

In this expository article we consider the problem of finding the structures of extremal graphs with preassigned diameters and connectivities. More specifically, we consider the problem of (1) determining the minimum possible number of edges for a graph of order n and diameter at most k having the property that, by deleting any s or fewer vertices (edges), the resulting subgraph (partial graph) has a diameter at most λ, $\lambda \geq k$, and (2) characterizing the graphs with the minimum number of edges. We present a comprehensive bibliography of all pertinent developments.

Acknowledgments

I wish to express my gratitude to Dr. K. Vijayan not only for introducing me to graph theory, but also for many useful and stimulating discussions on the subject. Thanks are also due to Professor F. Harary for suggesting that I write this article.

References

1. Bollobás, B. 1968. A problem in the theory of communications networks. Acta. Math. Acad. Sci. Hung. **19:** 75–80.
2. Bollobás, B. 1968. Graphs of given diameter. Theory of Graphs. Proc. Colloq. (held at Tihany, Hungary). G. Katona & P. Erdös, Eds. Academic Press, New York, N.Y. pp. 29–36.
3. Bollobás, B. 1969. Graphs with given diameter and maximal valency. Comb. Math. Appl. D. J. A. Welsh, Ed. Academic Press, New York, N.Y. pp. 25–37.
4. Bollobás, B. 1976. Strongly two connected graphs. Congr. Numerantium XVII. Proc. Seventh Southeastern Conf. Combinatorics, Graph Theory, Comput. Baton Rouge, La. F. Hoffman *et al.*, Eds. Utilitas Math., Winnipeg, Manitoba, Canada. 161–170.
5. Bollobás, B. 1976. Graphs with given diameter and minimal degree. Ars Combinatoria. **2:** 3–9.
6. Bollobás, B. & S. E. Eldridge. 1976. On graphs with diameter 2. J. Comb. Theory (B). **21:** 201–205.
7. Bollobás, B. & P. Erdös. 1975. An extremal problem of graphs with diameter 2. Math. Mag. **48:** 281–283.
8. Bollobás, B. & F. Harary. 1976. Extremal graphs with given diameter and connectivity. Ars Combinatoria. **1:** 281–296.
9. Bondy, J. A. & U. S. R. Murty. 1972. Extremal graphs of diameter 2 with prescribed minimum degree. Stud. Sci. Math. Hung. **7:** 239–241.

10. Bondy, J. A. & U. S. R. Murty. 1976. Graph Theory with Applications. Macmillan, London, England.
11. Caccetta, L. 1976. Some problems in extremal graph theory. University of Western Australia. Doctoral dissertation.
12. Caccetta, L. 1976. Extremal graphs of diameter 4. J. Comb. Theory (B). **21:** 104–115.
13. Caccetta, L. 1976. A problem in extremal graph theory. Ars Combinatoria. **2:** 33–56.
14. Caccetta, L. 1977. On extremal graphs. Proc. Eighth Southeastern Conf. Combinatorics, Graph Theory, Comput. Baton Rouge, La. F. Hoffman *et al.*, Eds. Utilitas Math. Winnipeg, Manitoba, Canada. Congr. Numerantium XIX: 125–137.
15. Caccetta, L. 1979. Extremal graphs of diameter 3. J. Austral. Math. Soc. (A), to appear.
16. Caccetta, L. 1977. Characterization of extremal graphs of diameter 4, I. Ars Combinatoria. **4:** 187–224.
17. Caccetta, L. 1977. Characterization of extremal graphs of diameter 4, II. Ars Combinatoria, to appear.
18. Erdös, P., A. Rényi & V. T. Sós. 1966. On a problem in graph theory. Stud. Sci. Math. Hung. **1:** 215–235.
19. Harary, F. 1962. The maximum connectivity of a graph. Proc. Nat. Acad. Sci. **48:** 1142–1146.
20. Harary, F. 1969. Graph Theory. Addison-Wesley, Reading, Mass.
21. Murty, U. S. R. 1966. Extremal graph-theoretic problems with applications. Doctoral dissertation. Indian Statistical Institute. Calcutta.
22. Murty, U. S. R. 1968. Critical graphs of diameter 2. Math. Mag. **41:** 138–140.
23. Murty, U. S. R. 1968. On some extremal graphs. Acta. Math. Acad. Sci. Hung. **19:** 69–74.
24. Murty, U. S. R. 1969. Extremal nonseparable graphs of diameter 2. *In* Proof Techniques in Graph Theory. Proc. Second Ann Arbor Graph Theory Conf., Ann Arbor, Mich. F. Harary, Ed. Academic Press, New York, N.Y. pp. 111–117.
25. Vijayan, K. & U. S. R. Murty. 1964. On accessibility in graphs. Sankhya Ser. A. **26:** 299–302.

SURVEY OF EXTENSIONS OF BROOKS' GRAPH COLORING THEOREM

Paul A. Catlin

Department of Mathematics
Wayne State University
Detroit, Michigan 48202

Following the notation of [11], we consider a graph G of maximum degree $\Delta(G) = h$ and denote its chromatic number by $\chi(G)$. We present here several extensions of the well-known theorem of Brooks which relates the chromatic number of a graph to its maximum degree.

THEOREM 1 (Brooks [3]). For any connected graph G,

$$\chi(G) \leq \Delta(G) + 1 \tag{1}$$

with equality if and only if either $\Delta(G) = 2$ and G is an odd cycle; or $\Delta(G) \geq 3$ and G is complete.

A clique $K_{\Delta(G)+1}$ is necessarily a component of a graph G, and if $\Delta(G) = 2$, then odd cycles are components. We refer to such components as *Brooks components* in order to use a collective name.

First, we shall consider extensions of THEOREM 1 in which one color class has a certain property, i.e., it is a maximum independent set, or its vertices all have degree $\Delta(G)$. Then we give related results of Dirac [9] in which $\deg_G(v) \leq h$ for all but a few exceptional vertices, and yet G still has an h-coloring. Finally, we present results which assert that if the size of the largest clique of G is small, or if the girth of G is large, then the bound $\chi(G) \leq \Delta(G)$ of Brooks' Theorem can be improved considerably.

THEOREM 2 (Catlin [5]). If Brooks' Theorem implies $\chi(G) \leq \Delta(G)$, then one of the color classes may be chosen to be some maximum independent set in a $\Delta(G)$-coloring of G.

Let $\beta(G)$ be the point independence number of G.

Let G_1 and G_2 be the graphs of FIGURE 1. Thus, for $h = 4$ and $h = 5$, G_{h-3} has no K_h and $|V(G_{h-3})| = 2h$, $\Delta(G_{h-3}) = h$, $\beta(G_{h-3}) = 2 = |V(G_{h-3})|/\Delta(G_{h-3})$.

THEOREM 3 (Albertson *et al.* [1]). Except for the two graphs G_1 and G_2 of FIGURE 1, if G has no K_h, where $\Delta(G) = h$, then $\beta(G) > |V(G)|/h$.

Of course, $\beta(G) \geq |V(G)|/h$ follows immediately from Brooks' Theorem, and so equality only holds for these two graphs.

Albertson, Bollobás, and Tucker conjectured that in THEOREM 3 a coloring of G in h colors could be found such that not all color classes have the same size, provided G is neither G_1 nor G_2. The truth of this conjecture follows immediately from THEOREMS 2 and 3.

Mitchem [13] gives a short proof of THEOREM 2. We obtained THEOREM 2 from a stronger result, presented next. Let $\Delta(G) = h$.

0077-8923/79/0328-0095$1.75/1 © 1979, NYAS

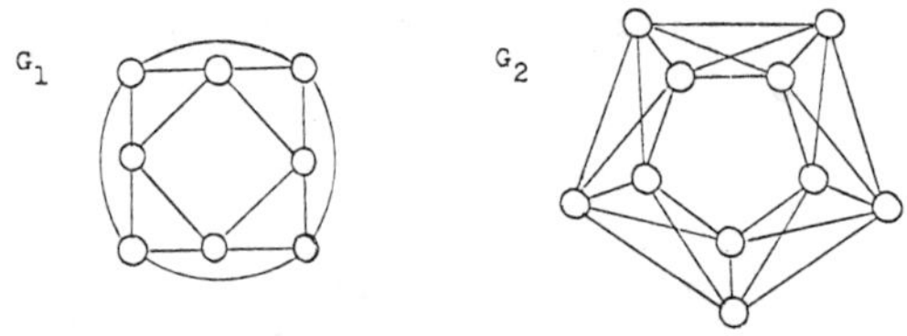

FIGURE 1. The Albertson-Bollobás-Tucker graphs.

Consider the following two questions: Is there an h-coloring of G in which:

(a) One color class consists entirely of vertices of degree h?

(b) One color class is a maximum independent subset of the set of vertices of degree h?

The answer to both questions (a) and (b) is negative for the following class of graphs: If $h = 3$, let G consist of pairwise disjoint odd cycles with edges joining different odd cycles, but in which the pairwise disjoint odd cycles are the only cycles of G (an example is shown in FIGURE 2). If $h \geq 4$, let G consist of pairwise disjoint cliques K_h with edges joining different cliques (see FIGURE 3). Of course the joining edges must also be bridges, with no two bridges incident with a common vertex, in order that $\Delta(G) = h$. If such graphs are connected, they have a tree-type structure, where the odd cycles or the cliques are "vertices." We call such connected graphs *Brooks trees of degree h.* (Of course, they are not trees.)

THEOREM 4 (Catlin [5]). Brooks trees are the only connected graphs which are exceptions to questions (a) and (b).

It is easy to verify that a Brooks tree G is an exception: each vertex in an independent set S of vertices of degree h is a vertex on a bridge. Since G has a tree-type structure, there are fewer bridges than there are cliques ($h \geq 4$) or odd cycles ($h = 3$). Hence, when S is deleted from G, a clique K_h, $h \geq 4$, or an odd cycle ($h = 3$) remains, and $G - S$ cannot be colored in $h - 1$ colors.

It is also easy to show that THEOREM 4 implies THEOREM 2. Suppose G is a graph and we wish to find an h-coloring in which a maximum independent set is monochromatic. Merely join to each vertex v of G exactly $h - \deg_G v$ edges which each link v with a new vertex outside $V(G)$ of degree 1. The resulting graph, $G^{\#}$, having no vertices of degree $h - 1$, is not a Brooks tree and, by THEOREM 4, for question (b), it follows that there is a maximum independent subset S of vertices of degree h in $G^{\#}$, where S is one color class in an h-coloring. Thus, S is a maximum independent set in G and S is monochromatic in some h-coloring of G.

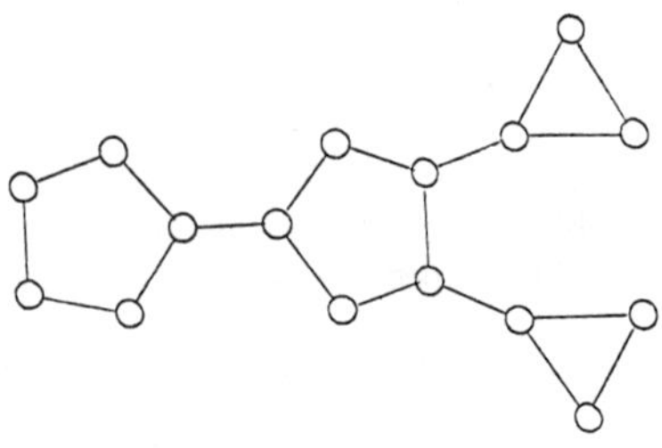

FIGURE 2. A Brooks tree of degree 3.

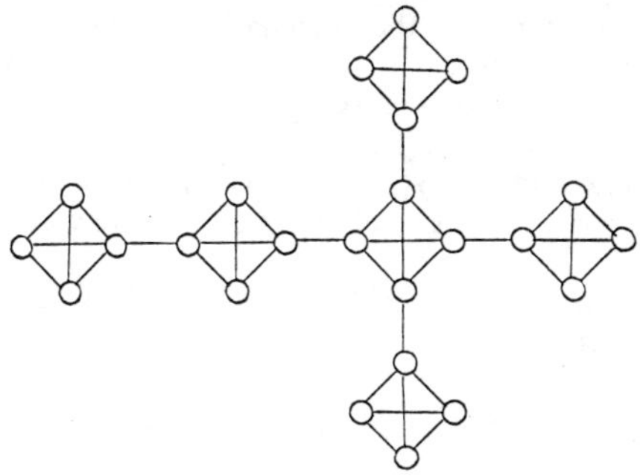

FIGURE 3. A Brooks tree of degree 4.

THEOREM 5 (Dirac [9, Theorems 4 and 10]). Let G be a connected graph with a distinguished vertex v, and suppose that $\deg_G x \leq h$, for all $x \in V(G) - v$. If $\chi(G) > h$ and if G is not K_{h+1}, then $G - v$ is a Brooks tree of degree h, and each vertex of degree $h - 1$ in $G - v$ is adjacent in G to v.

THEOREM 5 is easily seen to be equivalent to THEOREM 4 for question (a). If $\chi(G) > h$ and if G is a minimal graph satisfying the hypotheses of THEOREM 5, then the color class containing v is a color class of vertices of degree h in $G - v$.

Let

$$(a - b)^+ = \begin{cases} a - b, \text{ if } a - b \geq 0 \\ 0, \text{ if } a - b < 0 \end{cases}$$

THEOREM 6 (Dirac [9, Theorem 15]). Let $h \geq 3$. If

$$\sum_{v \in V(G)} (\deg_G x - h)^+ \leq h - 3 \tag{2}$$

then either $\chi(G) \leq h$ or G contains a K_{h+1}.

This relaxes the hypothesis $\deg_G(x) \leq h$ of Brooks' Theorem slightly. To see that this is best possible, consider the graph G on $2h + 1$ vertices formed by two disjoint cliques K_h joined by a single edge e (i.e., a Brooks tree of degree h on $2h$ vertices), together with a vertex v of degree $2h - 2$ joined to each vertex of the two K_h's except the two ends of e (see FIGURE 4). Note that $\chi(G) = h + 1$, G has no K_{h+1}, and the left side of (**2**), known as the *excess* of G, is

$$\deg_G v - h = h - 2$$

As far as we know, this single example is the only edge-minimal counterexample if the right side of (**2**) is replaced by $2h - 5$.

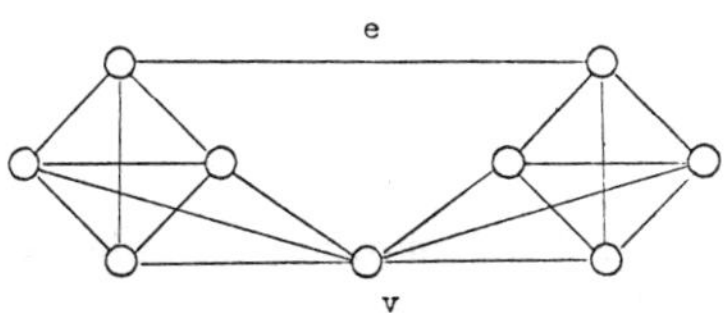

FIGURE 4. The extremal graph for Dirac's Theorem, $h = 4$.

Since Brooks' Theorem relates the chromatic number to the maximum degree and the size of the largest clique, it is natural to ask if the bound on $\chi(G)$ can be improved if there are no large cliques. Grünbaum [10] conjectured that such improvements were impossible: for all $h \geq 3$, it was conjectured that there exists a graph G with no triangles and with $\chi(G) = \Delta(G) = h$. Although there are triangle-free graphs with $\chi(G)$ arbitrarily large, Grünbaum's conjecture was recently proved false:

THEOREM 7 (Borodin and Kostochka [2], Catlin [4]). If G contains no K_{r+1}, where $3 \leq r \leq \Delta(G)$, then

$$\chi(G) \leq \frac{r}{r+1}(\Delta(G) + 2)$$

Although this result was obtained independently by Borodin and Kostochka [2] and by Catlin [4], in both proofs the following decomposition theorem of Lovász was used, in which for $X_i \subseteq V(G)$, the subgraph induced by X_i is written $\langle X_i \rangle$.

THEOREM 8 (Lovász [12]). Let G be a graph and n be a natural number. For any partition

$$\Delta(G) - (n-1) = h_1 + h_2 + \cdots + h_n$$

there is a decomposition of $V(G)$ into n sets $X_1, X_2, \ldots, X_n$ such that, for all i, $\Delta(\langle X_i \rangle) \leq h_i$.

One can obtain THEOREM 7 from THEOREMS 1 and 8 by setting $h_1 = h_2 = \cdots = h_{n-1} = r$.

A graph is *h-degenerate* if it has a subgraph of minimum degree h, but no subgraph with minimum degree $h + 1$ or more. The generalization of THEOREM 7 to h-degenerate graphs was obtained independently by Borodin and Kostochka [2] and by Mitchem [14]. Mitchem [15] had earlier generalized Brooks' Theorem to h-degenerate graphs.

Unfortunately, we know of no example that would make THEOREM 7 best possible. There are several examples with $\Delta(G)$ small in which $\chi(G)$ is substantially larger than the largest clique size, which constitutes a lower bound on $\chi(G)$. Chvátal [7] gives the smallest graph of girth 4 with $\chi(G) = \Delta(G) = 4$. The graph $G = C_5 + C_5$ is mentioned by Dirac [8, p. 90]. Obviously G has no K_5, and $\chi(G) = 6$, $\Delta(G) = 7$. This is of interest, because the least value of $\Delta(G)$ for which THEOREM 7 improves Brooks' Theorem is $\Delta(G) = 7$: if there is no K_4, then the chromatic number is at most 6.

There are two other results similar to THEOREM 7, both given in the last two theorems, which we state.

THEOREM 9 (Catlin [6]). Let G be a graph. If G has no $K_{r+2} - e$ as a subgraph, for some $r \geq 3$, then

$$\chi(G) \leq \frac{r}{r+1}(\Delta(G) + 3)$$

If also G has no 4-cycle, then

$$\chi(G) \leq \tfrac{2}{3}(\Delta(G) + 3)$$

In a communication from Borodin and Kostochka, the following unpublished result was announced:

THEOREM 10 (Kostochka). If G is a graph of girth at least $2\Delta(G)^2$, then $\chi(G) \leq (\Delta(G) + 4)/2$.

REFERENCES

1. ALBERTSON, M., B. BOLLOBÁS & S. TUCKER. 1976. The independence ratio and maximum degree of a graph. Proc. Seventh Southeastern Conference on Combinatorics, Graph Theory and Computing, LSU.
2. BORODIN, O. V. & A. V. KOSTOCHKA. 1977. On an upper bound of the graph's chromatic number depending on graph's degree and density. J. Comb. Theory Ser. B. **23:** 247–250.
3. BROOKS, R. 1941. On colouring the nodes of a network. Proc. Cambridge Phil Soc. **37:** 194–197.
4. CATLIN, P. A. 1978. A bound on the chromatic number of a graph. Discrete Math. **22:** 81–83.
5. CATLIN, P. A. Brooks' graph-coloring theorem and the independence number. J. Comb. Theory Ser. B. To appear.
6. CATLIN, P. A. 1978. Another bound on the chromatic number of a graph. **24:** 1–6.
7. CHVÁTAL, V. 1970. The smallest triangle-free-4-chromatic 4-regular graph. J. Comb. Theory Ser. B. **9:** 93–94.
8. DIRAC, G. A. 1952. A property of 4-chromatic graphs and some remarks on critical graphs. J. London Math. Soc. **27:** 85–92.
9. DIRAC, G. A. 1957. A theorem of R. L. Brooks and a conjecture of H. Hadwiger. Proc. London Math. Soc. **7:** (3) 161–195.
10. GRÜNBAUM, B. 1970. A problem in graph coloring. Am. Math. Mon. **77:** 1088–1092.
11. HARARY, F. 1969. Graph Theory. Addison-Wesley, Reading, Mass.
12. LOVÁSZ, L. 1966. On decomposition of graphs. Stud. Sc. Math. Hung. **1:** 237–238.
13. MITCHEM, J. A short proof of Catlin's extension of the theorem of Brooks. To appear.
14. MITCHEM, J. On the genus of graphs with Lick-White number k. Proc. Am. Math. Soc. To appear.
15. MITCHEM, J. 1977. An extension of Brooks' theorem to *n*-degenerate graphs. Discrete Math. **17:** 291–298.

SOME OPEN (AND SOME CLOSED) QUESTIONS ON SYMMETRIC BIBD'S AND PBIBD'S

Allan Gewirtz

Department of Mathematics
Brooklyn College
City University of New York
Brooklyn, New York 11210

Introduction and Background

The purpose of this paper is to show how two apparently distinct mathematical entities can be so interrelated that knowledge of the existence or nonexistence of either can lead directly to equivalent knowledge of the other. In addition, since the methods used to study the entities were different, it means that new methods can be brought to bear on the old problems, perhaps leading to solutions of the remaining problems, or indeed, creating new ones.

A. The first of these entities are graphs with maximal even girth studied in some detail in [1, 2]. These can be viewed as the class G of simple, undirected, connected graphs of diameter $d > 1$ and girth $2d$ such that if a pair of nodes (vertices) are at a distance d from each other then that pair of nodes is connected by $t > 1$ distinct paths of length d. (The girth of a graph is the length of the smallest circuit. For terms not explicitly defined see [3].) Such graphs clearly contain no triangles

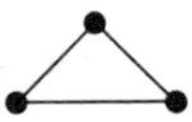

because then the girth of the graph would be $3 \neq 2d$ for any diameter $d > 1$. If the diameter is fixed at $d = 2$ then FIGURE 1 illustrates a graph with all the required properties and with $t = 2$.

If $d = 2$ and $t = 2$ there are graphs other than the diamond of FIGURE 1 which meet all the conditions. A necessary condition for such graphs is that they be regular of degree (valence) v. It is known that the next largest graph has valence 5 and 16 nodes. Up to isomorphism it is unique and illustrated in FIGURE 2.

All such graphs have the property that they can be displayed as a hierarchy [4] with any vertex of the graph selected to be identified as 0 and considered as being on level 0 of the hierarchy. The number of levels of the hierarchy is the diameter d of the graph and the girth $2d$ implies that no vertices on the same level are connected together except for those on level d which are connected together by edges called *reentry arcs*. FIGURE 3 illustrates the hierarchy of the unique graph with $d = 2$, $t = 2$, valence $v = 10$, and which has $n = 56$ nodes. Each of the forty-five vertices on level 2 are named for the two vertices of level 1 which are connected to them. The reentry arcs are too numerous to draw in the figure, but it should be noted that no vertex of level 2 is connected to any other vertex of level 2 that has the same digit in its name. For instance, vertex 12 is not connected to 13, 14, ..., 1t, 23, 24, ..., 2t, since to do so violates the triangle condition.

For further discussion it is of benefit to adopt the notation $g(d, t, v, n)$ for graphs

0077-8923/79/0328-0100$1.75/1 © 1979, NYAS

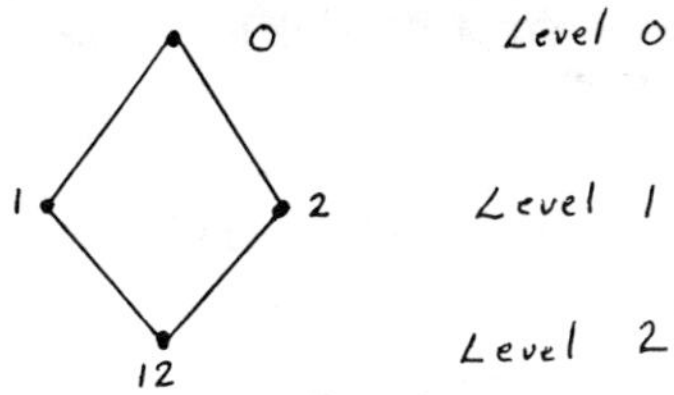

FIGURE 1.

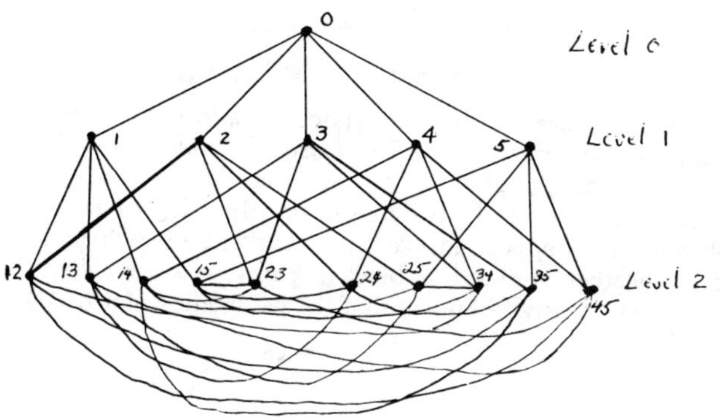

FIGURE 2.

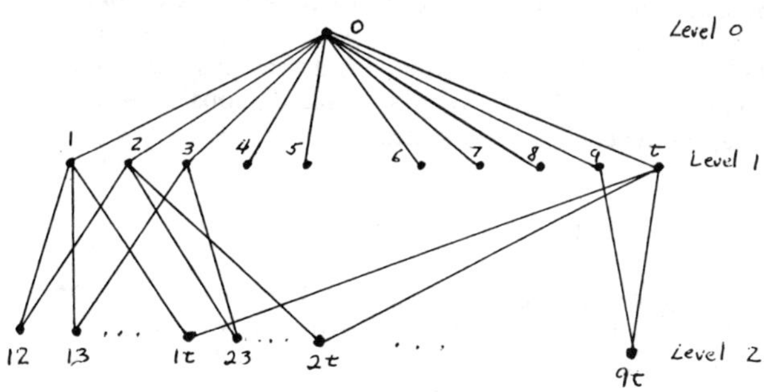

FIGURE 3.

g that meet the conditions and have diameter d, t (as described), valence v, and n vertices. For the purposes of this paper $d = 2$. The known graphs are:

$g(2, 2, 2, 4)$ (FIGURE 1)

$g(2, 2, 5, 16)$ (FIGURE 2)

$g(2, 2, 10, 56)$ (FIGURE 3)

$g(2, 6, 22, 100)$

$g(2, 4, 16, 77)$

All are known to be unique except possibly $g(2, 4, 16, 77)$. If $d = 2$ and $t = 2$, a necessary condition for g to exist is that:

$$v^2 + v + 2 = 2n$$

$$v = a^2 + 1, \text{ with integral } a \not\equiv 0 \ (4)$$

If $d = 2$ and $t = 4$, the equivalent formulas are:

$$v^2 + 3v + 4 = 4n$$

$$v = a^2, \ a > 1$$

If $d = 2$ and $t = 6$, the equivalent formulas are:

$$v^2 + 5v + 6 = 6n$$

$$v = a^2 - 3, \ a \geq 3, \ a \not\equiv 0, 4, 8 \ (12)$$

B. The second of these entities are symmetric *balanced incomplete block designs* (BIBD's) and symmetric *partially balanced incomplete block designs* (PBIBD's). A BIBD can be thought of as a collection of b sets (blocks) with k elements (varieties) in each set, the varieties to be selected from a set with v elements, each variety to appear in exactly r blocks, and each pair of varieties to appear together in exactly λ blocks; v, b, k, r, λ are called the parameters of the BIBD. It is well known that the parameters of a BIBD satisfy

$$vr = bk$$

$$r(k - 1) = \lambda(v - 1)$$

We write $B(v, b, k, r, \lambda)$.

A BIBD is called symmetric if $v = b$ and $r = k$. In that case, we write $B(b, k, \lambda)$. In [5] using computer methods and with the knowledge that $v = 1 + k(k - 1)/2$ must be true for a symmetric BIBD with $\lambda = 2$, it was shown that there exists a $B(56, 11, 2)$. In [6] it was pointed out that this particular $B(56, 11, 2)$ was inherent in the $g(2, 2, 10, 56)$ of [1]. This relationship will be explained in the broader context of the next section.

Graphs and Symmetric BIBD's

To start, we show how to create a $B(56, 11, 2)$ from $g(2, 2, 10, 56)$ using Figure 3 as a model. The varieties are the fifty-six vertices of Figure 3. Each of the fifty-six vertices also names a block; thus $b = v = 56$. Each block contains the vertex of its name and the ten vertices adjacent to the vertex of its name. Each vertex is in the block of its name and the blocks named by the ten vertices adjacent to it, thus $k = r = 11$.

Under the assignment of vertices to blocks, it is clear that if two vertices are adjacent, they are in two blocks together, the ones named by them. If two vertices are not adjacent, they are in the $t = 2$ blocks named by the vertices to which they are both adjacent. That this number is exactly 2 is clear, because if they are not adjacent, they must be at distance 2 from each other (otherwise $d > 2$), giving one common adjacency, and now the $t = 2$ condition asserts that there is exactly one more adjacency.

It now becomes clear that, for every $g(2, 2, v, n)$, there is an associated $B(n, v + 1, 2)$ obtained in exactly the same manner as just described. It is not clear, nor necessarily true that, given a $B(n, v + 1, 2)$, there is an associated $g(2, 2, v, n)$.

We return to this question in the final section. We do have:

$g(2, 2, v, n)$	$B(n, v + 1, 2)$
$g(2, 2, 2, 4)$	$B(4, 3, 2)$
$g(2, 2, 5, 16)$	$B(16, 6, 2)$
$g(2, 2, 10, 56)$	$B(56, 11, 2)$

Graphs and Symmetric PBIBD's

In the description of BIBD's let b, v, k, r be as described, and suppose that there exist a set of integers $\lambda_1, \ldots, \lambda_m$ such that each pair of varieties are in λ_i $(i = 1, \ldots, m)$, blocks together. Then the system with parameters v, b, r, k, $\lambda_1, \ldots, \lambda_m$ is called a *partially balanced incomplete block design* (PBIBD). If $v = b$, $r = k$, the PBIBD is *symmetric*. With this definition we can now show that, for any graph $g(2, t, v, n)$, $t > 2$, there exists an associated symmetric PBIBD with $\lambda_1 = 2$ and $\lambda_2 = t$. We write for symmetric PBIBD's with two λ's $PB(n, k, \lambda_1, \lambda_2)$.

To see the existence of such designs, we proceed to describe the construction which is almost exactly as that used to get symmetric BIBD's from $g(2, 2, v, n)$ graphs. This is true because the hierarchical structure of $g(2, t, v, n)$ graphs with $t > 2$ is the same as for the case $t = 2$ except that each node of level 2 now connects to $t > 2$ nodes of level 1 instead of connecting to $t = 2$ nodes.

The number and description of blocks, varieties, block size, and number of blocks containing each variety is exactly as in the previous section. Also, if two vertices are adjacent, they are in $\lambda_1 = 2$ blocks together, the ones they name. However, if two vertices are not adjacent, they are in t blocks together, those to which they are both adjacent. Therefore, we have:

$g(2, t, v, n)$	$PB(n, v + 1, 2, t)$
$g(2, 4, 16, 77)$	$PB(77, 17, 2, 4)$
$g(2, 6, 22, 100)$	$PB(100, 23, 2, 6)$

Open Problems

A question still to be answered (private correspondence—J. H. van Lint to A. Gewirtz) is does a unique graph $g(2, t, v, n)$ imply a unique $B(n, v + 1, 2)$ or $PB(n, v + 1, 2, t)$? In general, this question is still open. For $g(2, 2, 2, 4)$, clearly there is exactly one $B(4, 3, 2)$. It is clear that the structure of the graphs in question is such that no associated design can have two blocks both containing the same three varieties.

There are many open questions dealing with these matters; here are some (in no particular order):

(1) It is an open question whether or not there exists a $g(2, 3, 21, 162)$ (no other $t = 3$ graph can exist). The existence or nonexistence of a $PB(162, 22, 2, 3)$ may resolve this question. Certainly nonexistence would resolve it.

(2) Same as (1) for $g(2, 5, 55, 650)$ (no other $t = 5$ graph can exist).

(3) Same as (1) for $g(2, 8, 136, 2432)$ (no other $t = 8$ graph can exist).
(4) Are the associated designs, discussed above, unique?
(5) If $d > 2$, are there any associated designs?
(6) Are there other ways to assign the vertices to blocks to create further designs?

REFERENCES

1. GEWIRTZ, A. 1969. Graphs with maximal even girth. Can. J. Math. **21:** 915–934.
2. GEWIRTZ, A. 1969. The uniqueness of $g(2, 2, 10, 56)$ Trans. N.Y. Acad. Sci. **31:** 656–675.
3. HARARY, F. 1969. Graph Theory. Addison-Wesley, Reading, Mass.
4. HOFFMAN, A. J. & R. SINGLETON. 1960. On Moore graphs with diameters 2 and 3. IBM J. Res. **4:** 497–504.
5. KINGSLEY, R. A. & R. G. STANTON. 1971. On a block design (56, 56, 11, 11, 2). Proc. 25th Summer Meet. Can. Math. Cong. Thunder Bay, Ont. pp. 231–244.
6. GEWIRTZ, A. 1975. Review No. 1913. Math. Reviews. **50:** 267–268.
7. HALL, JR., M. 1964. Applied Combinatorial Mathematics. E. E. Bechenbach, Ed. Wiley, New York, N.Y. p. 372.

EXPOSITION OF THE CONSTRUCTION OF 103 IRREDUCIBLE GRAPHS FOR THE PROJECTIVE PLANE

Gary Haggard

Department of Mathematics
University of Maine at Orono
Orono, Maine 04473

INTRODUCTION

In 1930 Kuratowski [6] proved that a graph can be imbedded in the plane if and only if it does not contain a subgraph homeomorphic to either K_5 or $K_{3,3}$. A major unsolved problem in graph theory is to determine whether or not there is a finite set of graphs characterizing imbeddability on any other surface—orientable or nonorientable. The object of this paper is to give an exposition of some work of Glover *et al.* [5] which is a first step in finding a Kuratowski theorem for the projective plane P. This first step involves constructing a family of 103 graphs which are irreducible for P. In subsequent work it will have to be shown that in fact these are all the irreducible graphs for P. This construction of irreducible graphs for P is the first major advance towards a Kuratowski type theorem for any surface since 1930 except for a result of Milgram [7], and Glover and Huneke [1] which characterized the cubic irreducible graphs for P.

FUNDAMENTAL OPERATION

Let G be a graph with vertex v. Let V_1 and V_2 be a partition of the vertices of G adjacent to v. The graph $S_v G$, called the *split of G at v relative to V_1 and V_2*, is formed from G as follows:

(1) Replace v by v_1 and v_2.
(2) Join v_1 and v_2 by an edge.
(3) Join v_i to each vertex in V_i, where $i = 1, 2$.

An example should help clarify this construction (see FIGURE 1). When we can list the vertices in V_1, we denote the split graph as $G_{v(a,\ldots,b)}$, where $V_1 = \{a, \ldots, b\}$. As we are interested in imbeddings of graphs, we need to know how the imbeddability of this split graph is related to the imbeddability of the original graph.

LEMMA 1 (Glover, Huneke, and Wang). Let G be a graph and let $S_v G$ be a split of G at v. If $G \not\subset M$ (read G does not imbed on M), then $S_v G \not\subset M$.

Although the splitting process preserved the nonimbeddability property of the original graph, we are really interested in irreducible graphs for a surface which are a special family of graphs that do not imbed on a surface. A graph G is *irreducible for a surface M* if and only if $G \not\subset M$, but $G - e \subset M$, for each $e \in E(G)$. We call an edge e of a graph G *essential for a surface* M if $G \not\subset M$ but $G - e \subset M$. Thus G is irreducible for M if and only if each edge of G is essential for M. Since the split of a graph that does not imbed on M itself does not imbed on M, there is an irreducible graph for M contained in the graph formed by the splitting process. Rather than starting with an arbitrary collection of nonprojective planar graphs and performing

0077-8923/79/0328-0105$1.75/1

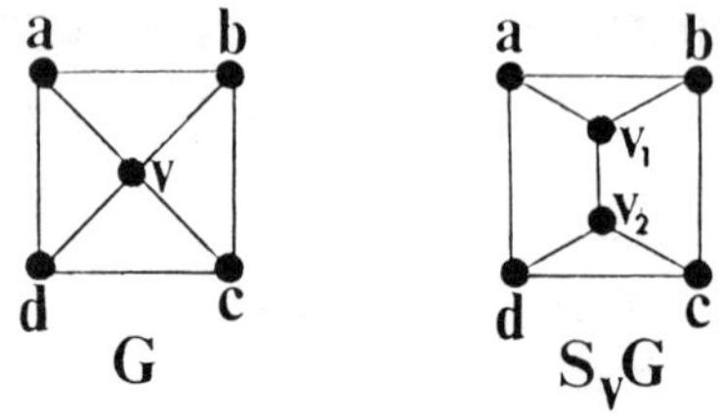

FIGURE 1. Split of G at V.

vertex splittings on these graphs in the hopes of finding irreducible graphs, Glover, Huneke, and Wang start with only the five graphs shown in FIGURE 2 and then proceed to find their 103 irreducible graphs by splitting vertices appropriately. Because of the splittings they choose, their graphs are found from splittings without any edge deletions. In another paper, Glover *et al.* [4] prove that *all irreducible graphs for P can be derived by splittings without any edge deletions from the five graphs in* FIGURE 2. Thus an important contribution of their work is the indication of a possibility of solving the Kuratowski problem for other surfaces without actually constructing *all* the irreducible graphs for the surface, but only finding the few from which the others can be derived using vertex splittings. In [2, 3] it is established that there are finitely many irreducible graphs for the projective plane.

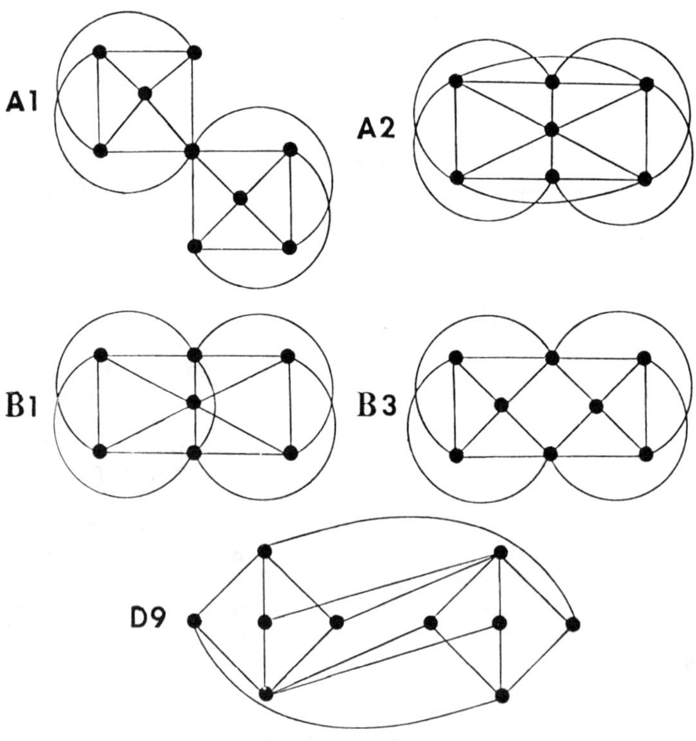

FIGURE 2. Generators of the 103 graphs.

Outline of the Proof

The key result for organizing the proof that these 103 graphs are irreducible for P is the following:

THEOREM 1 (Glover, Huneke, and Wang [5]). Let G be a graph and let $H = S_v G$ for some split of G at v. If $G \not\subset P$ and H is irreducible for P, then G is irreducible for P.

We can now describe the steps in the proof that these 103 graphs are irreducible for P.

(1) Construct a family of graphs which do not imbed on P.
(2) Find ways to split these graphs and the splits of these graphs so that all 103 graphs are realized.
(3) Prove that any graph which is at the end of a sequence of splittings performed in (2) is irreducible.
(4) Apply THEOREM 1 to prove that all 103 graphs are irreducible.
(5) Verify that each of these graphs is distinct from each of the others.

Typical Arguments

Besides a familiar Euler formula type argument for showing that a graph does not imbed on P, it is sometimes possible to remove an edge from a split of a graph that does not imbed on P and still have a graph which does not imbed on P. For example, if $G \not\subset P$ and $S_{v(w,z)}G$ contains a triangle (v, w, z), then both $S_{v(w,z)}G \not\subset P$ and $S_{v(w,z)}G - (w, z) \not\subset P$, since (w, z) can be imbedded once (v, w) and (v, z) are imbedded as indicated in FIGURE 3 by the dashed line.

In FIGURE 4 we see how this result can be used to find a graph which does

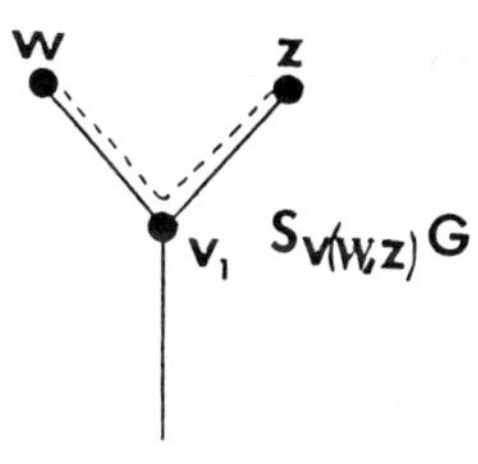

FIGURE 3. Redundancy of (w, z).

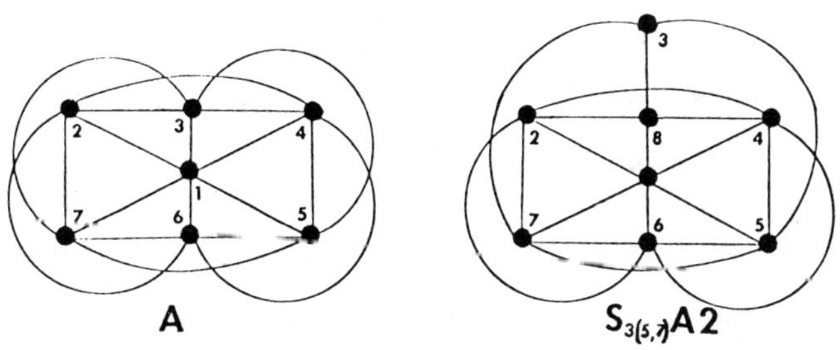

FIGURE 4. Application of THEOREM 1 (Glover, Huneke, and Wang).

not imbed on P. If we apply the previous result to the triangle (3, 5, 7) in $S_{3(5,7)}A2$ we find we can form $S_{3(5,7)}A2 - (5, 7) = B7$ and have $B7 \not\subset P$.

The next part of the proof involves finding sequences of splits such that each of the 103 graphs occurs in one of the chains. A typical chain is shown in FIGURE 5. Constructions of this sort lead to thirty-eight graphs which are at the ends of chains of splittings derived from the thirty-five graphs found in (1).

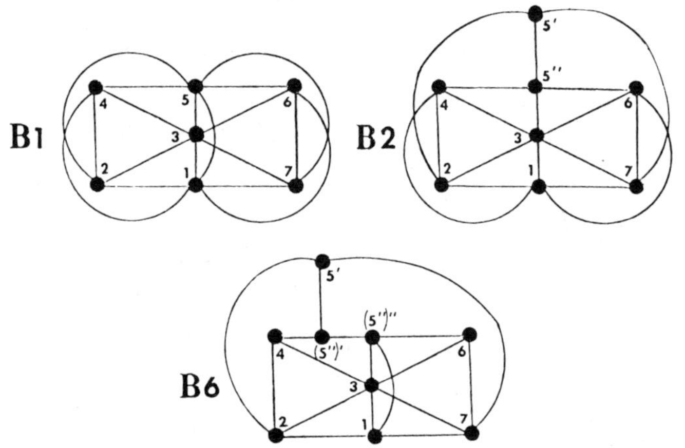

FIGURE 5. Chain of splits.

To accomplish (3) for the thirty-eight graphs in question involves finding an imbedding of $G - e$ on P for each edge e of each graph in the family. One chief tool in doing this is the set of symmetries of a graph. If e is an essential edge of G for P and Σ is a symmetry of G, then $\Sigma(e)$ is an essential edge of G for P. A second, more topological, tool is the following theorem.

THEOREM 2 (Glover, Huneke, and Wang). Let f: $G \to P$ be an immersion with exactly two crossing edges. Then $f(G - (v_1, v_2))$ is an imbedding where (v_1, v_2) is one of the crossing edges. Further, if e is a common edge for regions R and S, where $v_1 \in R$ and $v_2 \in S$, then e is essential in G for P.

FIGURE 6 helps make this result easier to understand.

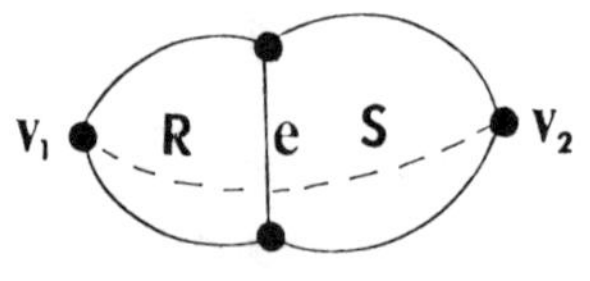

FIGURE 6. Edge e is essential.

We shall now give a proof that $E37$ is irreducible for P. We need three different immersions of $E37$ in P which are shown in FIGURE 7. By removing one of the crossing edges at a time, we are able to apply THEOREM 2 to the graphs shown in FIGURE 8.

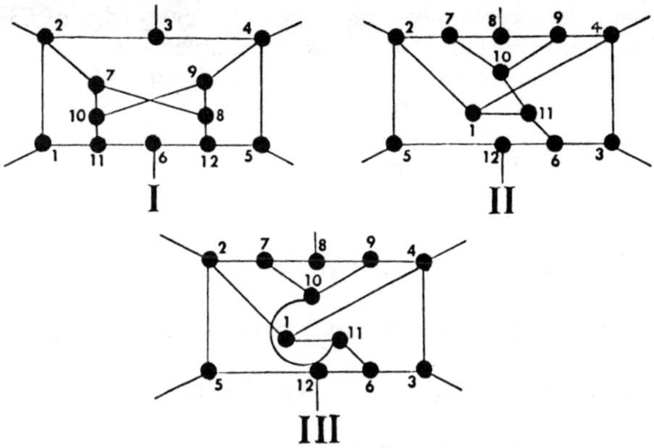

FIGURE 7. Immersions of $E37$.

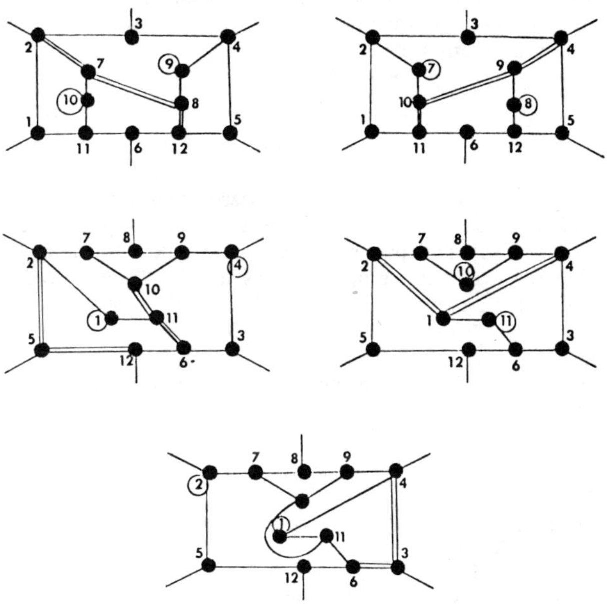

FIGURE 8. Double lines are essential.

There are two symmetries of the graph we would now like to apply to certain of the essential edges. The symmetries are:

$$\Sigma_1: (2, 4)(7, 9)$$

$$\Sigma_2: (8, 10)(11, 12)(1, 5)$$

Applying Σ_1 to (9, 10), (7, 8), and (3, 4) we get (7, 10), (7, 8), and (3, 4) to be essential. By applying Σ_2 to (5, 12), (1, 4), and (6, 11), we get (1, 11), (4, 5), and

(6, 12) to be essential. Since we have now proved that every edge of $E37$ is essential, and since $E37 \not\subset P$, we conclude that $E37$ is irreducible for P.

To complete the proof that these 103 graphs are irreducible for P requires that we show they are all distinct. In the appendix there is a listing of the edges of each of these 103 graphs and a listing of their cycle vectors as generated using Tiernan's algorithm [8]. Since all the cycle vectors are distinct, the graphs are all distinct.

Questions

A number of natural questions arise as a result of this work of Glover, Huneke, and Wang.

(1) Can one find a characterization of these 103 graphs other than in the context of splittings?
(2) Are there any special properties of the imbeddings of these graphs on P? Does each of these graphs have an imbedding for which the boundaries of the faces are cycles of the graph?
(3) Which results about P can be generalized to other surfaces?
(4) Can other tests for irreducibility be found which would give easier proofs than the ones required here?

Summary

Nearly fifty years after Kuratowski characterized the graphs which do not imbed in the plane, major results of a similar nature have been proven for the projective plane by Glover *et al.* [4]. This paper introduces the reader to the method of proof they used to find 103 irreducible graphs for this surface. The key to these results is the operation of splitting a vertex. Let G be a graph with vertex set V. Let $v \in V$ and let V_1 and V_2 be any partition of the vertices of G adjacent to v. The split of G at v relative V_1 and V_2 is the graph with vertex set $(V - \{v\}) \cup \{v_1, v_2\}$. In the new graph v_1 and v_2 are adjacent and v_i is adjacent to each vertex in V_i, for $i = 1, 2$.

References

1. Glover, H. H. & J. P. Huneke. 1975. Cubic irreducible graphs for the projective plane. Discrete Math. **13:** 341–355.
2. Glover, H. H. & J. P. Huneke. Graphs with bounded valency that do not embed in the projective plane. Discrete Math. To be published.
3. Glover, H. H. & J. P. Huneke. The set of irreducible graphs for the projective plane is finite. To appear.
4. Glover, H. H., J. P. Huneke & C. S. Wang. A Kuratowski theorem for the projective plane. To appear.
5. Glover, H. H., J. P. Huneke & C. S. Wang. 103 graphs which are irreducible for the projective plane. Submitted for publication.
6. Kuratowski, K. 1930. Sur le problème des courbes gauches en topologie. Fund. Math. **15:** 271–283.
7. Milgram, M. 1973. Irreducible graphs-2. J. Comb. Theory. **14:** 7–45.
8. Tiernan, J. C. 1970. An efficient search algorithm to find the elementary circuits of a graph. Commun. ACM. **13:** 722–726.

Appendix

LINE 1. Name of graph, number of vertices, number of edges.
LINE 2. List of edges.
LINE 3. List of edges.
LINE 4. Cycle vector beginning with the number of triangles.

A 1 9 20
1, 2; 1, 3; 1, 4; 1, 5; 1, 6; 1, 7; 1, 8; 1, 9; 2, 3; 2, 4; 2, 5; 3, 4; 3, 5;
4, 5; 6, 7; 6, 8; 6, 9; 7, 8; 7, 9; 8, 9;
20 20 30 24 0 0 0 0
A 2 7 18
1, 2; 1, 3; 1, 4; 1, 5; 1, 6; 1, 7; 2, 3; 2, 4; 2, 6; 2, 7; 3, 4; 3, 5; 3, 7;
4, 5; 4, 6; 5, 6; 5, 7; 6, 7;
18 20 51 108 160 120
A 3 10 21
1, 2; 1, 3; 1, 4; 1, 5; 1, 6; 1,10; 2, 3; 2, 4; 2, 5; 3, 4; 3, 5; 4, 5; 6, 7;
6, 8; 6, 9; 7, 8; 7, 9; 7,10; 8, 9; 8,10; 9,10;
21 17 27 24 6 0 0 0 0
A 4 11 22
1, 3; 1, 4; 1, 5; 1,11; 2, 3; 2, 4; 2, 5; 2,11; 3, 4; 3, 5; 4, 5; 6, 7; 6, 8;
6, 9; 6,10; 7, 8; 7, 9; 7,10; 8, 9; 8,10; 9,11;10,11;
22 14 24 24 12 0 0 0 0 0
A 5 10 20
1, 2; 1, 3; 1, 4; 1, 5; 2, 3; 2, 4; 2, 5; 3, 4; 3, 5; 4, 5; 6, 7; 6, 8; 6, 9;
6,10; 7, 8; 7, 9; 7,10; 8, 9; 8,10; 9,10;
20 20 30 24 0 0 0 0 0
B 1 7 17
1, 2; 1, 3; 1, 4; 1, 5; 1, 6; 1, 7; 2, 3; 2, 4; 2, 5; 2, 6; 2, 7; 3, 4; 3, 5;
4, 5; 5, 6; 5, 7; 6, 7;
17 19 42 72 84 48
B 2 8 18
1, 2; 1, 3; 1, 4; 1, 5; 1, 6; 1, 7; 2, 3; 2, 5; 2, 7; 2, 8; 3, 4; 3, 5; 4, 5;
4, 8; 5, 6; 5, 7; 6, 7; 6, 8;
18 13 32 56 79 88 54
B 3 8 18
1, 2; 1, 3; 1, 5; 1, 6; 1, 7; 1, 8; 2, 3; 2, 4; 2, 5; 3, 4; 3, 5; 4, 5; 4, 6;
4, 7; 4, 8; 6, 7; 6, 8; 7, 8;
18 14 27 48 72 72 36
B 4 8 18
1, 2; 1, 3; 1, 4; 1, 5; 1, 6; 1, 7; 2, 3; 2, 5; 2, 8; 3, 4; 3, 5; 4, 5; 4, 8;
5, 6; 5, 7; 6, 7; 6, 8; 7, 8;
18 12 31 52 84 96 52
B 5 8 18
1, 2; 1, 3; 1, 4; 1, 5; 1, 6; 1, 7; 2, 3; 2, 7; 2, 8; 3, 4; 3, 5; 4, 5; 4, 8;
5, 6; 5, 7; 5, 8; 6, 7; 6, 8;
18 12 29 56 87 92 58
B 6 9 19
1, 2; 1, 3; 1, 4; 1, 5; 1, 6; 1, 7; 2, 3; 2, 5; 2, 8; 3, 4; 3, 5; 4, 5; 4, 9;
5, 6; 5, 7; 6, 7; 6, 9; 7, 8; 8, 9;
19 11 23 41 60 87 88 48

B 7 8 18
1, 2; 1, 3; 1, 4; 1, 5; 1, 6; 1, 7; 2, 3; 2, 7; 2, 8; 3, 4; 3, 5; 3, 7; 4, 5;
4, 8; 5, 6; 5, 7; 6, 7; 6, 8;
18 13 30 57 85 96 66
B 8 9 19
1, 2; 1, 3; 1, 5; 1, 6; 1, 7; 1, 9; 2, 3; 2, 4; 2, 5; 3, 4; 3, 5; 4, 5; 4, 8;
6, 7; 6, 8; 6, 9; 7, 8; 7, 9; 8, 9;
19 14 18 21 36 72 72 36
B 9 9 19
1, 2; 1, 3; 1, 4; 1, 6; 1, 7; 1, 8; 2, 3; 2, 6; 2, 9; 3, 4; 3, 5; 4, 5; 4, 9;
5, 6; 5, 8; 6, 7; 7, 8; 7, 9; 8, 9;
19 7 21 45 66 108 113 59
B10 10 20
1, 2; 1, 3; 1, 4; 1, 5; 2, 3; 2, 4; 2, 9; 3, 4; 3, 5; 4, 5; 5, 6; 6, 7; 6, 8;
6,10; 7, 8; 7, 9; 7,10; 8, 9; 8,10; 9,10;
20 14 18 12 9 36 72 72 36
B11 10 20
1, 2; 1, 5; 1, 8; 1, 9; 2, 3; 2, 4; 2, 9; 3, 4; 3, 8; 3,10; 4, 5; 4, 7; 5, 6;
5,10; 6, 7; 6, 8; 6,10; 7, 8; 7, 9; 9,10;
20 4 12 40 56 88 149 136 60
C 1 10 19
1, 2; 1, 4; 1, 6; 1, 7; 1, 8; 1, 9; 1,10; 2, 3; 2, 5; 3, 4; 3, 6; 4, 5; 5, 6;
7, 8; 7, 9; 7,10; 8, 9; 8,10; 9,10;
19 10 24 12 6 0 0 0 0
C 2 9 18
1, 2; 1, 4; 1, 6; 1, 7; 1, 8; 1, 9; 2, 3; 2, 5; 3, 4; 3, 6; 3, 7; 3, 8; 3, 9;
4, 5; 5, 6; 7, 8; 7, 9; 8, 9;
18 7 27 24 42 36 36 0
C 3 9 18
1, 2; 1, 6; 1, 8; 1, 9; 2, 3; 2, 6; 2, 9; 3, 4; 3, 7; 4, 5; 4, 6; 4, 9; 5, 6;
5, 8; 5, 9; 6, 7; 6, 9; 7, 8;
18 8 18 30 49 52 60 36
C 4 9 18
1, 2; 1, 6; 1, 8; 1, 9; 2, 3; 2, 7; 2, 9; 3, 4; 3, 8; 3, 9; 4, 5; 4, 7; 4, 9;
5, 6; 5, 8; 5, 9; 6, 7; 6, 9;
18 6 18 30 52 66 54 42
C 5 10 19
1, 2; 1, 4; 1, 6; 1, 7; 2, 3; 2, 5; 3, 4; 3, 6; 3, 8; 3, 9; 3,10; 4, 5; 5, 6;
7, 8; 7, 9; 7,10; 8, 9; 8,10; 9,10;
19 7 18 15 24 36 36 36 0
C 6 11 20
1, 2; 1, 6; 1,11; 2, 3; 2, 5; 3, 4; 3, 6; 4, 5; 4,11; 5, 6; 7, 8; 7, 9; 7,10;
7,11; 8, 9; 8,10; 8,11; 9,10; 9,11;10,11;
20 10 20 16 2 4 0 0 0 0
C 7 8 17
1, 2; 1, 3; 1, 4; 1, 8; 2, 3; 2, 4; 2, 8; 3, 4; 3, 5; 3, 6; 4, 7; 5, 6; 5, 7;
5, 8; 6, 7; 6, 8; 7, 8;
17 10 18 36 64 72 40
C 8 11 20
1, 2; 1, 4; 1, 6; 1,10; 1,11; 2, 3; 2, 5; 3, 4; 3, 6; 4, 5; 5, 6; 7, 8; 7, 9;
7,10; 7,11; 8, 9; 8,10; 8,11; 9,10; 9,11;
20 7 21 12 12 0 0 0 0 0

C 9 11 20
1, 2; 1, 4; 1, 6; 1, 7; 2, 3; 2, 5; 3, 4; 3, 6; 3, 9; 4, 5; 5, 6; 7, 8; 7,10;
7,11; 8, 9; 8,10; 8,11; 9,10; 9,11;10,11;
20 7 18 6 15 18 36 36 36 0
C10 12 21
1, 2; 1, 4; 1, 6; 2, 3; 2, 5; 3, 4; 3, 7; 4, 5; 5, 6; 6, 7; 7, 8; 7,11; 8, 9;
8,10; 8,12; 9,10; 9,11; 9,12;10,11;10,12;11,12;
21 7 17 16 8 4 0 0 0 0 0
C11 11 19
1, 2; 1, 4; 1, 6; 2, 3; 2, 5; 3, 4; 3, 6; 4, 5; 5, 6; 7, 8; 7, 9; 7,10; 7,11;
8, 9; 8,10; 8,11; 9,10; 9,11;10,11;
19 10 24 12 6 0 0 0 0 0
D 1 10 18
1, 2; 1, 4; 1, 6; 1, 7; 1, 9; 1,10; 2, 3; 2, 5; 3, 4; 3, 6; 3, 7; 3, 9; 3,10;
4, 5; 5, 6; 7, 8; 8, 9; 8,10;
18 0 27 0 48 0 36 0 0
D 2 10 18
1, 2; 1, 7; 1, 9; 2, 3; 2,10; 3, 4; 3, 7; 4, 5; 4, 8; 5, 6; 5, 7; 5,10; 6, 7;
6, 9; 6,10; 7, 8; 7,10; 8, 9;
18 4 12 15 28 30 33 36 18
D 3 8 16
1, 2; 1, 4; 1, 6; 1, 7; 1, 8; 2, 3; 2, 5; 3, 4; 3, 6; 3, 7; 3, 8; 4, 5; 4, 7;
4, 8; 5, 6; 7, 8;
16 7 22 26 38 36 20
D 4 9 17
1, 2; 1, 5; 1, 7; 2, 3; 2, 6; 2, 8; 2, 9; 3, 4; 3, 8; 3, 9; 4, 5; 4, 7; 4, 8;
4, 9; 5, 6; 6, 7; 8, 9;
17 7 14 18 26 36 24 24
D 5 9 17
1, 2; 1, 5; 1, 7; 2, 3; 2, 6; 2, 8; 2, 9; 3, 4; 3, 8; 3, 9; 4, 5; 4, 7; 5, 6;
6, 7; 7, 8; 7, 9; 8, 9;
17 5 16 18 34 36 40 12
D 6 9 17
1, 2; 1, 5; 1, 7; 1, 8; 1, 9; 2, 3; 2, 6; 3, 4; 3, 8; 4, 5; 4, 7; 4, 9; 5, 6;
5, 8; 5, 9; 6, 7; 8, 9;
17 5 14 24 32 39 38 24
D 7 9 17
1, 2; 1, 5; 1, 7; 2, 3; 2, 8; 2, 9; 3, 4; 3, 6; 4, 5; 4, 7; 4, 8; 4, 9; 5, 6;
5, 8; 5, 9; 6, 7; 8, 9;
17 5 14 24 30 42 36 20
D 8 9 17
1, 2; 1, 4; 1, 7; 1, 8; 1, 9; 2, 3; 2, 6; 3, 4; 3, 7; 3, 8; 3, 9; 4, 5; 4, 9;
5, 6; 5, 8; 6, 7; 8, 9;
17 4 17 24 24 48 40 16
D 9 10 18
1, 2; 1, 4; 1, 8; 1, 9; 2, 3; 2, 7; 3, 4; 3, 6; 3, 9; 3,10; 4, 5; 5, 6; 5, 8;
5,10; 6, 7; 7, 8; 7, 9; 7,10;
18 0 19 0 64 0 88 0 0
D10 10 18
1, 2; 1, 4; 1, 8; 2, 3; 2, 6; 2, 9; 2,10; 3, 4; 3, 8; 3, 9; 3,10; 4, 5; 5, 6,
5,10; 6, 7; 7, 8; 7, 9; 9,10;
18 4 10 19 28 32 39 38 22

D11 10 18
1, 2; 1, 6; 1, 8; 1, 9; 1,10; 2, 3; 2, 7; 3, 4; 3, 9; 4, 5; 4,10; 5, 6; 5, 8; 6, 7; 6, 9; 6,10; 7, 8; 9,10;
18 4 11 16 30 28 47 30 23
D12 9 17
1, 2; 1, 5; 1, 7; 2, 3; 2, 6; 2, 8; 2, 9; 3, 4; 3, 8; 3, 9; 4, 5; 4, 7; 5, 6; 5, 9; 6, 7; 7, 8; 8, 9;
17 4 14 19 40 44 46 12
D13 10 18
1, 2; 1, 6; 1, 8; 2, 3; 2, 7; 2, 9; 2,10; 3, 4; 3, 9; 3,10; 4, 5; 4, 9; 4,10; 5, 6; 5, 8; 6, 7; 7, 8; 9,10;
18 7 14 6 14 24 36 24 24
D14 10 18
1, 2; 1, 5; 1, 8; 1, 9; 1,10; 2, 3; 2, 7; 3, 4; 3, 9; 3,10; 4, 5; 4, 8; 5, 6; 5,10; 6, 7; 6, 9; 7, 8; 9,10;
18 3 9 23 24 35 50 40 19
D15 10 18
1, 2; 1, 5; 1, 8; 1, 9; 2, 3; 2, 6; 3, 4; 3, 9; 4, 5; 4, 7; 4,10; 5, 6; 5, 9; 5,10; 6, 7; 7, 8; 8,10; 9,10;
18 3 10 19 27 41 39 41 25
D16 11 19
1, 2; 1, 6; 1, 9; 1,10; 1,11; 2, 3; 2, 8; 3, 4; 3,10; 4, 5; 4,11; 5, 6; 5, 9; 6, 7; 6,11; 7, 8; 7,10; 8, 9;10,11;
19 2 8 14 25 28 41 48 34 22
D17 8 16
1, 2; 1, 3; 1, 4; 1, 8; 2, 3; 2, 4; 2, 7; 3, 4; 3, 6; 4, 5; 5, 6; 5, 7; 5, 8; 6, 7; 6, 8; 7, 8;
16 8 12 24 48 48 30
D18 11 19
1, 2; 1, 7; 1, 9; 2, 3; 2, 8; 3, 4; 3,10; 3,11; 4, 5; 4,10; 4,11; 5, 6; 5,10; 5,11; 6, 7; 6, 9; 7, 8; 8, 9;10,11;
19 7 14 6 2 12 24 36 24 24
D19 11 19
1, 2; 1, 6; 1, 9; 2, 3; 2,10; 2,11; 3, 4; 3, 7; 4, 5; 4, 9; 4,10; 5, 6; 5,11; 6, 7; 7, 8; 8, 9; 8,11; 9,10;10,11;
19 2 4 20 26 24 47 60 40 16
E 1 11 18
1, 2; 1, 4; 1, 6; 1, 7; 1, 9; 1,11; 2, 3; 2, 5; 3, 4; 3, 6; 4, 5; 5, 6; 7, 8; 7,10; 8, 9; 8,11; 9,10;10,11;
18 0 18 0 12 0 0 0 0 0
E 2 11 18
1, 2; 1, 8; 1,10; 2, 3; 2,11; 3, 4; 3, 8; 4, 5; 4, 9; 5, 6; 5, 8; 6, 7; 6,11; 7, 8; 7,10; 8, 9; 8,11; 9,10;
18 0 12 0 28 0 39 0 36 0
E 3 8 15
1, 2; 1, 4; 1, 6; 1, 7; 1, 8; 2, 3; 2, 5; 3, 4; 3, 6; 3, 7; 3, 8; 4, 5; 5, 6; 5, 7; 5, 8;
15 0 30 0 60 0 0
E 4 9 16
1, 2; 1, 4; 1, 6; 1, 8; 2, 3; 2, 5; 3, 4; 3, 6; 3, 7; 3, 9; 4, 5; 5, 6; 5, 7; 5, 9; 7, 8; 8, 9;
16 0 18 12 24 36 12 0

E 5 9 16
1, 2; 1, 4; 1, 6; 1, 7; 1, 9; 2, 3; 2, 5; 3, 4; 3, 6; 3, 7; 3, 9; 4, 5; 4, 8;
5, 6; 7, 8; 8, 9;
16 0 22 0 44 0 36 0
E 6 10 17
1, 2; 1, 5; 1, 7; 2, 3; 2, 6; 2, 8; 2, 9; 3, 4; 3,10; 4, 5; 4, 7; 4, 8; 4, 9;
5, 6; 6, 7; 8,10; 9,10;
17 0 14 12 8 36 0 24 0
E 7 10 17
1, 2; 1, 5; 1, 7; 2, 3; 2, 8; 3, 4; 3, 6; 4, 5; 4, 7; 4, 9; 4,10; 5, 6; 6, 7;
6, 9; 6,10; 8, 9; 8,10;
17 0 14 8 20 24 28 16 0
E 8 10 17
1, 2; 1, 5; 1, 7; 1, 8; 1,10; 2, 3; 2, 6; 3, 4; 3, 7; 3, 8; 3,10; 4, 5; 4, 9;
5, 6; 6, 7; 8, 9; 9,10;
17 0 14 8 20 24 16 24 0
E 9 10 17
1, 2; 1, 4; 1, 6; 1, 7; 2, 3; 2, 5; 3, 4; 3, 6; 3, 9; 4, 5; 5, 6; 5, 8; 5,10;
7, 8; 7,10; 8, 9; 9,10;
17 0 12 12 12 36 18 24 0
E10 10 17
1, 2; 1, 5; 1, 7; 1, 8; 2, 3; 2, 6; 3, 4; 3, 9; 4, 5; 4, 7; 4,10; 5, 6; 6, 7;
6, 9; 6,10; 8, 9; 8,10;
17 0 9 18 14 26 37 12 6
E11 10 17
1, 2; 1, 5; 1, 7; 2, 3; 2, 6; 2, 8; 2, 9; 3, 4; 3,10; 4, 5; 4, 7; 5, 6; 5, 9;
6, 7; 7, 8; 8,10; 9,10;
17 0 12 8 24 24 24 28 0
E12 10 17
1, 2; 1, 5; 1, 7; 1, 8; 1,10; 2, 3; 2, 6; 3, 4; 3,10; 4, 5; 4, 7; 4, 8; 5, 6;
5, 9; 6, 7; 8, 9; 9,10;
17 0 12 10 22 22 25 22 12
E13 11 18
1, 2; 1, 6; 1, 8; 2, 3; 2, 7; 2, 9; 2,10; 3, 4; 3,11; 4, 5; 4, 9; 4,10; 5, 6;
5, 8; 6, 7; 7, 8; 9,11;10,11;
18 0 14 0 20 0 36 0 24 0
E14 11 18
1, 2; 1, 6; 1, 8; 1, 9; 2, 3; 2, 7; 3, 4; 3,10; 4, 5; 4, 8; 5, 6; 5,11; 6, 7;
7, 8; 7,10; 7,11; 9,10; 9,11;
18 0 6 14 14 18 29 32 14 8
E15 11 18
1, 2; 1, 5; 1, 8; 2, 3; 2, 9; 3, 4; 3, 6; 4, 5; 4, 7; 4, 9; 4,10; 5, 6; 5,11;
6, 7; 7, 8; 8,10; 9,11;10,11;
18 0 8 8 20 16 37 24 24 12
E16 12 19
1, 2; 1, 4; 1, 6; 1,11; 1,12; 2, 3; 2, 5; 3, 4; 3, 6; 4, 5; 5, 6; 7, 8; 7,10;
7,12; 8, 9; 8,11; 9,10; 9,12;10,11;
19 0 14 4 8 4 0 0 0 0 0
E17 12 19
1, 2; 1, 6; 1, 9; 2, 3; 2, 7; 3, 4; 3,11; 4, 5; 4, 8; 5, 6; 5,12; 6, 7; 7, 8;
7,11; 7,12; 8, 9; 9,10;10,11;10,12;
19 0 4 8 20 8 27 24 36 16 8

E18 8 15
1, 2; 1, 4; 1, 6; 2, 3; 2, 5; 2, 7; 3, 4; 3, 6; 3, 8; 4, 5; 4, 7; 5, 6; 5, 8;
6, 7; 7, 8;
15 0 27 0 60 0 36
E19 9 16
1, 2; 1, 6; 1, 8; 2, 3; 2, 4; 2, 9; 3, 4; 3, 7; 3, 9; 4, 5; 4, 9; 5, 6; 5, 8;
6, 7; 7, 8; 7, 9;
16 5 8 12 20 30 24 20
E20 9 16
1, 2; 1, 6; 1, 8; 2, 3; 2, 4; 2, 9; 3, 4; 3, 7; 3, 9; 4, 5; 4, 9; 5, 6; 5, 8;
6, 7; 6, 9; 7, 8;
16 4 9 12 24 30 24 18
E21 9 16
1, 2; 1, 4; 1, 7; 2, 3; 2, 6; 2, 8; 3, 4; 3, 7; 3, 9; 4, 5; 4, 8; 5, 6; 5, 9;
6, 7; 7, 8; 8, 9;
16 0 17 10 28 32 20 24
E22 9 16
1, 2; 1, 4; 1, 6; 1, 9; 2, 3; 2, 7; 2, 8; 3, 4; 3, 6; 4, 5; 4, 8; 5, 6; 5, 9;
6, 7; 7, 9; 8, 9;
16 0 18 0 52 0 54 0
E23 10 17
1, 2; 1, 6; 1, 9; 2, 3; 2, 4; 2,10; 3, 4; 3, 8; 3,10; 4, 5; 4,10; 5, 6; 5, 9;
6, 7; 7, 8; 7,10; 8, 9;
17 4 5 10 18 26 26 28 16
E24 10 17
1, 2; 1, 5; 1, 8; 1,10; 2, 3; 2, 6; 3, 4; 3, 9; 4, 5; 4, 7; 4,10; 5, 6; 5, 9;
6, 7; 7, 8; 8, 9; 9,10;
17 0 9 16 16 28 33 22 16
E25 10 17
1, 2; 1, 6; 1, 8; 1, 9; 2, 3; 2,10; 3, 4; 3, 7; 4, 5; 4,10; 5, 6; 5, 8; 5, 9;
6, 7; 6,10; 7, 8; 9,10;
17 0 12 6 26 26 22 36 8
E26 10 17
1, 2; 1, 6; 1, 8; 2, 3; 2, 9; 2,10; 3, 4; 3, 7; 4, 5; 4, 9; 4,10; 5, 6; 5, 8;
6, 7; 6,10; 7, 8; 8, 9;
17 0 10 8 28 24 30 32 0
E27 10 17
1, 2; 1, 5; 1, 7; 2, 3; 2, 6; 2, 9; 3, 4; 3, 8; 3,10; 4, 5; 4, 7; 5, 6; 5,10;
6, 7; 7, 8; 8, 9; 9,10;
17 0 10 12 16 36 22 32 4
E28 10 17
1, 2; 1, 5; 1, 8; 1, 9; 2, 3; 2, 7; 3, 4; 3,10; 4, 5; 4, 8; 4, 9; 5, 6; 5,10;
6, 7; 6, 9; 7, 8; 9,10;
17 0 11 12 16 32 24 28 18
E29 11 18
1, 2; 1, 4; 1, 7; 1, 8; 2, 3; 2, 6; 3, 4; 3, 7; 3,10; 4, 5; 5, 6; 5,11; 6, 7;
6, 9; 8, 9; 8,11; 9,10;10,11;
18 0 6 12 14 22 30 30 24 4
E30 11 18
1, 2; 1, 6; 1, 8; 2, 3; 2, 9; 3, 4; 3, 7; 4, 5; 4,10; 4,11; 5, 6; 5, 8; 6, 7;
6,11; 7, 8; 8,10; 9,10; 9,11;
18 0 6 10 18 22 26 34 26 4

E31 11 18
1, 2; 1, 5; 1, 8; 1, 9; 2, 3; 2, 6; 3, 4; 3,10; 4, 5; 4, 8; 4,11; 5, 6; 6, 7;
6,11; 7, 8; 7,10; 9,10; 9,11;
18 0 3 18 16 12 39 38 12 12
E32 11 18
1, 2; 1, 6; 1, 9; 1,10; 2, 3; 2, 8; 3, 4; 3,11; 4, 5; 4,10; 5, 6; 5, 9; 6, 7;
6,11; 7, 8; 7,10; 8, 9;10,11;
18 0 6 13 12 24 26 33 24 14
E33 11 18
1, 2; 1, 6; 1, 8; 1, 9; 2, 3; 2,11; 3, 4; 3, 7; 4, 5; 4,11; 5, 6; 5, 8; 5, 9;
6, 7; 6,10; 7, 8; 9,10;10,11;
18 0 8 8 14 28 20 32 26 12
E34 11 18
1, 2; 1, 5; 1, 8; 1, 9; 2, 3; 2, 6; 3, 4; 3, 9; 4, 5; 4, 7; 4,10; 5, 6; 5,11;
6, 7; 7, 8; 8,10; 9,11;10,11;
18 0 6 12 14 24 26 28 24 16
E35 11 18
1, 2; 1, 6; 1, 9; 2, 3; 2,11; 3, 4; 3, 7; 4, 5; 4,11; 5, 6; 5, 8; 5,10; 6, 7;
6,11; 7, 8: 8, 9; 9,10;10,11;
18 0 7 10 13 28 23 26 33 10
E36 12 19
1, 2; 1, 7; 1, 9; 2, 3; 2, 8; 3, 4; 3,10; 3,11; 4, 5; 4,12; 5, 6; 5,10; 5,11;
6, 7; 6, 9; 7, 8; 8, 9;10,12;11,12;
19 0 14 0 8 12 0 36 0 24 0
E37 12 19
1, 2; 1, 4; 1, 8; 1, 9; 2, 3; 2, 6; 3, 4; 3, 8; 3,11; 4, 5; 5, 6; 5,12; 6, 7;
7, 8; 7,10; 9,10; 9,12;10,11;11,12;
19 0 4 8 14 20 18 36 32 20 8
E38 12 19
1, 2; 1, 6; 1,10; 1,11; 2, 3; 2, 9; 3, 4; 3,12; 4, 5; 4,11; 5, 6; 5,10; 6, 7;
7, 8; 7,12; 8, 9; 8,11; 9,10;11,12;
19 0 4 8 16 12 29 24 40 20 12
E39 12 19
1, 2; 1, 6; 1, 9; 2, 3; 2,12; 3, 4; 3, 7; 4, 5; 4,12; 5, 6; 5, 8; 5,10; 6, 7;
6,11; 7, 8; 8, 9; 9,10;10,11;11,12;
19 0 4 10 10 18 28 24 30 30 10
E40 13 20
1, 2; 1, 5; 1,10; 1,13; 2, 3; 2, 7; 3, 4; 3, 6; 4, 5; 4, 7; 5, 6; 6, 7; 8, 9;
8,11; 8,13; 9,10; 9,12;10,11;11,12;12,13;
20 0 10 8 4 8 0 0 0 0 0 0
E41 13 20
1, 2; 1, 6; 1, 9; 2, 3; 2,10; 3, 4; 3, 7; 4, 5; 4,10; 5, 6; 5,13; 6, 7; 6,12;
7, 8; 8, 9; 8,13; 9,11;10,12;11,12;11,13;
20 0 2 8 12 8 24 32 20 40 26 8
E42 12 18
1, 2; 1, 4; 1, 6; 2, 3; 2, 5; 3, 4; 3, 6; 4, 5; 5, 6; 7, 8; 7,10; 7,12; 8, 9;
8,11; 9,10; 9,12;10,11;11,12;
18 0 18 0 12 0 0 0 0 0 0
F 1 9 15
1, 2; 1, 5; 1, 7; 1, 8; 2, 3; 2, 9; 3, 4; 3, 6; 3, 8; 4, 5; 4, 7; 4, 9; 5, 6;
6, 7; 8, 9;
15 0 10 12 8 28 8 12

F 2 10 16
1, 2; 1, 6; 1, 8; 1, 9; 2, 3; 2,10; 3, 4; 3, 7; 3, 9; 4, 5; 4,10; 5, 6; 5, 8;
6, 7; 7, 8; 9,10;
16 0 8 6 10 18 16 16 8
F 3 10 16
1, 2; 1, 6; 1, 8; 1, 9; 2, 3; 2,10; 3, 4; 3, 7; 4, 5; 4, 9; 5, 6; 5, 8; 6, 7;
7, 8; 7,10; 9,10;
16 0 5 13 7 17 24 10 10
F 4 10 16
1, 2; 1, 6; 1, 8; 2, 3; 2, 9; 3, 4; 3, 7; 3,10; 4, 5; 4, 9; 5, 6; 5, 8; 6, 7;
6,10; 7, 8; 9,10;
16 0 7 8 8 24 12 16 10
F 5 10 16
1, 2; 1, 6; 1, 8; 1, 9; 2, 3; 2,10; 3, 4; 3, 7; 4, 5; 4, 9; 5, 6; 5, 8; 6, 7;
6,10; 7, 8; 9,10;
16 0 6 10 10 18 18 14 12
F 6 10 16
1, 2; 1, 6; 1, 8; 2, 3; 2, 7; 2, 9; 3, 4; 3,10; 4, 5; 4, 9; 5, 6; 5, 8; 5,10;
6, 7; 7, 8; 9,10;
16 0 10 0 20 0 32 0 16
F 7 11 17
1, 2; 1, 7; 1, 9; 2, 3; 2, 8; 2,10; 3, 4; 3,11; 4, 5; 4,10; 5, 6; 5,11; 6, 7;
6, 9; 7, 8; 8, 9;10,11;
17 0 10 0 4 16 0 32 0 16
F 8 11 17
1, 2; 1, 7; 1, 9; 2, 3; 2,10; 3, 4; 3, 8; 4, 5; 4,11; 5, 6; 5,10; 6, 7; 6, 9;
6,11; 7, 8; 8, 9;10,11;
17 0 6 4 8 12 20 16 16 8
F 9 11 17
1, 2; 1, 6; 1, 9; 2, 3; 2,11; 3, 4; 3, 7; 4, 5; 4,10; 5, 6; 5, 8; 6, 7; 7, 8;
7,11; 8, 9; 9,10;10,11;
17 0 2 11 9 11 21 19 14 10
F10 11 17
1, 2; 1, 7; 1, 9; 2, 3; 2,10; 3, 4; 3,11; 4, 5; 4, 8; 5, 6; 5,10; 6, 7; 6, 9;
7, 8; 7,11; 8, 9;10,11;
17 0 4 8 6 16 20 12 20 8
F11 12 18
1, 2; 1, 8; 1,10; 2, 3; 2, 9; 3, 4; 3,11; 4, 5; 4,12; 5, 6; 5,11; 6, 7; 6,12;
7, 8; 7,10; 8, 9; 9,10;11,12;
18 0 10 0 4 0 16 0 32 0 16
F12 12 18
1, 2; 1, 8; 1,10; 2, 3; 2,11; 3, 4; 3, 9; 4, 5; 4,12; 5, 6; 5,11; 6, 7; 6,10;
7, 8; 7,12; 8, 9; 9,10;11,12;
18 0 2 8 4 8 24 16 12 24 6
F13 12 18
1, 2; 1, 7; 1,10; 2, 3; 2,11; 3, 4; 3, 8; 4, 5; 4,12; 5, 6; 5, 9; 6, 7; 6,11;
7, 8; 8, 9; 9,10;10,12;11,12;
18 0 0 9 9 9 18 19 18 18 9
F14 12 18
1, 2; 1, 6; 1,10; 2, 3; 2,12; 3, 4; 3, 8; 4, 5; 4,11; 5, 6; 5, 9; 6, 7; 7, 8;
7,12; 8, 9; 9,10;10,11;11,12;
18 0 0 8 12 8 13 24 20 16 8

G 10 15
1, 2; 1, 6; 1, 8; 2, 3; 2,10; 3, 4; 3, 9; 4, 5; 4,10; 5, 6; 5, 8; 6, 7; 7, 8;
7, 9; 9,10;
15 0 6 0 12 0 24 0 12

AN INTRODUCTION TO THE CATEGORY OF GRAPHS*

Pavol Hell

Department of Mathematics
Rutgers University
New Brunswick, New Jersey 08903

1. Motivation

In recent years the category-theoretical approach to graph theory has gained considerably in acceptance, mainly because of its most spectacular successes [41, 47, 48]. It is our intention here to motivate further interest in this fruitful area by describing and illustrating the general approach in a reasonably self-contained yet elementary and (as much as possible) nontechnical way. A more extensive survey of the area, with a comprehensive bibliography, is being planned by D. Waller and the author and will be submitted to the *Journal of Graph Theory*. We would welcome any papers on, or references to, graph-theoretic results involving homomorphisms or using categorical ideas (explicitly or implicitly).

What advantages does the category-theoretical approach to graph theory have to offer? In relating, in a unified setting, graph theory to the other modern mathematical theories, it helps to identify concepts and ideas in graph theory common to most of mathematics and to clarify the reasons behind many results. By taking such a global view, the graph theorist is led to interesting and deep combinatorial problems; Many powerful techniques, results, and constructions become available to him; and his work leads to nice applications in and outside of pure mathematics. Of course, the purpose of this approach is not to replace traditional graph theory, but rather to supplement it with a certain way of thinking, a framework. At its best, it makes use especially of that framework to pose and answer problems, to make and generalize constructions, and to apply results from and to other fields.

2. Categories

What is a category? Consider the following example: let A and B be two groups, say A a two-element group and B a trivial group. What is the structure of all group-homomorphisms involving only A and B? Figure 1 is one way of summarizing the information.

Clearly, there are only five homomorphisms: the two identities 1_A and 1_B, and the three constants: c, mapping A to the unit of A; f, mapping A to the unit of B; and g, mapping B to the unit of A. The composition of these mappings, when defined, is shown in the accompanying table.

The information contained in Figure 1 is an example of a category. It is no longer possible to tell what kind of objects A and B were, and what were the corresponding morphisms 1_A, 1_B, c, f, g. The figure could equally well depict, say, two rooted topological spaces and their five root-preserving continuous mappings. The category-theorist passed from the concrete case of the two groups to an abstract situation where A, B are some "objects" and 1_A, 1_B, c, f, g some "morphisms" for

* Support from the N.S.F. (Grant MCS 04949) is gratefully acknowledged.

0077-8923/79/0328-0120$1.75/1 © 1979, NYAS

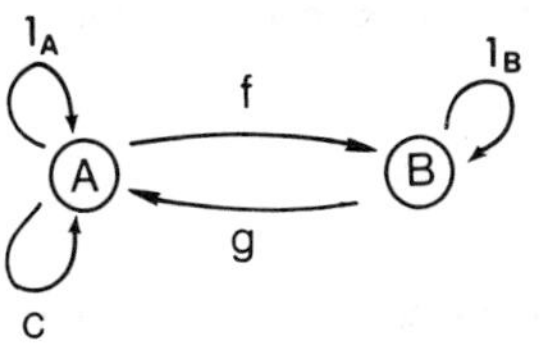

	1_A	c	f	g	1_B
1_A	1_A	c		g	
c	c	c		g	
f	f	f		1_B	
g			c		g
1_B			f		1_B

FIGURE 1. Two objects, all homomorphisms and their composition.

which "composition" is sometimes defined and given. In fact, "morphisms" are not necessarily mappings, and even if they are, they are treated as indivisible entities, rather than on the element-by-element basis.

Formally, a *category* $\mathscr{C}$ consists of a class of *objects* $O_{\mathscr{C}}$, a class of *morphisms* $M_{\mathscr{C}}$, and a partial binary operation called *composition* on $M_{\mathscr{C}}$. The class $M_{\mathscr{C}}$ is partitioned into sets $M_{\mathscr{C}}(X, Y)$ (for $X, Y \in O_{\mathscr{C}}$), and the following axioms are satisfied:

A1: For any $X, Y, Z \in O_{\mathscr{C}}$, the composition of any element of $M_{\mathscr{C}}(Y, Z)$ and any element of $M_{\mathscr{C}}(X, Y)$ is defined and belongs to $M_{\mathscr{C}}(X, Z)$; the composition is undefined in all other cases.

A2: The composition is associative; i.e., for any $X, Y, Z, W \in O_{\mathscr{C}}$ and $h \in M_{\mathscr{C}}(X, Y)$, $g \in M_{\mathscr{C}}(Y, Z), f \in M_{\mathscr{C}}(Z, W)$, we have $f \circ (g \circ h) = (f \circ g) \circ h$.

A3: For any $X \in O_{\mathscr{C}}$, the set $M_{\mathscr{C}}(X, X)$ contains a special morphism 1_X with the property that $1_X \circ f = f$ and $g \circ 1_X = g$ whenever the composition is defined.

	a	c	f	g	b
a	a	c		g	
c	c	c		g	
f	f	f		b	
g			c		g
b			f		b

FIGURE 2. The category of FIGURE 1 defined solely by its morphisms.

When defining a category, it is possible to suppress the objects completely, and define a category simply as a class of morphisms with a partial binary operation (composition). The axioms are constructed so as to assure that this definition agrees with the previous one [44, 55]. This approach is illustrated in FIGURE 2, where the category of FIGURE 1 is presented in this way.

First note that one can tell from FIGURE 2, which morphisms are the identities. By A3, an identity morphism i will satisfy $ix = x$ and $yi = i$ whenever the composition is defined. In this case, there are exactly two identities, a and b, hence there are exactly two objects A,B. Then observe that for each morphism it is possible to determine its beginning and end just from knowing when the composition is

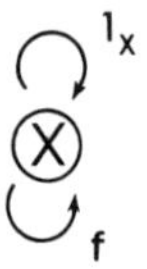

	1_X	f
1_X	1_X	f
f	f	1_X

FIGURE 3. A one-object category (monoid).

defined; for instance f goes from A to B because it can be composed with a on the right and with b on the left.

Another example of a category appears in FIGURE 3. There is only one object here, i.e., $|O_{\mathscr{C}}| = 1$; therefore all compositions are defined. Hence a one-object category is nothing but a semigroup with identity (also called a *monoid*).

A category is a generalization of a monoid in which the operation is not always defined and so there may be more than one identity (compare the table in FIGURE 3 with that in FIGURE 2). Note that the table in FIGURE 3 is the operation table of a group! In a way then, *a group is generalized to a monoid, and a monoid to a category*. This will be further illustrated in Section 5, where the study of full embeddings turns out to be a generalization of the study of graphs with given group.

Both of our examples above are small categories. This statement has a precise meaning, as a category is defined to be *small* if its class of objects $O_{\mathscr{C}}$ is a set (of any finite or infinite cardinality). A category which is not small is called *large*; its class of objects $O_{\mathscr{C}}$ is not a set, i.e., it is a proper class. (See Reference 55 (p. 16) or 11 (p. 18) on the distinction between sets and classes.) Large categories are typified by the category **Set**. The objects of **Set** are all sets, the morphisms all mappings. Since $O_{\mathbf{Set}}$ is the class of all sets, it is not a set itself. Other large categories include: **Group**, in which the objects are all groups and the morphisms all group-homomorphisms; **Vect**, in which the objects are all real vector spaces and the morphisms all linear transformations; **Top**, in which the objects are all topological spaces and the morphisms all continuous mappings; and **Simp**, in which the objects are all abstract simplicial complexes and the morphisms are all simplicial mappings.

A collection of large categories (so-called *relational categories*) may be defined as follows: Let Δ be a finite sequence of positive integers, $\Delta = (n_1, n_2, \ldots, n_k)$. The objects of the category **Rel**(Δ) are sets with k relations, the ith relation being n_i-ary ($i = 1, 2, \ldots, k$); the morphisms are mappings compatible with all relations. (A mapping of sets $f\colon V \to V'$ is *compatible* with the n-ary relations $r \subset V^n$, $r' \subset V'^n$ if $(x_1, x_2, \ldots, x_n) \in r$ implies $(f(x_1), f(x_2), \ldots, f(x_n)) \in r'$.) Note that the category **Rel**(Δ) with $\Delta = (2)$ can be thought of as the *category of digraphs*, and with $\Delta = (2, 2, \ldots, 2)$ as the *category of* (arc-) *colored digraphs*. A similar definition can be made for so-called *operational categories* **Op**(Δ): simply replace "relations" by "operations." (A mapping of sets $f\colon V \to V'$ is *compatible* with the n-ary operations $\sigma\colon V^n \to V$, $\sigma'\colon V'^n \to V$ if $f(\sigma(x_1, x_2, \ldots, x_n)) = \sigma'(f(x_1), f(x_2), \ldots, f(x_n))$ for all $x_1, x_2, \ldots, x_n \in V$.) The generalizations of **Rel**(Δ) and **Op**(Δ) to infinitely many relations and operations of infinite arities are evident [24]. All these categories have one fact in common, namely, the objects are sets with some additional structure and the morphisms are mappings compatible with the structure; a category of this type (or isomorphic to one of this type) is called *concrete*. Any small category is concrete; it may be shown isomorphic to a relational category with sufficiently many binary relations, **Rel**(2, 2, ...) (cf. [24, 50]). (The idea is similar to constructing the Cayley color-graph of a given group [17, p. 168] (cf. also [12]).) There exist categories that are not concrete. They are typified by the category **Toph** defined as follows (cf. [44, p. 16]). The objects of **Toph** are all topological spaces. The morphisms of **Toph** are all homotopy classes of continuous mappings with composition defined as the usual composition of homotopy classes [27, p. 30]. The fact that **Toph** is not concrete follows from [14]. Note that **Toph** is also an example of a category in which morphisms *are not* mappings.

We shall say that a category $\mathscr{D}$ is a *subcategory* of the category $\mathscr{C}$ if $O_{\mathscr{D}} \subset O_{\mathscr{C}}$, $M_{\mathscr{D}} \subset M_{\mathscr{C}}$, and the identities and composition in $M_{\mathscr{D}}$ are the same as in $M_{\mathscr{C}}$. The subcategory is *full* if $M_{\mathscr{D}}(X, Y) = M_{\mathscr{C}}(X, Y)$ whenever $X, Y, \subset O_{\mathscr{D}}$. Observe that **Group** is a full subcategory of **Op**(2). Furthermore, observe that **Op**(2) can be thought of as a full subcategory of **Rel**(3) (by viewing each binary operation σ as a ternary relation $\{(x, y, \sigma(x, y)): x, y \in V\}$).

3. Homomorphisms

We want to define, in the spirit of Section 2, the category **Graph**. It is more or less clear what the objects are, but what would the appropriate morphisms be? Several definitions have been proposed [8, 9], but the one accepted today was pioneered by Sabidussi [52, 54], Hedrlín and Pultr [22, 23], and Hedetniemi [18]. They are called compatible mappings, graph-homomorphisms, homomorphisms, or just morphisms, and are defined as follows: if G, H are graphs, a *homomorphism* $f: G \to H$ is a mapping of the vertex-sets $f: V(G) \to V(H)$ such that $f(g)$ is adjacent to $f(g')$ whenever g is adjacent to g'. While a homomorphism $f: G \to H$ is a mapping $f: V(G) \to V(H)$, it is clear that it also induces a mapping $f^{\#}: E(G) \to E(H)$. If both f and $f^{\#}$ are surjective, we say that f is a *full* homomorphism, and H a *homomorphic image* of G.

Now we form the *category of graphs*, **Graph**, by taking for its objects all graphs, and for its morphisms all homomorphisms. (Different interpretations of the notion of graph will give rise to different categories; we shall, for the most part, consider only graphs [17] without loops or multiple edges, but will admit both finite and infinite graphs.) Note that **Graph** can be thought of as a full subcategory of **Rel**(2), by viewing each edge $[x, y]$ as the union of the two arcs (x, y), (y, x). That is to say, when graphs are regarded as special digraphs, the compatible mappings among them are exactly the homomorphisms.

Homomorphisms are closely related to many traditional graph-theoretic concepts: An isomorphism of G to H is nothing but a full bijective homomorphism $G \to H$; an n-coloring of G is just a homomorphism $G \to K_n$ (the so-called multicolorings of recent interest [7, 41, 56] are also certain homomorphisms); a complete n-coloring [17, p. 143] of G is a full homomorphism $G \to K_n$. Hence the chromatic number $\chi(G)$ is the smallest n such that G admits a homomorphism $G \to K_n$, and the achromatic number $\psi(G)$ is the largest m such that G admits a full homomorphism $G \to K_m$. (The fact that this definition is meaningful even for infinite m was observed by P. Erdös and the author.) Therefore, *if there exists a homomorphism* $G \to H$ *then* $\chi(G) \leq \chi(H)$, *and if there exists a full homomorphism* $G \to H$ *then* $\psi(G) \geq \psi(H)$. A full homomorphism $P_n \to G$ is a walk in G containing each edge of G at least once; similarly, a full homomorphism $C_n \to G$ is a closed walk in G containing each edge of G at least once. Thus any finite connected graph is the homomorphic image of some path and of some cycle. From these remarks (and Euler's Königsberg theorem) it follows without much difficulty that

$$\psi(P_n) = \max \left\{k: \left[\frac{k+2}{2}\right] \cdot (k-2) \leq n\right\}$$

and

$$\psi(C_n) = \max \left\{k: k \cdot \left[\frac{k}{2}\right] \leq n\right\} - s(n),$$

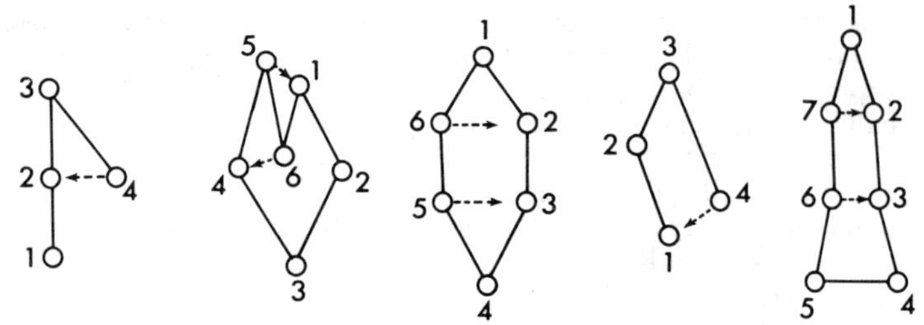

FIGURE 4. Homomorphisms $P_4 \to P_3$, $C_6 \to C_4$, $C_6 \to P_4$, $P_4 \to C_3$, $C_7 \to C_3 \cdot P_3$.

where $s(n)$ is the number of positive integers x satisfying $n = 2x^2 + x + 1$ (see [1, 16, 35, 40] for independent calculations of these numbers).

In some cases homomorphisms between graphs are easily found. For example, there are evident homomorphisms $P_{n+1} \to P_n$, $C_{n+2} \to C_n$, $C_{2n} \to P_{n+1}$, $P_{n+1} \to C_n$, and $C_{n+2k} \to C_n \cdot P_{k+1}$. (Here $C_n \cdot P_{k+1}$ is the graph obtained by identifying a vertex of C_n with an endpoint of P_{k+1}.)

FIGURE 4 implies, in view of the preceding paragraph, that *any finite connected graph G is a homomorphic image of each sufficiently long path and each sufficiently long cycle (respectively sufficiently long even cycle, if G is bipartite)* (cf. FIGURE 5).

In other cases the question of the existence of a homomorphism $G \to H$ is very difficult. The two graphs of FIGURE 6 appear very similar from many points of view (they have the same chromatic number, clique number, etc.), yet there is no homomorphism of G into H, or of H into G [29]. (Furthermore, the only homomorphisms $G \to G$, $H \to H$ are the identities; cf. also FIGURE 12.)

The general problem of determining if there exists a homomorphism between two finite graphs (the "homomorphism problem") is NP-complete [38] since it contains as a special case the problem of determining if a finite graph is n-colorable. It is an important problem, which frequently occurs in applications [26, 49, 58] and to which many other homomorphism problems may be polynomially transformed. In other words *an algorithm solving the homomorphism problem for graphs can be*

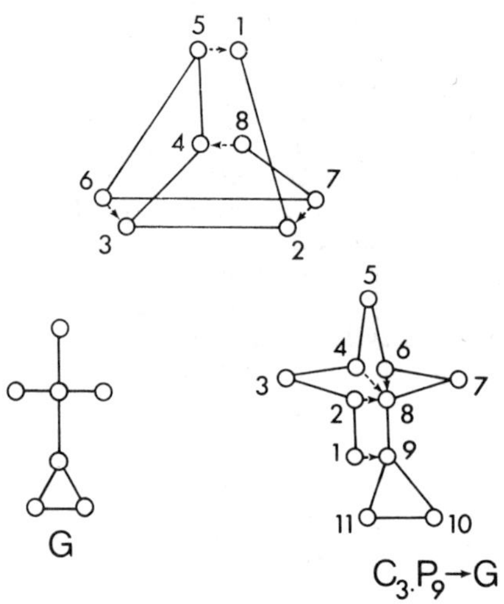

FIGURE 5. For all $n \geq 8$, $P_n \to P_8 \to K_4$; for all $n \geq 9$, $C_{2n+1} \to C_3 \cdot P_9 \to G$.

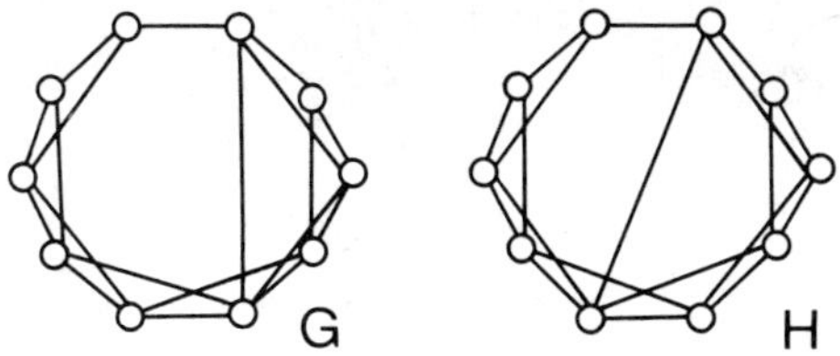

FIGURE 6. Two graphs with no homomorphisms (except identities).

used to check the existence of a morphism between two multigraphs, finite relational structures (such as digraphs, colored digraphs, or sets with several, not necessarily binary, relations), finite operational structures (including groups, rings, and lattices), or finite objects of certain other categories. This will be further explained in Section 5.

Another well-known NP-complete problem is that of finding the independence number of a graph [38]. It follows that there exists a polynomial reduction of any problem in NP to the independence number problem. A direct polynomial transformation of the homomorphism problem to the independence number problem was constructed by V. Chvátal and the author as follows (cf. [5] for a similar transformation for colorings): Let G, H be finite graphs. Define the graph $G * H$ on the vertex-set $V(G) \times V(H)$ to consist of all edges of the form $[(g, h), (g', h')]$ where ($g = g'$ and $h \neq h'$) or ($[g, g'] \in E(G)$ and $[h, h'] \notin E(H)$). It is now easy to see that a homomorphism $G \to H$ exists if and only if the independence number of $G * H$ is equal to $|V(G)|$. Hence *every algorithm that finds the largest independent set of a graph can be used to determine if there exists a homomorphism $G \to H$, and to find one if its exists.* Specifically, if the independence algorithm applied to a graph with p vertices and q edges takes $f(p, q)$ steps to find a maximum independent set, then applied to a graph G_1 with p_1 vertices and q_1 edges and a graph G_2 with p_2 vertices and q_2 edges, it takes

$$f\left(p_1 \cdot p_2, \binom{p_2}{2} \cdot (p_1 + 2q_1) + q_1 \cdot (p_2 - 2q_2)\right)$$

steps to find a homomorphism $G_1 \to G_2$ or to conclude that none exists.

In practical applications, a homomorphism $G \to H$ may correspond to an optimal arrangement of objects, or schedule of activities. If there does not exist a homomorphism $G \to H$ (or if it is too difficult to find one), an approximative homomorphism will be of interest; in particular we may want to find a vertex-mapping f: $V(G) \to V(H)$ which is as nearly a homomorphism as possible. Let $d_+(f)$ denote the minimum number of edges that can be added to H so that f is a homomorphism (and let $d_+(f) = \infty$ if all such additions fail to make f a homomorphism); let $d_-(f)$ denote the minimum number of edges that can be deleted from G so that f is a homomorphism; and let $d(f)$ denote the minimum number of edges each of which can be either added to H or deleted from G so that f is a homomorphism. The quantity $d(f)$ is called the *defect*; $d_+(f)$, the *positive defect*; and $d_-(f)$, the *negative defect* of f. Let p denote the number of vertices and n the chromatic numbers of G. It is not difficult to prove that the minimum positive defect among all vertex-mappings f: $G \to \bar{K}_p$ is $\binom{n}{2}$; the minimum negative defect among all vertex-mappings f: $K_p \to G$ seems to be an interesting parameter which has not been much studied [20]. The former parameter is the number of edges in a "minimum range" of G, i.e., a graph H with p vertices and the minimum number of edges for which there is a homomorphism $G \to H$. Similarly, the latter parameter is the number of edges in a "maximum domain" of G, i.e., a graph F with p vertices and

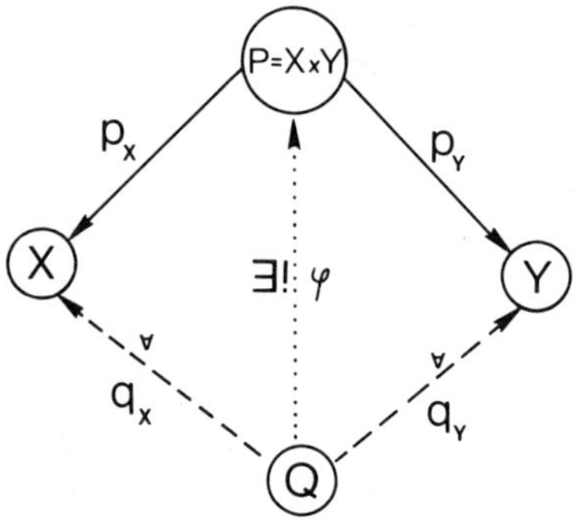

FIGURE 7. The product of X and Y.

the maximum number of edges for which there is a homomorphism $F \to G$. Each minimum range of a graph is a complete graph with n vertices together with $p - n$ isolated vertices. The situation is more interesting for digraphs (without loops): some (but not each) minimum range of a digraph is always a union of isolated vertices and a supertournament (digraph with a spanning subtournament) which may have more than n vertices [20]. Approximative homomorphisms with given negative defect were used in an inclusion-exclusion counting argument by V. Müller [47], to deduce that the *edge-reconstruction conjecture holds for all graphs with at least* $n \cdot \log_2 n$ *edges.* (Since the reconstruction conjecture is known to hold for disconnected graphs, Müller's result comes very near to settling the edge-reconstruction problem.)

4. Category-Theoretical Concepts

Many notions that appear, at first sight, to depend on the structure of the objects for which they are defined, turn out to be category-theoretical concepts, i.e., defined in a universal fashion for all categories.

The concept of *isomorphism* is a case in point. Is it possible to define isomorphism without the customary reference to the elements of the objects, i.e., not on an element-by-element basis? Yes: it suffices to realize that f is an isomorphism if and only if for some g both fg and gf are identities. The preceding sentence serves as the definition of isomorphism in any category. Naturally, this definition agrees with the traditional one in **Graph**.

Consider the concept of a *product*. There are well-known products of objects like groups, topological spaces, and so forth. Is product a category-theoretical concept (and if so what is the "correct" product in **Graph**)?

Let $\mathscr{C}$ be a category, X and Y objects of $\mathscr{C}$. The *product* of X and Y is defined to be an object P together with morphisms $p_X \in M(P, X)$ and $p_Y \in M(P, Y)$, such that for any object Q and morphisms $q_X \in M(Q, X)$ and $q_Y \in M(Q, Y)$ there is a unique morphism $\varphi \in M(Q, P)$ for which $p_X \circ \varphi = q_X$ and $p_Y \circ \varphi = q_Y$ (FIGURE 7).

The definition can, of course, be adapted for the product of a family of objects, X_α, $\alpha \in A$. It is easily proved [55] that a product, if it exists, is unique (up to an isomorphism). We shall use the symbol $X \times Y$ for the product of objects X and Y, and the symbol $\prod_{\alpha \in A} X_\alpha$ for the product of a family X_α, $\alpha \in A$.

It is, of course, possible that for a pair of objects X, Y of $\mathscr{C}$, there is no object P satisfying the definition of $X \times Y$. However, in **Graph** *any pair of objects has a product*, as we shall illustrate below. (In fact, any nonempty family of objects of

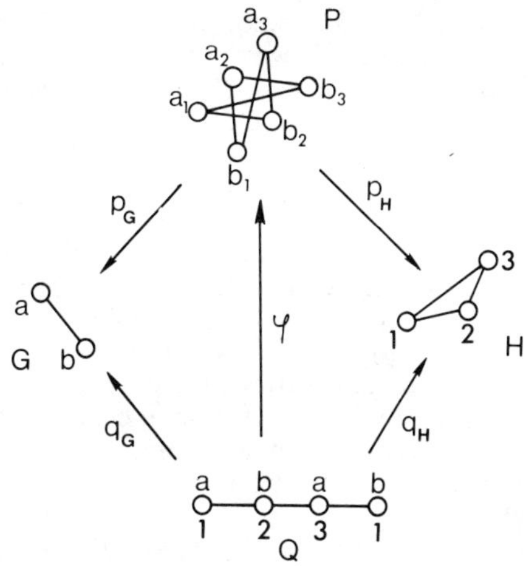

FIGURE 8. $K_2 \wedge K_3 \cong C_6$ is the product of K_2 and K_3 in **Graph**.

Graph has a product. If loops are allowed, then the resulting category has all products, coproducts, pullbacks, and pushouts, and so all limits and colimits [28]; such a category is called complete and cocomplete [55]. A different proof of this fact is given in [42].) It is perhaps interesting to find that *the conjunction* [17, p. 25], *assumes the role of the product in* **Graph**. (This is the reason that the conjunction is often referred to as the *categorical product* [45], or just *the* product.) The *conjunction* of G and H, $G \wedge H$, is the graph with the vertex-set $V(G) \times V(H)$, and all edges $[(g, h), (g', h')]$ for $[g, g'] \in E(G)$ and $[h, h'] \in E(H)$. In FIGURE 8 we illustrate the fact that $G \wedge H$ is the product of G and H in the category **Graph**.

Taking $G = K_2$ and $H = K_3$, we find $P = G \wedge H \cong C_6$. The homomorphisms $p_G: P \to G$ and $p_H: P \to H$ are evident from the picture (p_G takes a_i to a and b_i to b; p_H takes a_i and b_i to i). If $Q = P_4$ and if $q_G: Q \to G$ and $q_H: Q \to H$ are the homomorphisms depicted in FIGURE 8 (i.e., a vertex of Q with the labels x, i is taken to x by q_G and to i by q_H), then there is precisely one homomorphism $\varphi: Q \to P$ for which $p_G \circ \varphi = q_G$ and $p_H \circ \varphi = q_H$. Namely, it must take the vertex of Q labeled x, i to x_i. Since the conjunction $G \wedge H$ satisfies the definition of the product $G \times H$ (and the product is unique), we have $G \times H \cong G \wedge H$.

An interesting conjecture concerning the categorical product/conjunction was made by Hedetniemi [18]. It states that *the chromatic number of $G \wedge H$ is equal to the smaller of the chromatic numbers of G and H*; i.e., $\chi(G \wedge H) = \min(\chi(G), \chi(H))$. It is clear that $\chi(G \wedge H)$ cannot exceed $\chi(G)$ or $\chi(H)$ because there are homomorphisms $G \wedge H \to G$ and $G \wedge H \to H$ (namely, p_G and p_H). A moment's reflection shows that it is sufficient to prove that $\chi(G) = n$ and $\chi(H) = n$ implies $\chi(G \wedge H) \geqq n$. This has been verified [18] for $n \leqq 3$, for the case when both G and H have clique number n, for $G \cong H$, and most recently [4], for the case when each vertex of G belongs to a clique with $n - 1$ vertices.

An equivalent formulation of the Hedetniemi conjecture, that takes into account the categorical properties of the conjunction, is the following: *If $\chi(G) = n$ and $\chi(H) = n$ then some Q with $\chi(Q) = n$ admits homomorphisms into both G and H*. To prove the equivalence of the two conjectures, note that if Hedetniemi's conjecture is true, we may take $Q = G \wedge H$; on the other hand, if some Q with $\chi(Q) = n$ admits homomorphisms into G and H, then by the definition of the product in **Graph**, it admits a homomorphism into $G \wedge H$, and hence $\chi(G \wedge H) \geqq n$. In this more categorical formulation all the solved special cases (with the notable exception of the result in [4]) are immediate: For $n = 1$ or 2 take $Q = K_n$; for $n = 3$ let Q be an odd cycle of length at least as large as the length of some odd cycle in G and some odd cycle in H; when both G and H have clique number n take $Q = K_n$ (this is, of course, a generalization of the case $n \leqq 2$); take $Q = G$ if $G \cong H$, or more generally if there is a homomorphism $G \to H$.

Another problem in which the categorical properties of the conjunction may be exploited is the problem of isometric embeddings. An *isometric* (or distance preserving) *embedding* $f: G \to H$ is an injective homomorphism in which the distance of vertices is preserved, i.e., $d_H(f(g), f(g')) = d_G(g, g')$ for all $g, g' \in V(G)$. Isometric embeddings of bipartite graphs into cartesian products of n copies of K_2 were investigated by Djoković [10], and others. A natural companion question is one of isometric embeddings into categorical products/conjunctions. The conjunction of n copies of K_2 is not too interesting, being just the disjoint union of the copies. But replacing the single edges by paths, we obtain the perhaps unexpected result, that *any bipartite graph can be isometrically embedded into a conjunction of finite paths*. This embedding was used in [32] to establish a subdirect representation theorem for bipartite graphs, and it can be constructed as follows: Let P_n have the (*consecutive*) vertices 1, 2, ..., n and P_ω the vertices 1, 2, ... The *eccentricity* of a vertex g in a connected graph G is defined as $\text{ecc}(g) = \sup \{d_G(g, g'): g' \in V(G)\}$ (cf. [17, p. 35] for finite graphs). If G is a connected bipartite graph and g a vertex of G, we define $p_g: G \to P_{\text{ecc}(g)+1}$ by $p_g(g') = d_G(g, g') + 1$. It is easy to see that each p_g is a homomorphism, and so by the categorical properties of the conjunction there exists a homomorphism $\varphi: G \to \prod_{g \in V(G)} P_{\text{ecc}(g)+1}$. It can be shown [32] that φ is an isometric embedding, that the construction can be extended to include *all* bipartite graphs, and that all copies of P_ω occurring in $\prod P_{\text{ecc}(g)+1}$ can be replaced by products of finite paths.

The definition of a coproduct in a category $\mathscr{C}$ is dual to that of a product [55]. We shall not bother to state it, as the coproducts in **Graph** are just the disjoint unions, and the situation is not too interesting for a graph-theorist. However, the concept of a colimit, which generalizes the coproduct, is extremely useful and corresponds in the category **Graph** to a construction that appears under various guises. Intuitively speaking, a colimit of a family of graphs according to a scheme (or diagram) is what one obtains from the disjoint union of the graphs by identifying certain vertices, edges, or subgraphs as the scheme specifies.

A familiar instance of the colimit construction in graph theory is the replacement of each arc (of a given color) of a (colored) directed graph by a specified graph with two distinguished vertices one of which is identified with the beginning of the arc and the other with the end of the arc. Such constructions occur, for example, in the study of graphs with given group [15, 17, p. 169], or the study of full embeddings [6, 22, 30, 31, 36, 43]. Clearly, first the disjoint union of all the replacing graphs is taken, and then some identification of vertices is performed according to the structure of the given digraph. This point of view of such constructions (also known as šip-constructions) [43] was taken in [30, 31] and further details on the colimit construction in graph theory may be found there. We limit ourselves here to an illustration of one further use of the colimit construction, namely, in finding graphs in which all neighborhoods are isomorphic to a given graph, FIGURE 9. (This method was used both in [2, 3] and in [34].)

Limits in a category $\mathscr{C}$ are defined as certain generalizations of products; some of the properties of limits in **Graph** are studied in [28, 42, 57].

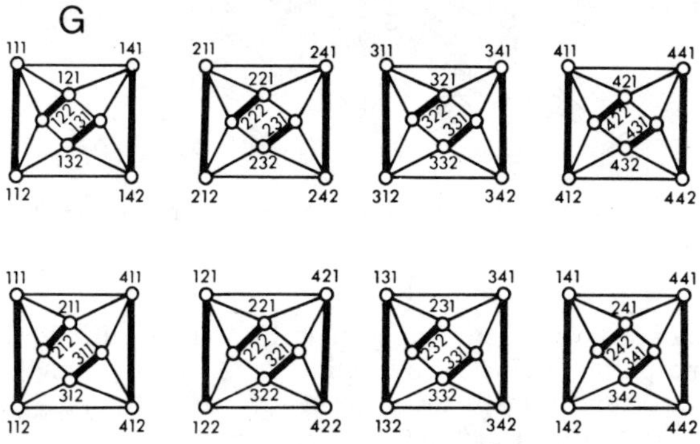

FIGURE 9. The neighborhood of each vertex of G is isomorphic to P_4; after carrying out the indicated colimit construction (i.e., identifying vertices of equal labels), the neighborhood of each vertex is isomorphic to P_7.

5. FUNCTORS

If a category is a generalization of a group (or monoid), then a functor is the corresponding generalization of a homomorphism. Let $\mathscr{C}$, $\mathscr{D}$ be categories defined in terms of their morphisms only. Then a functor $\Phi: \mathscr{C} \to \mathscr{D}$ is a mapping of the morphisms which preserves identities and satisfies $\Phi(fg) = \Phi(f)\Phi(g)$ whenever fg is defined. In terms of the formal definition of categories given in Section 2 (i.e., as classes of objects and morphisms), a *functor* Φ is a mapping $O_{\mathscr{C}} \to O_{\mathscr{D}}$ and a mapping $M_{\mathscr{C}} \to M_{\mathscr{D}}$ such that $\Phi(1_X) = 1_{\Phi(X)}$ and $\Phi(fg) = \Phi(f)\Phi(g)$ if fg is defined; it is easy to deduce that $f \in M_{\mathscr{C}}(X, Y)$ implies $\Phi(f) \in M_{\mathscr{D}}(\Phi(X), \Phi(Y))$. If a functor $\Phi: \mathscr{C} \to \mathscr{D}$ is injective on the objects of $\mathscr{C}$, we define the category $\Phi(\mathscr{C})$ as the subcategory of $\mathscr{D}$ having the objects $\Phi(X)$, $X \in O_{\mathscr{C}}$, and the morphisms $\Phi(f)$, $f \in M_{\mathscr{C}}$.

An *isomorphism* of categories is a functor which is bijective on both objects and morphisms. A *full embedding* is an isomorphism onto a full subcategory; i.e., $\Phi: \mathscr{C} \to \mathscr{D}$ is a full embedding if and only if $\Phi: \mathscr{C} \to \Phi(\mathscr{C})$ is an isomorphism and $\Phi(\mathscr{C})$ is a full subcategory of $\mathscr{D}$.

Consider the following example [13] of a functor Φ: **Graph** $\to$ **Graph**. For a graph G we let $\Phi(G) = G \wedge K_2$, for a homomorphism $f: G \to H$ we let $\Phi(f): G \wedge K_2 \to H \wedge K_2$ be the mapping taking (g, i) to $(f(g), i)$. It can be easily verified that $\Phi(f)$ is indeed a homomorphism, that $\Phi(1_G) = 1_{\Phi(G)}$, and that $\Phi(fg) = \Phi(f)\Phi(g)$ whenever fg is defined. Note that Φ is an isomorphism of **Graph** onto Φ(**Graph**), but that Φ is not a full embedding as we illustrate in FIGURE 10. Finally note that Φ(**Graph**) consists of only bipartite graphs, so that Φ is a functor from the category of all graphs to the category of all bipartite graphs.

A full embedding Φ: **Rel**(3) $\to$ **Rel**(2, 2, 2) may be defined as follows: If $X = (V; r)$ is an object of **Rel**(3), i.e., V a set and $r \subset V \times V \times V$, we take a set of new elements $E = \{e_{u,v,w}: (u, v, w) \in r\}$ and define $\Phi(X) = (V \cup E; r_1, r_2, r_3)$ where

$$r_1 = \{(e_{u,v,w}, u): (u, v, w) \in r\},$$

$$r_2 = \{(e_{u,v,w}, v): (u, v, w) \in r\},$$

$$r_3 = \{(e_{u,v,w}, w): \in (u, v, w) \in r\}.$$

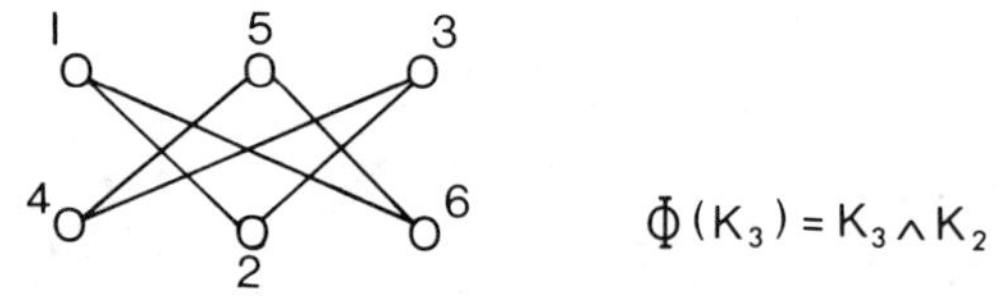

FIGURE 10. The mapping $i \to i + 1$ (mod 6) is a homomorphism $\Phi(K_3) \to \Phi(K_3)$ which is not of the form $\Phi(f)$ for any f: $K_3 \to K_3$.

(An illustration of this construction appears in FIGURE 11.) If f: $X = \{V, r) \to X' = (V'; r')$ is a morphism in **Rel**(3), we define $\Phi(f)$: $\Phi(X) \to \Phi(X')$ by extension: $\Phi(f)$ equals f on V and it maps the element $e_{u,v,w}$ of E to $e_{f(u),f(v),f(w)}$ (note that $(u, v, w) \in r$ implies $(f(u), f(v), f(w)) \in r'$).

It is easy to verify that Φ is injective and that the only morphisms from $\Phi(X)$ to $\Phi(X')$ in **Rel**(2, 2, 2) are of the form $\Phi(f)$ where f: $X \to X'$ is a morphism in **Rel**(3), i.e., Φ is an isomorphism of **Rel**(3) onto a full subcategory of **Rel**(2, 2, 2). Similarly, one can obtain a full embedding **Rel**(Δ) $\to$ **Rel**(2, 2, ...). This was first done in [24], and the present construction is only a minor simplification of the original method. (A further simplification presented in passing in [31, p. 121, part b] is not correct.)

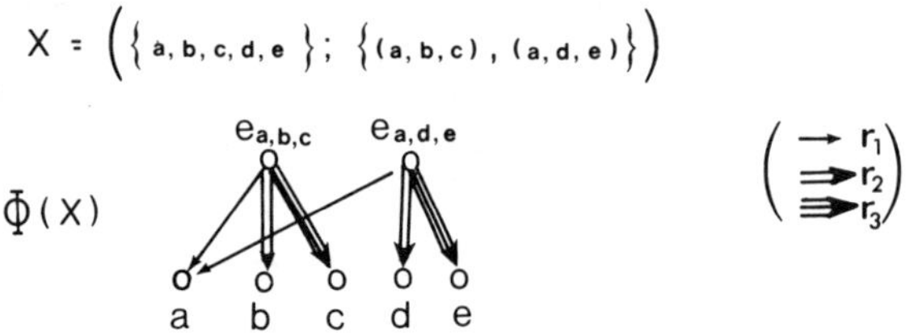

FIGURE 11. An illustration of the functor Φ: **Rel**(3) $\to$ **Rel**(2, 2, 2).

A full embedding Φ: **Rel**(2, 2, 2) $\to$ **Graph** (and in general **Rel**(2, 2, ...) $\to$ **Graph**) may be found using the colimit construction discussed in Section 4 (cf. [31, 33]). Roughly speaking, graphs G_1, G_2, G_3 with very special properties are constructed and then, for $X = (V; r_1, r_2, r_3)$, the graph $\Phi(X)$ is obtained from X by replacing each arc of the relation r_i by a copy of G_i (identifying specified vertices with the beginning and end of the arc). Each homomorphism f: $X \to X'$ induces in a natural way a homomorphism $\Phi(X) \to \Phi(X')$ which is then called $\Phi(f)$. The special properties of the graphs G_i [22, 31, 43] are required to assure that each homomorphism $\Phi(X) \to \Phi(X')$ is of the form $\Phi(f)$; i.e., Φ is a full embedding. In other words, the graphs G_i have to imitate the homomorphism properties of the arcs; in the similar construction of R. Frucht [15, 17 p. 169], the graphs G_i only have to imitate the automorphism properties of the arcs—these are easier to construct. Arbitrarily large families of suitable replacement graphs G_i are constructed in [30, 31] (cf. FIGURE 12). (The original constructions of a full embedding Φ: **Rel**(2, 2, ...) $\to$ **Graph** [22, 50, 51] were based on a more complicated method, but required only one replacement graph.)

We remark that the composition of two full embeddings is a full embedding, and that the identity functor $\mathscr{C} \to \mathscr{D}$ is a full embedding if and only if $\mathscr{C}$ is a full

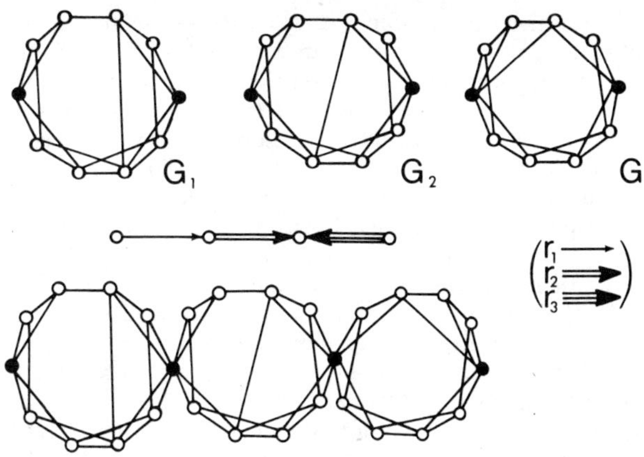

FIGURE 12. Three graphs which can be used in the colimit construction full embedding Φ: **Rel**(2, 2, 2) → **Graph**. The vertices labelled *b* (respectively, *e*) are identified with the beginning (respectively, end) of the replaced arc, as illustrated.

subcategory of $\mathscr{D}$. Using these two observations, we get the full embeddings **Rel**(3) → **Rel**(2, 2, 2) → **Graph** (and in general **Rel**(Δ) → **Graph**), **Op**(2) → **Rel**(3) → **Graph** (and in general **Op**(Δ) → **Graph**), and **Group** → **Op**(2) → **Graph**. It is becoming apparent that many categories admit a full embedding into **Graph**, i.e., that the category of graphs is sufficiently rich to model many other categories within itself. A powerful theorem proved by Z. Hedrlín and L. Kučera [19] asserts that *any concrete category admits a full embedding into* **Graph**. (The proof uses results of Goralčík, de Groot, Isbell, Lambek, Pultr, Sichler, Trnková, and Vopěnka, and involves a set-theoretic assumption which in its simplest form requires that each ultrafilter closed under countable intersections be principal [19, 25].) Intuitively speaking, this means that for any concrete category $\mathscr{C}$ there is a piece of **Graph** isomorphic to $\mathscr{C}$. A further result of L. Kučera [19] (and, independently, of P. Freyd) states that *any category can be obtained from a concrete category by identification of morphisms*, i.e., the same way **Toph** is obtained from **Top**. Hence *any category is a morphism identification of a suitable full subcategory of* **Graph**.

There are, of course, other categories in which all concrete categories can be fully embedded. Such categories are called binding, or *universal*. Note that a category $\mathscr{C}$ is universal if and only if there is a full embedding **Graph** → $\mathscr{C}$. *Universal categories include* **Op**(2) [24] *and many algebraic categories* (*such as the category of semigroups* [21], *the category of digraphs* (**Rel**(2), cf. [24]), *of strongly connected digraphs* ([21]), *the category of u-uniform hypergraphs* ($u \geqq 2$) ([36]), *and many full subcategories of* **Graph** *such as the category of k-chromatic graphs* ($k \geqq 3$), *the category of l-connected graphs, the category of graphs of girth g, and the category of graphs with a given subgraph* ([6, 31, 36, 43]). Since in the category **Graph** there are objects X, Y with $M(X, Y) = \varnothing$ (cf. FIGURE 6), the same must be true of any universal category $\mathscr{C}$ (if Φ: **Graph** → $\mathscr{C}$ is a full embedding, then any morphism in $M_{\mathscr{C}}(\Phi(X), \Phi(Y))$ would have to be of the form $\Phi(f)$). Consequently **Top**, **Group**, *and many other categories are not universal. In particular, the category of all bipartite graphs and all their homomorphisms is not universal*, since there always are homomorphisms between bipartite graphs.

Let us now relate the theory of full embeddings and universal categories to the study of graphs with given group. Since a group is a special one-object category (Section 2), and so a concrete category, it may be fully embedded into **Graph**, or any other universal category. Let us interpret the full embedding in this particular situation. To the group X, we assign one object $\Phi(X)$ of **Graph**, so that the full subcategory of **Graph** induced by $\Phi(X)$ is isomorphic to X, i.e., each homomorphism of $\Phi(X)$ corresponds to a member of X and composition of homomorphisms corresponds to the multiplication in X. Therefore the homomorphisms are in fact automorphisms of X, and $\Phi(X)$ is a graph whose automorphism group is isomorphic to X. Consequently, *for any group X and universal category $\mathscr{C}$ there is an object of $\mathscr{C}$ whose automorphism group is isomorphic to X*—we obtain u-uniform hypergraphs ($u \geqq 2$), k-chromatic graphs ($k \geqq 3$), l-connected graphs, graphs of girth g, and so forth, with given group (compare with [15, 37, 53]). Universality of $\mathscr{C}$ is a much stronger property than merely the existence of an object with given group. It also implies that for each monoid X there is an object Y with the set of homomorphisms $Y \to Y$ (endowed by composition) isomorphic to X; that any two-object category "exists" in $\mathscr{C}$ (cf. FIGURE 13, FIGURE 1), and so forth.

	1_X	1_Y
1_X	1_X	—
1_Y	—	1_Y

FIGURE 13. A two-object category $\mathscr{C}$. (A full embedding Φ: $\mathscr{C} \to$ **Graph** is obtained by letting $\Phi(X)$, $\Phi(Y)$ be the graphs G, H of FIGURE 6.)

As another application of the theory of full embeddings, we turn to some questions of algorithmic complexity. For this purpose we restrict all categories mentioned below to objects with finitely many elements; i.e., **Graph** now consists of finite graphs, **Top** of finite topological spaces, and so forth.

Consider again the sequence of full embeddings **Group** $\to$ **Op**(2) $\to$ **Rel**(3) $\to$ **Rel**(2, 2, 2) $\to$ **Graph**, and their composition Φ: **Group** $\to$ **Graph**. Since Φ is a full embedding, a homomorphism of groups $G \to G'$ exists if and only if there exists a homomorphism of graphs $\Phi(G) \to \Phi(G')$. Furthermore, Φ involves only a polynomial increase in the size of the objects: no increase occurs in **Group** $\to$ **Op**(2) $\to$ **Rel**(3); in **Rel**(3) $\to$ **Rel**(2, 2, 2) the size of an object increases by the number of triples in its relation; in **Rel**(2, 2, 2) $\to$ **Graph** the size of a resulting graph is the sum, over the three relations, of the number of arcs times the number of vertices of the corresponding replacement graph. Thus we have found *a polynomial transformation of the homomorphism problem for groups to the homomorphism problem for graphs.* To determine if there is a morphism $G \to G'$ in **Group**, we form $\Phi(G)$, $\Phi(G')$ and determine if there is a morphism $G \to G'$ in **Graph**. Naturally, the former will be an isomorphism if and only if the latter will (cf. Section 4). Hence we have, at the same time, *a polynomial transformation of the isomorphism problem for groups to the isomorphism problem for graphs.* (A different polynomial transformation is given in [46].)

Similar comments apply to many other full embeddings. To summarize some of the conclusions, let **Strong** be the full subcategory of **Rel**(2) consisting of strongly connected digraphs, **Graph** (P) the full subcategory of **Graph** consisting of all graphs with property P, where P is any of the following properties: $\chi(G) = k$ ($k \geq 3$ given),

$\kappa(G) = l$ (l given), G contains G_0 as an induced subgraph (G_0 given), G has girth g (g given), and so forth [6, 31, 43]. Also, let **Hyp**(u) be the category whose objects are u-uniform hypergraphs (objects (V, E) where V is a set and E a set of u-element subsets of V), and whose morphisms are vertex-mappings which preserve edges (mappings f: $V \to V'$ such that $e \in E$ implies $f(e) \in E'$), and **Mult** the category of multigraphs. All the full embeddings below

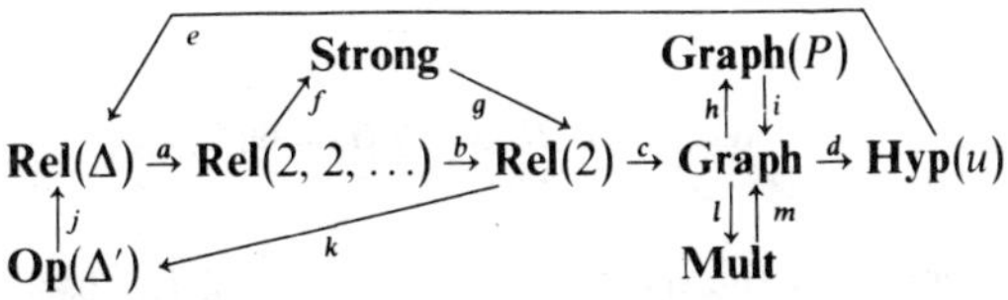

involve only a polynomial increase in the size of the objects. (We assume Δ and Δ' contain at least one 2, cf. [24]; $X \to Y(\Delta)$ means X admits a full embedding to $Y(\Delta)$ for some Δ, while $X(\Delta) \to Y$ means that a full embedding exists for each Δ.) The full embeddings a, $b \circ c$, j were illustrated here and the details may be found in [21, 24, 31]. The full embeddings g, i, l are evident. A full embedding **Hyp**(u) $\xrightarrow{e}$ **Rel**(u) is obtained by regarding each hypergraph (V, E) to be a u-ary relation (V, R) where $R = \{(v_1, \ldots, v_u): \{v_1, \ldots, v_u\} \in E\}$. The full embeddings d, m are described in [36, 33], f in [21], and k in [24].

It follows that *the morphism and the isomorphism problems for all of the above categories are polynomially equivalent* (any one can be polynomially transformed to any other). In particular, the morphism problem in each of the categories is *NP*-complete. The categories of groups, semigroups, rings, lattices, and so forth, are full subcategories of suitable categories **Op**(Δ), and hence their homomorphism and isomorphism problems may be polynomially reduced to the corresponding problems in **Graph** (or any other of the above categories). Furthermore, *the isomorphism problem in* **Top** *and* **Hyp** (*the category of all hypergraphs*) *may be polynomially transformed to the isomorphism problem in* **Rel**(2) (*and hence* **Graph**) by assigning to each topological space or hypergraph (V, E), the digraph with the vertex-set $V \cup E$ and all arcs (v, e) where $v \in e$. (We view a topological space as a set V together with a family E of open sets.) It is easy to verify that two hypergraphs or topological spaces are isomorphic (homeomorphic) if and only if the resulting digraphs are isomorphic.

As our last example of a functor we shall consider a functor Φ: **Graph** $\to$ **Simp** defined by Lovász [41]. If G is a graph, let $\Phi(G)$ be the simplicial complex whose vertices are the vertices of G and whose simplexes are those subsets of $V(G)$ which have a common neighbor; if f: $G \to H$ is a homomorphism, let $\Phi(f) = f$. Since a graph homomorphism takes a set of vertices with a common neighbor to a set of vertices with a common neighbor, $\Phi(f)$ is a simplicial mapping $\Phi(G) \to \Phi(H)$. Obviously, $\Phi(1_G) = 1_{\Phi(G)}$ and $\Phi(fg) = \Phi(f)\Phi(g)$ if fg is defined, i.e., Φ is a functor. In FIGURE 14, we illustrate the fact that Φ is not injective.

Let Ψ denote the composition of Φ with the well-known functor [27], $|\ |$: **Simp** $\to$ **Top** which assigns to each complex K the polyhedron $|K|$ determined by it.

Since Ψ: **Graph** $\to$ **Top** is not a full embedding, the existence of a continuous mapping $\Psi(G) \to \Psi(H)$ is not equivalent with the existence of a homomorphism $G \to H$. However, it was shown by Lovász [41] that *if* $\Psi(G)$ *is* $(n-2)$*-connected then* G *is not* n*-colorable*; i.e., $\chi(G) \geq n + 1$. Clearly, this is an extremely valuable result, as the proofs of noncolorability of a graph (i.e., lower bounds for χ) are quite difficult. For instance, it has been conjectured by Kneser

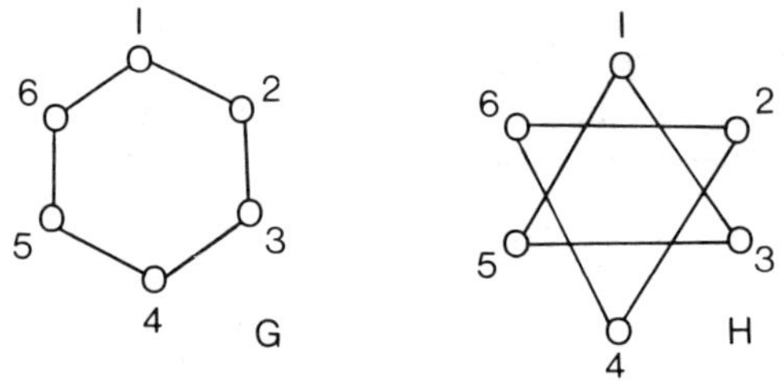

FIGURE 14. $G \neq H$, but $\Phi(G) = \Phi(H)$.

[39] that the complement of the intersection graph of all k-subsets of an n-set has chromatic number $n - 2k + 2$. It is not difficult to find an $(n - 2k + 2)$-coloring of the graph, but to prove that it does not admit an $(n - 2k + 1)$-coloring seems quite hard. Lovász [41] was able to use the above property of Ψ to prove Kneser's conjecture. Another natural context for an application of the above property of Ψ is Hedetniemi's conjecture. The conjecture follows for any pair of n-chromatic graphs G, H such that $\Psi(G \wedge H) = |\Phi(G) \times \Phi(H)|$ is $(n - 3)$-connected. (Note that, as Lovász remarks [41], an n-chromatic graph G does not necessarily yield an $(n - 3)$ connected space $\Psi(G)$.)

The help of the referee, who caught many errors and inconsistencies in the first draft, is gratefully acknowledged.

REFERENCES

1. BORIES, F. & J. L. JOLIVET. 1975. On complete colorings of graphs. *In* Recent Advances in Graph Theory. M. Fiedler, Ed. Academia, Prague, Czechoslovakia. pp. 75–88.
2. BROWN, M. & R. CONNELLY. 1973. On graphs with constant link I. *In* New Directions in the Theory of Graphs. F. Harary, Ed. Academic Press, New York, N.Y. pp. 19–51.
3. BROWN, M. & R. CONNELLY. 1975. On graphs with constant link II. Discrete Math. **11:** 199–232.
4. BURR, S. A., P. ERDÖS & L. LOVÁSZ. 1976. On graphs of ramsey type. Ars Combinatoria. **1:** 167–190.
5. CHVÁTAL, V. 1973. Edmonds polytopes and a hierarchy of combinatorial problems. Discrete Math. **4:** 305–337.
6. CHVÁTAL, V., P. HELL, L. KUČERA & J. NEŠETŘIL. 1971. Every finite graph is a full subgraph of a rigid graph. J. Comb. Theory. **11:** 284–286.
7. CLARKE, F. H. & R. E. JAMISON. 1976. Multicolorings, measures, and games on graphs. Discrete Math. **14:** 241–245.
8. ČULÍK, K. 1958. Zur Theorie der Graphen. Časopis Pěst. Mat. **83:** 133–135.
9. DIRAC, G. A. 1964. Homomorphism theorems for graphs. Math. Ann. **153:** 69–80.
10. DJOKOVIĆ, D. Ž. 1973. Distance-preserving subgraphs of hypercubes. J. Comb. Theory (B). **14:** 263–267.
11. DUGUNDJI, J. 1968. Topology. Allyn and Bacon, Boston, Mass.
12. EILENBERG, S. & S. MACLANE. 1945. General theory of natural equivalences. Trans. Am. Math. Soc. **58:** 231–294.
13. FARZAN, M. & D. A. WALLER. 1977. Kronecker products and local joins of graphs. Canad. J. Math. **29:** 255–269.
14. FREYD, P. 1970. On the concreteness of certain categories. Symposia Mathematica, Vol. IV (INDAM, Rome): 431–456. Academic Press, London.
15. FRUCHT, R. 1938. Herstellung von Graphen mit vorgegebener abstrakter Gruppe. Compositio Math. **6:** 239–250.

16. GELLER, D. & H. KRONK. 1974. Further results on the achromatic number. Fund. Math. **85:** 285–290.
17. HARARY, F. 1969. Graph Theory. Addison-Wesley, Reading, Mass.
18. HEDETNIEMI, S. T. 1966. Homomorphisms of graphs and automata. University of Michigan Technical Report 03105-44-T.
19. HEDRLÍN, Z. 1971. Extensions of structures and full embeddings of categories. Actes, Congrès Intern. Math., Nice, Tome **1:** 319–322. Gauthier-Villars, Paris.
20. HEDRLÍN, Z. & P. HELL. 1971. On approximation of morphisms of graphs and other structures. Unpublished manuscript.
21. HEDRLÍN, Z. & J. LAMBEK. 1969. How comprehensive is the category of semigroups? J. Algebra **11:** 195–212.
22. HEDRLÍN, Z. & A. PULTR. 1965. Symmetric relations (undirected graphs) with given semigroups. Monatsh. Math. **69:** 318–322.
23. HEDRLÍN, Z. & A. PULTR. 1966. On rigid undirected graphs. Canad. J. Math. **18:** 1237–1242.
24. HEDRLÍN, Z. & A. PULTR. 1966. On full embeddings of categories of algebras. Illinois J. Math. **10:** 392–409.
25. HEDRLÍN, Z. & P. VOPĚNKA. 1966. An undecidable theorem concerning full embeddings into categories of algebras. Comment. Math. Univ. Carolinae **7:** 401–409.
26. HEIL, G. H. 1973. Structure in social networks. Harvard University Department of Sociology Report.
27. HILTON, P. J. & S. WYLIE. 1960. Homology Theory. University Press, Cambridge.
28. HELL, P. 1972. Rétractions de graphes. Ph.D. thesis. Université de Montréal.
29. HELL, P. 1974. Absolute retracts in graphs. Graphs and Combinatorics. R. A. Bari & F. Harary, Eds.: 291–301. (Lecture Notes in Mathematics, Vol. 406). Springer-Verlag, New York, N.Y.
30. HELL, P. 1972. Full embeddings into some categories of graphs. Algebra Universalis **2:** 129–141.
31. HELL, P. 1974. On some strongly rigid families of graphs and the full embeddings they induce. Algebra Universalis **4:** 108–126.
32. HELL, P. 1975. Subdirect products of bipartite graphs. Infinite and Finite Sets. A. Hajnal, R. Rado & V. T. Sós, Eds. Vol. 2: 857–866. János Bolyai Math. Society. North Holland, Amsterdam.
33. HELL, P. 1975. On some independence results in graph theory. Proceedings of the Conference on Algebraic Aspects of Combinatorics. D. Corneil & E. Mendelsohn, Eds.: 89–122. Utilitas Math. Publ. Winnipeg.
34. HELL, P. Graphs with given neighborhoods I. Proc. Colloque International C.N.R.S. Orsay: to appear.
35. HELL, P. & D. J. MILLER. 1976. Graphs with given achromatic number. Discrete Math. **16:** 195–207.
36. HELL, P. & J. NEŠETŘIL. 1970. Graphs and *k*-societies. Canad. Math. Bull. **13:** 375–381.
37. IZBICKI, H. 1959. Unendliche Graphen endlichen Grades mit vorgegebenen Eigenschaften. Monatsh. Math. **63:** 298–301.
38. KARP, R. M. 1972. Reducibility among combinatorial problems. Complexity of Computer Computations. R. E. Miller & J. W. Thatcher, Eds.: 85–103. Plenum Press, New York, N.Y.
39. KNESER, M. 1955. Jber. Deut. Math. Ver. **58:** Aufgabe 300.
40. KO, C. S. 1977. The achromatic number of paths and cycles. Personal communication.
41. LOVÁSZ, L. Kneser's conjecture, chromatic number, and homotopy. J. Combinatorial Theory, to appear.
42. MENDELSOHN, E. 1968. Full embeddings and the category of graphs, with applications to topology and algebra. Ph.D. thesis. McGill University.
43. MENDELSOHN, E. 1972. On a technique for representing semigroups as endomorphism semigroups of graphs with given properties. Semigroup Forum **4:** 283–294.
44. MACLANE, S. 1971. Categories for the Working Mathematician. Springer-Verlag, New York, N.Y.
45. MILLER, D. J. 1968. The categorical product of graphs. Canad. J. Math. **20:** 1511–1521.
46. MILLER, G. L. 1977. Graph isomorphism, general remarks. Ninth Annual Symposium on Theory of Computing: 143–150. ACM. New York, N.Y.

47. MÜLLER, V. 1977. The edge reconstruction hypothesis is true for graphs with more than $n \cdot \log_2 n$ edges. J. Combinatorial Theory (B) **22:** 281–283.
48. NEŠETŘIL, J. & V. RÖDL. 1976. The ramsey property for graphs with given forbidden complete subgraphs. J. Combinatorial Theory (B) **20:** 243–249.
49. PFALTZ, J. L. 1972. Graph structures. J.A.C.M. **19:** 411–422.
50. PULTR, A. 1964. Concerning universal categories. Comment. Math. Univ. Carolinae **5:** 227–239.
51. PULTR, A. & Z. HEDRLÍN. 1965. Representation of small categories. Doklady Akad. Nauk SSSR. Tom 160, Vol. 6: 284–286.
52. SABIDUSSI, G. 1957. Graph Theory. Mimeographed notes. Tulane University, New Orleans.
53. SABIDUSSI, G. 1957. Graphs with given group and given graph-theoretic properties. Canad. J. Math. **9:** 515–525.
54. SABIDUSSI, G. 1961. Graph derivatives. Math. Z. **76:** 385–401.
55. SCHUBERT, H. 1972. Categories. Springer-Verlag, New York, N.Y.
56. STAHL, S. 1976. *n*-tuple colorings and associated graphs. J. Combinatorial Theory (B) **20:** 185–203.
57. WALLER, D. A. 1976. Pullbacks in the category of graphs. Proceedings of the Fifth British Combinatorial Conference. C. St. J. A. Nash-Williams & J. Sheehan, Eds.: 637–642. Utilitas Math. Publ. Winnipeg.
58. WALLER, D. A. 1976. Products of graph projections as a model for multi-stage communication networks. Electronic Letters **12:** 206–207.

SYMMETRIC EMBEDDINGS OF CAYLEY GRAPHS

Henry Levinson

Department of Mathematics
Rutgers University
Newark, New Jersey 07102

Historical Introduction

In 1878 Cayley formulated a method of presenting a group in terms of an edge colored digraph [3]. The method consisted of taking a set S of distinct points P_g corresponding to distinct elements $g \in G$, where $G \simeq \langle x_1, x_2, \ldots, x_n;\ R_1(x_i) = R_2(x_i) = \cdots = R_m(x_i) = 1 \rangle$ is some particular presentation of a group G on generators $x_1, \ldots, x_n$ and defining relators (words in the x_i's equal to the group identity) $R_1(x_i), \ldots, R_m(x_i)$.

Let g and g' be elements of G. If, as a consequence of the defining relators, $g' = gx_i$, for some generator x_i, join P_g to $P_{g'}$ by an arc (directed edge) colored "i". The result is a colored digraph giving a pictorial representation of the multiplication table for the group. In this digraph, a copy of each colored arc enters and leaves each point once, and any path corresponding to a word in the x_i which is equal to 1 as a consequence of the defining relators is a cycle. These colored digraphs are now known as *Cayley graphs*; previously they had been called "Cayley's color groups."

It should be noted that a Cayley graph is determined not only by the group in question, but also by the particular presentation taken. Thus a given group may have several nonisomorphic Cayley graphs. An example of this is given for S_3, the symmetric group of all permutations of three objects. This group may be presented on two generators a and b, colored by $\rightarrow$ and $\twoheadrightarrow$, respectively, in FIGURE 1a, and three defining relators: $a^2 = b^3 = (ab)^2 = 1$. S_3 may also be presented differently on two generators x and y, colored $\rightarrow$ and $\twoheadrightarrow$ in FIGURE 1b, with defining relators: $x^2 = y^2 = (xy)^3 = 1$.

Soon afterward, in 1882, Dyck [5] gave a method of presenting groups as tilings of Riemann surfaces by reflections and rotations of face-colored polygons. Dyck's tilings are the dualizations of Cayley graphs. By 1889, Cayley graphs had drawn the attention of Kempe (see [4]).

Burnside [2], in his famous book of 1897, had two remarkable chapters in which many of the principal questions in the topological theory of Cayley graphs were anticipated. The fundamental question in the area concerns a relationship between graph theory, group theory, and the topology of manifolds in which each field benefits from the others. A precise formulation of such a question has not yet been achieved.

One of the principal goals concerning Cayley graphs is to determine all those with given genus. The *genus of a Cayley graph* is the genus of its underlying (uncolored undirected) graph: the minimum of the genera of all orientable surfaces on which the graph has an embedding. Maschke [10] in 1896 determined all finite Cayley graphs which are planar (of genus 0). Maschke only considered "minimal" presentations, those in which no defining generator could be expressed in terms of the other defining generators.

We call an embedding of a Cayley graph *point-symmetric* (p-s for short) if the clockwise succession of colored arcs is the same, up to cyclic permutations, at each

0077-8923/79/0328-0137$1.75/1

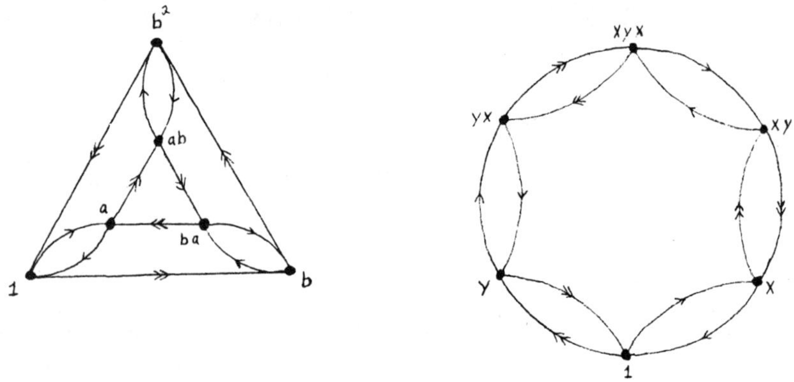

$a^2 = b^3 = (ab)^2 = 1$ $\quad\quad$ $x^2 = y^2 = (xy)^3 = 1$

(a) $\quad\quad$ (b)

FIGURE 1. Two Cayley graphs for the symmetric group S_3.

point. An embedding of a Cayley graph is *weakly point-symmetric* (w-p-s for short) if the succession of colored arcs, clockwise or counterclockwise, up to cyclic permutations, is the same at each point. An embedding of a Cayley graph is *symmetric* if it is p-s or w-p-s, and no p-s or w-p-s embedding of that Cayley graph exists on any surface of lower genus. Each of the Cayley graphs Maschke found have symmetric planar embeddings.

In 1931 Baker [1] characterized Cayley graphs with symmetric embeddings in the torus and in the projective plane. However, there are several toroidal groups he failed to list through an error of omission. His method presupposed the symmetry of the embedding, and indeed depended heavily on this assumption. Baker's techniques may be used to obtain all Cayley graphs with symmetric embeddings on surfaces of higher genus, but the number of possibilities which must be considered is computationally formidable. However, Baker utterly ignored the possibility that a symmetric embedding may not be an embedding of a Cayley graph on a surface of minimum genus. That is, the genus of a particular Cayley graph might only be realizable by an embedding which is neither p-s nor w-p-s.

There was a thirty-nine-year hiatus in the pursuit of these studies until 1970 when we proved in [7] that the genus of any Cayley graph of an infinite group is either 0 or ∞. In 1972 White [15] defined the *genus of a group* to be the minimum of the genera of all Cayley graphs for the group. He proceeded to enumerate certain classes of toroidal groups and to calculate the genera of certain abelian groups. He also gave upper bounds on the genera of symmetric groups and Hamiltonian groups. In terms of White's definition, we may set ourselves the problem of enumerating all groups of given genus.

Strasser and the author [13] determined all presentations of groups with p-s planar embeddings of their Cayley graphs in 1975. Kleinian groups are discrete subgroups of the set of all Möbius transformations $(az + b)/(cz + d)$ with complex coefficients a, b, c, d satisfying $\left|\begin{smallmatrix} a & b \\ c & d \end{smallmatrix}\right| = 1$. From a result of Maskit and the author [9] in 1973, it is known that the set of all groups having Cayley graphs with p-s planar embeddings coincides with the set of all Kleinian groups, each with an invariant component of discontinuity. It was also shown that any group with a w-p-s planar embedding for one of its Cayley graphs is an extension of a Kleinian group by a group of order 2.

Most recently, Proulx has determined in her doctoral dissertation [12] the set of all groups of genus 1. Although she did not assume the embeddings to be symmetric, this set of groups coincides with the set of all toroidal groups as characterized by Baker [1].

A *genus presentation of a group* is a presentation the genus of whose Cayley graph is minimal over all genera of Cayley graphs of presentations for that group. J. L. Gross (unpublished) asked if all genus presentations of groups have symmetric embeddings on surfaces of their genus types. We shall show that a sufficient condition for an affirmative answer to Gross's question is that all planar groups have symmetric planar embeddings of their planar Cayley graphs.

Embeddings

Lemma 1. Let G be a graph embedded in a 2-cell manner on a 2-manifold M. Then there exists a set of cycles of G which generate $\Pi_1(M)$, the fundamental group of M.

Proof: A *loop on M* is a closed curve on M. A *generating system of loops on M* is a subset of the set of all loops on M with a common initial endpoint, called the *base point*, in terms of which any loop on M may be expressed, modulo continuous deformations on M.

Let $L = \{l_1, l_2, \ldots, l_n\}$ be a generating system of loops on M which generate $\Pi_1(M)$. Without loss of generality, since M is presumed connected, we may assume the base point of the system is a point of G. Each loop l_i of L may be deformed, by a topological retraction, to a cycle c_i of G, since the embedding of G is presumed a 2-cell embedding and therefore G is a 1-skeleton of M. Thus the cycles $c_1, c_2, \ldots, c_n$ (not necessarily simple or edgewise disjoint) may each be deformed topologically into elements of L and hence generate $\Pi_1(M)$. □

The first consequence of Lemma 1 is a presentation theorem for finite groups.

Theorem 1. Each finite group is isomorphic to the quotient group of a Kleinian group by the normal closure of a set of $2n$ elements of that group.

Proof: Let G be a finite group and let P be one of its presentations. Let $C(P)$ denote the Cayley graph of P.

Using the embedding technique of Heffter [6], we may embed $C(P)$ on an orientable surface of genus n in a p-s manner (or in a w-p-s manner, if we so wish). By Lemma 1, there is a system S of cycles of $C(P)$ which generate $\Pi_1(M)$. If one were to cut along each edge of each cycle in S, the result would be to splay M into a simply connected (planar) region, a polygon with $4n$ polygonal edges, each of which is a path of (doubled) edges of $C(P)$.

For an example of this cutting and splaying in the case of a w-p-s embedding of a Cayley graph on the torus, see Figure 2. In Figure 2a the Cayley graph of the presentation

$$\langle a, b;\ (ab)^2 = (ab^{-1})^2 = a^4 = b^4 = 1\rangle$$

is embedded on a torus. Cutting along the cycles $(1, a, a^2, a^3)$ and $(1, b, b^2, b^3)$ and splaying the result into a square, one obtains Figure 2b. Note how the cut cycles have become doubled paths.

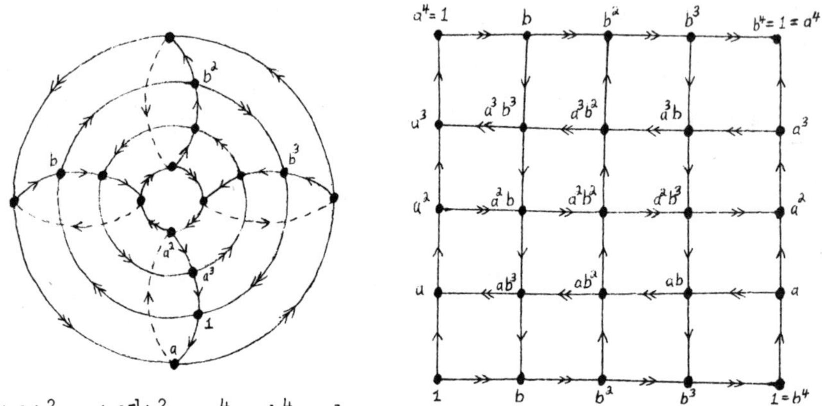

$(ab)^2 = (ab^{-1})^2 = a^4 = b^4 = 1$

(a) (b)

FIGURE 2. Cutting and splaying a torus with an embedded Cayley graph.

This polygonal region contains $C(P)$, p-s embedded (or w-p-s embedded, if we so wish), except for the necessity of identifying pairs of edges on the boundary of the region. If this region were used as the fundamental region for a (non-Euclidean) tiling of the plane, the Cayley graph $C(K)$ obtained would be p-s embedded (or w-p-s embedded) in the plane.

In FIGURE 3 such a tiling has been done using the region in FIGURE 2b as fundamental domain. The result is the Cayley graph of the presentation

$$K = \langle a, b; (ab)^2 = (ab^{-1})^2 = 1 \rangle$$

If $C(K)$ is p-s embedded in the plane, then K must be a presentation of a Kleinian group, by Theorem 4 of [9]. Let $B_1(x_i), \ldots, B_{2n}(x_i)$ denote the *distinct*

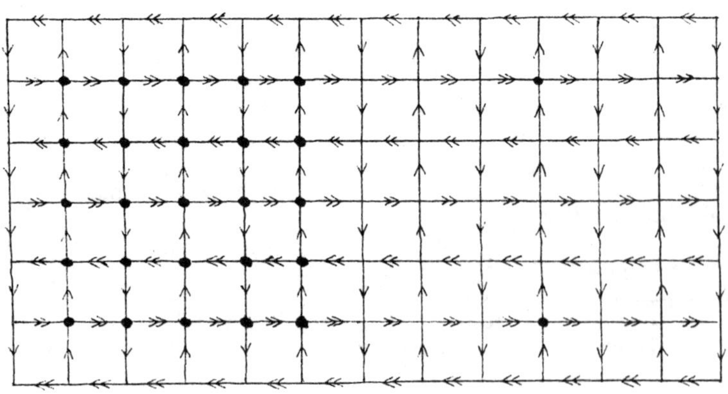

$(ab)^2 = (ab^{-1})^2 = 1$

FIGURE 3. Tiling of the plane with copies of FIGURE 2b.

edges (cycles in $C(P)$) of the polygonal region used to tile the plane. If

$$K = \langle x_1, \ldots, x_k;\ R_1(x_i) = \cdots = R_l(x_i) = 1\rangle$$

denotes the presentation of which $C(K)$ is the Cayley graph, then

$$G \simeq K/gp[B_1(x_i), \ldots, B_{2n}(x_i)]$$

since

$$G \simeq \langle x_1, \ldots, x_k;\ R_1(x_i) = \cdots = R_l(x_i) = 1 = B_1(x_i) = \cdots = B_{2n}(x_i)\rangle$$

which was our original assertion. □

The notions of *covering spaces* and *liftings* are defined in standard works in algebraic topology (see [11]). In these terms, a p-s embedding of a Cayley graph on a surface may be lifted to a p-s embedding of the Cayley graph of a Kleinian group in the plane (a universal covering space of the surface). Thus any Cayley graph of a presentation of a finite group may be covered by the Cayley graph of some presentation of a Kleinian group. A characterization of all p-s planar Cayley graphs of Kleinian groups is given in [13].

It is not known if there exists a finite group with a Cayley graph of lowest genus which embeds on a surface M of that genus, but *not symmetrically*. Should such a group G exist, consider a Cayley graph C for G of minimum genus γ. Assuming C does not have a symmetric embedding on a surface M of genus γ, an embedding of C on M lifts to an embedding of some planar Cayley graph K by the method used in the proof of THEOREM 1. The induced embedding of K in the plane is neither p-s nor w-p-s. Were K to have another p-s embedding in the plane, it would of necessity, by [9], be the Cayley graph of a Kleinian group. An inspection of all Kleinian groups in [13] which admit p-s planar embeddings of their Cayley graphs reveals that either the Cayley graph is 3-connected or it has cut points deriving from the possibility of presentations which are free products without amalgamations. In the former case, the embedding is unique, due to the well-known result of Whitney [15]. In the latter case, passing from the non-p-s to the p-s embedding is effected by reflections of "blocks" in the plane, keeping certain cut points fixed. A cycle is *simple* if it does not intersect itself. Since embeddings of simple cycles are essentially the same under such reflections, the fundamental region from which the non-p-s embedding of K was constructed deforms into a new fundamental region for the p-s embedding of K. Since the boundaries of the old and new regions consist of the same edges in the same orders about each simple cycle, the new p-s embedded region may have its appropriate boundary edges identified to form a surface homeomorphic to M on which the original Cayley graph C for G is p-s embedded, contrary to our initial assumption.

Thus the existence of a group such as G must necessarily imply the existence of an infinite planar Cayley graph which admits no p-s planar embedding. There are presentations of groups whose Cayley graphs are planar but which admit only w-p-s planar embeddings. All such presentations are similar to those in [13] and again are either 3-connected or have cut points deriving from presentations which are free products without amalgamations. The details of these presentations will be given elsewhere.

This proves the following theorem.

THEOREM 2. If each planar Cayley graph has a symmetric planar embedding, then each finite Cayley graph has a symmetric embedding which is its genus embedding.

Open Problems

Since all Cayley graphs of finite positive genus arise from finite groups, each of which is the result of adding $2n$ relators to a presentation of a Kleinian group, it may well be that such relators, for a given genus, fit some discernable pattern. Such a pattern, if its exists, would serve to characterize all Cayley graphs with given genus. Examples are needed to form a basis from which to generalize.

Given a presentation of some group, one may wish to determine whether or not the group in question is finite. Of course a complete answer to this problem is impossible due to the recursive unsolvability of the word problem in general. However a classification of finite groups by their genus might give us insight about classes of presentations with solvable word problems. If one could show the genus of a group to be finite and positive, then of course that group must be finite, and the determination of its entire structure is algorithmically possible.

Although all groups with toroidal graphs are known and many groups have been found whose genus exceeds 2, no group is known of genus 2. A. T. White has conjectured that there is no group of genus 2.

A presentation is called *minimal* if no defining generator may be expressed in terms of the rest of the defining generators. It is possible to have two minimal presentations of the same group with different numbers of defining generators. A minimal presentation on a minimum number of defining generators is called a *minimum* presentation. The genus of a group must be equal to the genus of a Cayley graph of one of its minimal presentations since the deletion of edges corresponding to superfluous generators can only serve to reduce the genus of a Cayley graph. J. L. Gross questions whether the genus of a group is always realized as the genus of the Cayley graph of one of its minimum presentations. Although he believes the answer to be negative, no one has yet found an example of a group all of whose minimum presentations have Cayley graphs with higher genera than that of the Cayley graph of some minimal, but not minimum, presentation for that group.

Finally, it is possible to embed an arbitrary graph in a Cayley graph in various ways, preserving certain properties of the original graph and destroying others. An example of a planarity-preserving embedding is given in [8]. Topology has been described as the study of properties which remain invariant under classes of mappings. Which properties of arbitrary graphs may always be preserved under *some* embedding into a Cayley graph? Although a complete answer to such a general question may never be given, partial answers may elucidate the relationships between graph theory, group theory, and the topology of manifolds to the mutual benefit of all three areas.

Summary

It is shown that embeddings of planar, infinite Cayley graphs in which the sequence of colored arcs about each point is the same (modulo cyclic permutations) give rise to embeddings of finite Cayley graphs on surfaces having the same property. If each infinite planar Cayley graph has such a "symmetric" planar embedding, then so do finite Cayley graphs on surfaces of their genus types.

Acknowledgment

The author wishes to acknowledge and thank J. L. Gross for his kind help and thoughtful criticism of the material and terminology in this paper.

References

1. Baker, R. P. 1931. Cayley diagrams on the anchor ring. Am. J. Math. **53:** 645–669.
2. Burnside, W. 1897. Theory of Groups of Finite Order. Cambridge University Press, Cambridge, England.
3. Cayley, A. 1878. Desiderata and suggestions, No. 2: Theory of groups: Graphical representations. Am. J. Math. **1:** 174–176.
4. Cayley, A. 1889. On the theory of groups. Am. J. Math. **11:** 139–157.
5. Dyck, W. 1882. Gruppentheoretische Stuedien. Math. Ann. **20:** 1–44.
6. Heffter, L. 1891. Uber das Problem der Nachbargebiete. Math. Ann. **38:** 477–508.
7. Levinson, H. 1970. On the genera of graphs of group presentations. Ann. N.Y. Acad. Sci. **175:** 277–284.
8. Levinson, H. 1972. On the genera of graphs of group presentations III. J. Comb. Theory (B). **13:** 298–302.
9. Levinson, H. W. & B. Maskit. 1975. Special embeddings of Cayley graphs. J. Comb. Theory (B). **18:** 12–17.
10. Maschke, H. 1896. The representation of finite groups, especially of the rotation groups of the regular bodies of three and four-dimensional space, by Cayley's color diagrams. Am. J. Math. **18:** 156–194.
11. Massey, W. S. 1967. Algebraic Topology: An Introduction. Harcourt, Brace & World, New York, N.Y.
12. Proulx, V. 1978. Classification of the toroidal groups. J. Graph Theory **2:** 269–273.
13. Strasser, E. R. & H. W. Levinson. 1975. Planarity of Cayley diagrams: Planar presentations. Proc. Sixth S-E Conf. Comb., Graph Theory, Comput. pp. 567–593.
14. White, A. T. 1972. On the genus of a group. Trans. Am. Math. Soc. **173:** 203–214.
15. Whitney, H. 1933. Planar graphs. Fund. Math. **22:** 73–84.

THE CONSTRUCTION OF MINIMAL-LINE GRAPHS WITH GIVEN AUTOMORPHISM GROUP

Donald J. McCarthy

Department of Mathematics and Computer Science
St. John's University
Jamaica, New York 11439

Louis V. Quintas*

Department of Mathematics
Pace University
New York, New York 10038

INTRODUCTION

A graph whose automorphism group is isomorphic (as an abstract group) to the group $\mathscr{G}$ will be termed a $\mathscr{G}$-graph, and a $(\mathscr{G}, n)$-graph is a $\mathscr{G}$-graph having n points. The object of this paper is to survey the evolution and current status of the following:

MINIMAL-LINE PROBLEM. Given a finite group $\mathscr{G}$, determine, for each positive integer n, the least number $e(\mathscr{G}, n)$ of lines possible in a $(\mathscr{G}, n)$-graph.

The problem was initially formulated for asymmetric graphs (i.e., for the case when $\mathscr{G} = id$, the identity group) in 1963 by P. Erdös and A. Rényi [1]. In 1967, $e(id, n)$ was determined by L. V. Quintas [2], who initiated the study of the general problem.

Shortly after, $e(\mathscr{G}, n)$ was determined for various groups. Specifically, the minimal-line problem was solved for the symmetric group $\mathscr{S}_k$, for each $k > 1$ (L. V. Quintas, 1968 [3]); for the cyclic group $\mathscr{C}_3$ of order three (R. Frucht *et al.*, 1970 [4]); for the dihedral group $\mathscr{D}_k$ of order $2k$, for each k (G. Haggard, 1973 [5]). In each case, a similar pattern seemed to hold. For small values of n the behavior of $e(\mathscr{G}, n)$ was somewhat erratic and the structure of the minimal-line graphs could fluctuate considerably. (FIGURE 1 in the Appendix displays the minimal-line $(\mathscr{C}_3, n)$-graphs for $9 \leq n \leq 26$; none exist for $n < 9$.) However, a certain regularity was apparent for large values of n. From some point on, $e(\mathscr{G}, n)$ was always attained by adding to some fixed $\mathscr{G}$-graph M a minor variant of the minimal-line (id, m)-graph constructed by Quintas. This "stability" phenomenon was formalized by McCarthy and Quintas [6], who showed that the pattern was not confined to these examples, but was valid for every group $\mathscr{G}$. Their Stability Theorem, which provides an explicit characterization of the "stability graph" M and a precise description of the minimal-line $(\mathscr{G}, n)$-graphs for n sufficiently large, will be stated as soon as the necessary preliminaries have been dealt with. This will be followed by explicit illustrations of the theorem, a survey of known results, and brief remarks on some open questions related to the minimal-line problem.

* Partially supported by a Pace University Scholarly Research Committee grant.

0077-8923/79/0328-0144$1.75/1 © 1979, NYAS

Basic Results

For any graph G, let $e(G)$, $v(G)$, and $c(G)$ denote, respectively, the number of lines, points, and connected components of G. The cycle rank denoted by $r(G)$, is given by $r = e - v + c$. We define the *defect* $d(G)$ to be $r(G) - c(G)$. Since we also have $d = e - v$, it is clear that in any collection of graphs each having the same number of points, minimizing the number of lines amounts to minimizing the defect. Note that for a connected graph G, $d(G) \geq -1$, with equality holding precisely when G is a tree. It follows that if there exists a $(\mathscr{G}, n)$-graph which is a forest, then $e(\mathscr{G}, n)$ is attained by the $(\mathscr{G}, n)$-forest having the maximum number of components possible. With this in mind, and following Quintas [2], we present a construction of F_n, a standard maximal-component asymmetric forest on n points.

We start with an ordering

$$T_1, T_2, \ldots, T_i, \ldots \tag{1}$$

of the set of all nonisomorphic asymmetric trees, subject only to the condition that the trees are listed in order of increasing size, i.e., $v(T_i) \leq v(T_{i+1})$, for all i. We remark that $v(T_1) = 1$ and $v(T_2) = 7$; also, for each integer $k \geq 7$, we have $v(T_i) = k$, for some i.

The graph F_n, having n points, is obtained essentially by taking an initial segment of the list (**1**). Specifically, $F_1 = T_1$, $F_7 = T_2$, and for $n \geq 8$ define F_n as follows. Let c be maximal such that

$$v(T_1 \cup T_2 \cup \cdots \cup T_c) \leq n \tag{2}$$

(Throughout, $G \cup H$ denotes the disjoint union of the graphs G and H.) If equality holds in (**2**), define $F_n = T_1 \cup T_2 \cup \cdots \cup T_c$. Otherwise, take $F_n = T_1 \cup \cdots \cup T_{c-1} \cup T_i$, where T_i is the first tree in the list (**1**) having $n - v(T_1 \cup \cdots \cup T_{c-1})$ points. This defines F_n for $n = 1$ and $n \geq 7$; it is also convenient to let $F_0 = \phi$, the empty set.

In view of the earlier remarks, it follows that for $n \geq 7$, F_n is a minimal-line $(\imath d, n)$-graph, and $e(\imath d, n) = n - c(F_n)$. The same is true when $n = 1$. For $2 \leq n \leq 5$, there are no $(\imath d, n)$-graphs, and $e(\imath d, 6)$ is attained by the unicyclic graph U_1 depicted in Figure 3 of the Appendix.

This describes the solution to the minimal-line problem when $\mathscr{G} = \imath d$. In order to describe the minimal-line graphs for arbitrary $\mathscr{G}$, we introduce a slight modification of the construction of F_n. Namely, for $s > 0$, let $F_{n,s}$ be the asymmetric forest on n points obtained by precisely the same construction as earlier, but after first deleting the trees $T_1, T_2, \ldots, T_s$ from the list (**1**). Thus $F_{n,s}$ is defined for $n \geq v(T_{s+1})$; for convenience, let $F_{n,0} = F_n$.

For any finite group $\mathscr{G}$, let $s(\mathscr{G})$ denote the number of nontrivial direct factors in the standard decomposition of $\mathscr{G}$ (into a direct product of indecomposable groups) which are isomorphic to symmetric groups. A $\mathscr{G}$-graph G is said to be *semireduced* if the number of nonisomorphic asymmetric components of G does not exceed $s(\mathscr{G})$. We can now state the

Stability Theorem [6]. Given any finite group $\mathscr{G}$, let d_0 denote the least defect attainable by a graph in the set of semireduced $\mathscr{G}$-graphs. Among all semireduced $\mathscr{G}$-graphs having defect d_0 let M be one having the least number of points, v_0. Then for sufficiently large n

$$e(\mathscr{G}, n) = n + d_0 - c(F_{m,s})$$

where $m = n - v_0$ and $s = s(\mathcal{G})$. Furthermore, M can be chosen so that $e(\mathcal{G}, n)$ is attained by the $(\mathcal{G}, n)$-graph $M \cup F_{m,s}$.

Detailed illustrations of this theorem are provided in the next section.

Special Classes of Groups

In what follows, we survey the known results on the minimal-line problem for various classes of groups. We deal with the cases when $\mathcal{G}$ is symmetric, dihedral, hyperoctahedral, a direct product of symmetric groups, cyclic, or commutative. In each case, the emphasis is on exhibiting a stability graph M, and providing explicit descriptions of minimal-line "winners."

$\mathcal{G}$ is symmetric [3]. Suppose $\mathcal{G} \simeq \mathcal{S}_k$, the symmetric group of degree k, with $k \geq 2$. Since $\mathcal{S}_k$ is directly indecomposable, $s(\mathcal{S}_k) = 1$. Thus a semireduced $\mathcal{S}_k$-graph can contain at most one type of asymmetric component. It follows that there are two possibilities for such a graph G:

(a) $G = kC$, where C is a connected asymmetric graph. (Throughout, kC will denote the disjoint union of k copies of the graph C.)

(b) $G = S \cup A$, where S is a connected $\mathcal{S}_k$-graph, and A is a connected asymmetric graph or $A = \phi$.

In case (b), $d(G) \geq -2$, whereas in case (a) we can obtain $d(G) = -k$ by taking C to be a tree. Thus $d_0 = -k$, and by choosing $C = T_1$, we see that $v_0 = k$. The stability graph here is $M = kT_1$, and the minimal-line $(\mathcal{S}_k, n)$-graph will be $kT_1 \cup F_{n-k,1}$, for n "sufficiently large." In fact, this holds whenever $n \geq k + 7$.

When $n < k$, no $(\mathcal{S}_k, n)$-graphs exist, and when $n = k$ the minimal-line winner is kT_1. This leaves only the cases $k + 1 \leq n \leq k + 6$. When $k = 2$, the desired minimal-line graph is $P_{n-1} \cup T_1$, for $3 \leq n \leq 8$, where P_m denotes the path (of length $m - 1$) on m points. When $k > 2$, the minimal-line $(\mathcal{S}_k, n)$-graphs for $n = k + 1$ to $k + 6$ are, respectively, S_k, $S_k \cup T_1$, $S_{k,2}$, $S_{k,2} \cup T_1$, $S_{k,3} \cup T_1$, $kT_1 \cup U_1$. Here S_k denotes the star on $k + 1$ points, and $S_{k,j}$ is the tree on $k + j + 1$ points obtained by attaching (one endpoint of) a path of length j to the center of S_k.

$\mathcal{G}$ is dihedral [5]. Let $\mathcal{G} \simeq \mathcal{D}_k$, the dihedral group of order $2k$. For $1 \leq k \leq 10$, the stability graph for $\mathcal{D}_k$ is displayed in Figure 2 of the Appendix. Despite the apparent disorder, these stability graphs follow an entirely regular pattern, for $k > 6$. The deviations from this pattern, when $k \leq 6$, are due to irregular structural features of the small dihedral groups. The basic algebraic facts are as follows:

(a) $\mathcal{D}_k$ is never isomorphic to a symmetric group, except when $k = 1, 3$; $\mathcal{D}_1 \simeq \mathcal{S}_2$ and $\mathcal{D}_3 \simeq \mathcal{S}_3$

(b) $\mathcal{D}_k$ is directly indecomposable unless $k \equiv 2 \pmod 4$; when $k = 2m$, with m odd, $\mathcal{D}_k \simeq \mathcal{S}_2 \times \mathcal{D}_m$

(c) $\mathcal{D}_k$ is never isomorphic to a complete monomial group $\mathcal{S}_m[\mathcal{A}]$ with $m > 1$, $|\mathcal{A}| > 1$ except when $k = 4$; $\mathcal{D}_4 \simeq \mathcal{S}_2[\mathcal{S}_2]$.

Note that as a consequence of (a) and (b), we have $\mathcal{D}_2 \simeq \mathcal{S}_2 \times \mathcal{S}_2$ and $\mathcal{D}_6 \simeq \mathcal{S}_2 \times \mathcal{S}_3$. Later on, we will discuss the minimal-line problem for $\mathcal{G}$ a direct product of symmetric groups, hence defer $\mathcal{D}_2$ and $\mathcal{D}_6$ until then. Similarly, $\mathcal{D}_4$ will be subsumed under a treatment of the hyperoctahedral groups $\mathcal{S}_k[\mathcal{S}_2]$. Since $\mathcal{D}_1$ and $\mathcal{D}_3$ are symmetric, hence have already been discussed, we concern ourselves here with $\mathcal{D}_k$ when $k = 5$ and $k \geq 7$.

For this restricted range of values for k, it follows from (a) and (b) that $s(\mathcal{D}_k) = 1$ or 0 depending on whether or not $k \equiv 2 \pmod 4$. If $k \not\equiv 2 \pmod 4$, then $s(\mathcal{D}_k) = 0$

and, in view of (b) and (c), it turns out that a semireduced $\mathscr{D}_k$-graph must be connected, but not a tree. Thus we have $d_0 \geq 0$, and equality is attained here by C_k, the cycle on k points; indeed C_k is the stability graph. When $k \equiv 2 \pmod 4$ and $k > 6$, we see from (b) that $s(\mathscr{D}_k) = 1$. Letting $k = 2m$ with m odd, it is easy to verify that the stability graph is $2T_1 \cup C_m$, so that $d_0 = -2$ and $v_0 = m + 2$.

Until recently (indeed, until well after the original draft of the present manuscript) it was believed that, for $k > 6$, the minimum number of points possible in a $\mathscr{D}_k$-graph was v_0 (where, as above, $v_0 = k$ when $k \not\equiv 2 \pmod 4$, and $v_0 = (k/2) + 2$ when $k \equiv 2 \pmod 4$). Unfortunately, the proof of this result contained an oversight, and is valid only when k or $k/2$ is a prime power. It turns out [27] that for most other values of k, the minimum number of points possible in a $\mathscr{D}_k$-graph is not $v_0 = k$ (resp. $v_0 = (k/2) + 2$), but rather $v_1 = q' + 2p'$ (resp. $v_1 = 2 + q' + 2p'$), where p' is the sum of the distinct primes dividing k (resp. $k/2$) and q' is the sum of the highest powers of these primes which divide k. The values of $e(\mathscr{D}_k, n)$ for $v_1 \leq n < v_0$ are not yet known. Moreover, even when $n \geq v_0$ many of the $(\mathscr{D}_k, n)$-graphs previously believed to be minimal-line winners can be overthrown by combining a small "new" $\mathscr{D}_k$-graph with a large asymmetric forest; the larger $v_0 - v_1$ is, the more extensive the damage to earlier results. Thus, at present, the minimal-line problem for $\mathscr{D}_k$ is still open except when k is a prime power or twice a prime power. The results in the latter cases are described below.

Consider first the case when $k = 2m$, where $m > 3$ is a power of an odd prime. As remarked above, no $(\mathscr{D}_k, n)$-graphs exist for $n < m + 2$, and the minimal-line winners for $n \geq m + 2$ are easy to describe; namely, $2T_1 \cup C_m \cup F_{n-m-2,1}$ wins for $n = m + 2$ and $n \geq m + 9$; and in the six remaining cases, the winner is $P_{n-m-1} \cup C_m \cup T_1$.

Now suppose k is itself a prime power and $k > 4$. No $(\mathscr{D}_k, n)$-graphs exist for $n < k$, and $C_k \cup F_{n-k}$ is the minimal-line winner for $n = k$, $k + 1$ and when $n \geq k + 7$. The winner for $n = k + 2$ is $W_k \cup T_1$ and for $n = k + 6$ it is $C_k \cup U_1$, where W_k denotes the wheel on $k + 1$ points and (as before) U_1 is the unicyclic $\mathscr{id}$-graph on six points. This leaves the three cases $n = k + 3$, $k + 4$, and $k + 5$. In order to describe the minimal-line graphs here for various values of k, we need a little more notation. Let $W_{k,j}$ denote the graph on $k + 1 + j$ points and $2k + j$ lines obtained from the wheel W_k by attaching a path of length j at the center (hub) of the wheel. When j divides k define the graph $C_{k,j}$ on $k + j$ points and $2k$ lines, which is obtained from C_k as follows. Let the points of C_k be labelled $v_1, v_2, \ldots, v_k$ (with v_i adjacent to v_{i+1}); C_k will be a subgraph of $C_{k,j}$. Let the additional points of $C_{k,j}$ be $u_1, u_2, \ldots, u_j$; the additional lines are (v_s, u_t) for precisely those subscripts satisfying $s \equiv t \pmod j$.

When k is even, the minimal-line $(\mathscr{D}_k, n)$-graphs for $n = k + 3$, $k + 4$, $k + 5$ are $C_{k,2} \cup T_1$, $C_{k,4}$ and $C_{k,4} \cup T_1$. When k is odd, things are a little different. For $n = k + 3$ and $k + 4$, the winners are $C_{k,3}$ and $C_{k,3} \cup T_1$ when $k \equiv 0 \pmod 3$; and are $W_{k,1} \cup T_1$ and $W_{k,2} \cup T_1$ when $k \not\equiv 0 \pmod 3$. For $n = k + 5$, the winner is $C_{k,5}$ when $k \equiv 0 \pmod 5$ and $W_{k,3} \cup T_1$ when $k \not\equiv 0 \pmod 5$.

$\mathscr{G}$ is hyperoctahedral [7]. The hyperoctahedral groups are the complete monomial groups (wreath products) $\mathscr{H}_k = \mathscr{S}_k[\mathscr{S}_2]$, for $k > 1$. A detailed description of these groups is given in [7], where the minimal-line problem is solved. It turns out that $\mathscr{H}_k$ is never a direct product of symmetric groups except when $k = 3$; since $\mathscr{H}_3 \simeq \mathscr{S}_2 \times \mathscr{S}_4$, this case will be treated later. In what follows, we assume that $\mathscr{G} \simeq \mathscr{H}_k$, with $k > 1$, $k \neq 3$. We have $s(\mathscr{H}_k) = 0$ or 1 depending on whether k is even or odd. The stability graph is kP_2 when k is even, and $kP_2 \cup T_1$, when k is odd. However, since $T_1 \cup F_{m-1,1} = F_m$, for $m > 7$, the minimal-line $\mathscr{H}_k$-graphs are the same for k even as for k odd.

There are no $(\mathscr{H}_k, n)$-graphs, for $n < 2k$, and the minimal-line winner is $kP_2 \cup F_{n-2k}$ when $n = 2k$, $2k + 1$ and when $n \geq 2k + 7$. This leaves only the five values $2k + 2 \leq n \leq 2k + 6$, for each k. To describe the minimal-line graphs in these cases, we introduce the graph $P_{k,j}$ obtained by attaching each point of kP_2 to the same terminal point of P_j. Thus $P_{k,j}$ has $2k + j$ points and $3k + j - 1$ lines; the graph resembles a pinwheel. When $k > 5$, the minimal-line $(\mathscr{H}_k, n)$-graph is given by $P_{k,n-2k-1} \cup T_1$, for $2k + 2 \leq n \leq 2k + 5$, and by $kP_2 \cup U_1$, for $n = 2k + 6$. When $k \leq 5$, the minimal-line graphs are as above with the following exceptions. When $k = 2$, the winners for $n = 2k + j$, with $j = 2, 3, 4, 5$ are given by $2P_3$, $2P_3 \cup T_1$, $2P_4$, $2P_4 \cup T_1$, respectively. When $k = 4$, the winners for $n = 2k + 4$, $2k + 5$ are $4P_3$ and $4P_3 \cup T_1$. When $k = 5$, the winners for $n = 2k + 5$, $2k + 6$ are $5P_3$ and $5P_3 \cup T_1$.

$\mathscr{G}$ is a direct product of symmetric groups. Suppose $\mathscr{G}$ is the direct product of s symmetric groups of degrees $k_1, k_2, \ldots, k_s$ with $k_1 \geq k_2 \geq \cdots \geq k_s > 1$. If $k_s > 2$, then it is easy to see that the stability graph for $\mathscr{G}$ is

$$M = k_1 T_1 \cup k_2 T_2 \cup \cdots \cup k_s T_s \tag{3}$$

where the T_i are the trees in the list (**1**). If $k_s = 2$, however, this is not the case. For example, the stability graph for $\mathscr{S}_3 \times \mathscr{S}_2$ is not $3T_1 \cup 2T_2$, but rather, $3T_1 \cup P_2 \cup T_2$; the defect in each case is $d_0 = -5$, but the latter has only $v_0 = 12$ points, whereas the former has seventeen points. Keeping this example in mind, it is not hard to determine the stability graph in the general case. Changing notation slightly, suppose $\mathscr{G}$ is the direct product of $s + t$ symmetric groups whose degrees satisfy $k_1 \geq k_2 \geq \cdots \geq k_s > 2$, and the remaining t factors have degree 2. Then the stability graph for $\mathscr{G}$ is $M \cup N$, where M is given by (**3**) and N is obtained as follows. Let $R_1, R_2, R_3, \ldots$ be listing of all trees whose automorphism group has order 1 or 2, but excluding the trees $T_1, T_2, \ldots, T_s$. It is assumed that the listing is in order of increasing size, in the sense that $i < j$ implies $v(R_i) \leq v(R_j)$, and is compatible with the listing (**1**) in that the asymmetric R_i's are chosen sequentially from $T_{s+1}, T_{s+2}, \ldots$. Then

$$N = R_1 \cup \cdots \cup R_t \cup T_{s+1} \cup \cdots \cup T_{s+t} \tag{4}$$

Thus the stability graph is known whenever $\mathscr{G}$ is a direct product of symmetric groups. The complete solution to the minimal-line problem has not yet been obtained for these groups; however, the case when there are only two terms in the product has received some attention. Christopher [8] has solved the problem when $\mathscr{G} \simeq \mathscr{S}_k \times \mathscr{S}_2$, for $k \geq 2$. The results in this case are easily described. No $(\mathscr{G}, n)$-graphs exist when $n < k + 2$, and the minimal-line graphs are given by $kT_1 \cup P_{n-k}$, for $k + 2 \leq n \leq k + 8$, and by $kT_1 \cup P_2 \cup F_{n-k-2,1}$, for $n \geq k + 9$. Results can be readily obtained for $\mathscr{G} \simeq \mathscr{S}_k \times \mathscr{S}_3$, when $k \geq 3$. In this case, no $(\mathscr{G}, n)$-graphs exist for $n < k + 3$, and the minimal-line graph, for $n \geq k + 3$, is of the form $kT_1 \cup G_{n-k}$. The graphs G_j are described below, for $j \geq 3$; G_j is given by K_3, S_3, and $W_{3,1}$, respectively, when $j = 3$, 4, and 5, and by

$$S_{3,j-4}, \text{ for } 6 \leq j \leq 10$$

$$S_3 \cup F_{j-4,1}, \text{ for } 11 \leq j \leq 46$$

$$3T_2 \cup F_{j-21,2}, \text{ for } j \geq 47$$

$\mathscr{G}$ is cyclic. For simplicity, we start with the case when $\mathscr{G} \simeq \mathscr{C}_q$, where $q > 1$ is a power of a prime. Here $\mathscr{G}$ is directly indecomposable so that $s(\mathscr{G}) \leq 1$. In fact, $s(\mathscr{C}_q) = 1$ only when $q = 2$; and $\mathscr{C}_2 \simeq \mathscr{S}_2$. Thus we assume $q > 2$; we have $s(\mathscr{C}_q) = 0$, and since no tree has automorphism group $\mathscr{C}_q$, it is easy to see that $d_0 = 0$. It follows that the stability graph for $\mathscr{C}_q$ will be the connected unicyclic $\mathscr{C}_q$-graph

having the least number of points. It has been shown by Foy [9] that whenever $m > 2$ every connected unicyclic $\mathscr{C}_m$-graph is of the form $U(m)$, where U is a connected unicyclic $\mathscr{id}$-graph. Here $U(m)$ denotes the "m-fold replication" of U; if we regard U as a necklace, then $U(m)$ is obtained by unclasping U, taking m copies of this unclasped necklace and joining their clasps in the most natural manner. (Examples are provided in FIGURE 3 in the Appendix.) Since the smallest connected unicyclic $\mathscr{id}$-graph U_1 has six points, it follows that for $\mathscr{G} \simeq \mathscr{C}_q$, $v_0 = 6q$ and $M = U_1(q)$.

Assume now that $\mathscr{G}$ is cyclic of order $m > 2$. The direct product decomposition of $\mathscr{G}$ is determined by the prime factorization of m. If $m = q_1 q_2 \ldots q_t$, where the q's are powers of different primes, then $\mathscr{G} \simeq \mathscr{C}_{q_1} \times \mathscr{C}_{q_2} \times \cdots \times \mathscr{C}_{q_t}$. It follows easily from the above that if $m \not\equiv 2 \pmod 4$, then the stability graph for $\mathscr{G}$ is $U_1(q_1) \cup U_1(q_2) \cup \cdots \cup U_1(q_t)$. If $m \equiv 2 \pmod 4$, we can assume $q_1 = 2$. The stability graph here is just as above except that $U_1(q_1)$ should be replaced by $2T_1$ (cf. FIGURE 4).

Again, although the stability graph for $\mathscr{C}_m$ is known, the minimal-line problem has not been completely solved except in special cases. The case $m = 2$ was discussed with the other symmetric groups, and as mentioned earlier, the case $m = 3$ was solved by Frucht *et al.* [4]. The minimal-line $(\mathscr{C}_3, n)$-graph is $U_1(3) \cup F_{n-18}$, for $n \geq 25$, and the winners for $n < 25$ are given in FIGURE 1 of the Appendix. The case when m is prime has also been solved [10], but has not yet appeared in print. The minimum number of points possible in a $\mathscr{C}_m$-graph was determined by Meriwether [11].

$\mathscr{G}$ is commutative. Since every finite commutative group can be decomposed as the direct product of cyclic groups, it is not surprising that a good description can be given for the stability graph of every commutative group. As we shall see, however, there are some difficulties in obtaining a completely explicit description. At any rate, attention can be confined to the case when $\mathscr{G}$ has no direct factors of order 2. For suppose $\mathscr{G}$ is such a group, and let $\mathscr{G}_t$ be the direct product of t copies of the cyclic group $\mathscr{S}_2$. It turns out that if M is a stability graph for $\mathscr{G}$, and N is given by (**4**), with $s = 0$, then $M \cup N$ is a stability graph for $\mathscr{G} \times \mathscr{G}_t$.

Thus assume $\mathscr{G}$ is commutative with no direct factors of order 2. We have $s(\mathscr{G}) = 0$ and $d_0 = 0$. It follows that the stability graph M must be of the form

$$M_1 \cup M_2 \cup \cdots \cup M_k \tag{5}$$

where $k \geq 1$ and the M_i are nonisomorphic connected graphs, each of which is unicyclic (i.e., has rank 1) and has cyclic automorphism group of order $m_i > 1$. In addition, we have

$$\mathscr{G} \simeq \mathscr{C}_{m_1} \times \mathscr{C}_{m_2} \times \cdots \times \mathscr{C}_{m_k} \tag{6}$$

In fact, we know that M must be the graph of this type which has the least number of vertices possible. This leads to a more explicit description of M. Suppose we start with a decomposition (**6**); then it is easy to describe the minimal-point $\mathscr{G}$-graph of the form (**5**) in which each M_i is a connected unicyclic $\mathscr{C}_{m_i}$-graph. We know that $M_i = U(m_i)$, where U is a connected unicyclic $\mathscr{id}$-graph, depending on i. If all m_i are distinct, we simply take $M_i = U_1(m_i)$, for all i, yielding the minimum possible number of points: $6(m_1 + m_2 + \cdots + m_k)$. However, if $m_i = m_j$, with $i \neq j$, care must be taken so that M_i and M_j are not isomorphic (or (**5**) will not yield a $\mathscr{G}$-graph), hence, we cannot use $U = U_1$ in both cases. This leads to the following procedure. Let $U_1, U_2, U_3, \ldots$ be a listing of all nonisomorphic connected unicyclic $\mathscr{id}$-graphs, in order of increasing size. Suppose that among $m_1, m_2, \ldots, m_k$ there are j distinct integers $n_1, n_2, \ldots, n_j$ with n_i appearing r_i times. Then the minimal-point $\mathscr{G}$-graph corresponding to the decomposition $\mathscr{D}$ given in (**6**) is $M(\mathscr{D}) =$

$N_1 \cup N_2 \cup \cdots \cup N_j$, where $N_i = U_1(n_i) \cup U_2(n_i) \cup \cdots \cup U_{r_i}(n_i)$. The number of points of $M(\mathscr{D})$ is

$$n_1 u(r_1) + n_2 u(r_2) + \cdots + n_j u(r_j) \tag{7}$$

where $u(r) = v(U_1 \cup U_2 \cup \cdots \cup U_r)$. If $\mathscr{D}_0$ is a decomposition (6) for which (7) is a minimum, then $M = M(\mathscr{D}_0)$ is a stability graph for $\mathscr{G}$.

An explicit description of $\mathscr{D}_0$ and hence of M, can be given for specific groups $\mathscr{G}$, but seems difficult in full generality; that is, it is difficult to specify in the general case which decomposition of $\mathscr{G}$ yields a minimum for (7). The following simple example should provide some indication of the nature of the difficulty. Suppose $\mathscr{G}$ is the direct product of r copies of $\mathscr{C}_q$ and a single copy of $\mathscr{C}_p$, where p and q are powers of different primes (and q, $p > 2$). Then $\mathscr{G}$ has only two different decompositions as a direct product of cyclic groups: the one $(\mathscr{D}')$ given above, and another $(\mathscr{D}'')$ containing $r - 1$ copies of $\mathscr{C}_q$ and a single copy of $\mathscr{C}_{pq}$. Note that this has nothing to do with the actual values of q and p; however, the determination of which decomposition yields the stability graph does depend on these values. Direct comparison of the number of points in $M(\mathscr{D}')$ and $M(\mathscr{D}'')$ reveals that $M = M(\mathscr{D}')$ when $v(U_r) \leq 6p(q-1)/q$ and $M = M(\mathscr{D}'')$ otherwise. For each r, either case is realizable for appropriate choices of q, p.

When $\mathscr{G}$ is commutative, little is known about minimal-line $(\mathscr{G}, n)$-graphs for small values of n, but the minimum number of points possible in a $\mathscr{G}$-graph has been determined by Arlinghaus [12].

Remarks and Open Questions

It should be clear at this point that the minimal-line problem for a given group $\mathscr{G}$ can be broken down into several subproblems:

(a) Determining the stability graph M for $\mathscr{G}$;

(b) Determining the point at which stability begins, i.e., finding the smallest integer n_0 so that for $n \geq n_0$, $e(\mathscr{G}, n)$ is attained by $M \cup F_{n-v_0,s}$

(c) Determining the minimal-line $(\mathscr{G}, n)$-graphs for $n < n_0$.

The previous section contains an ample discussion of (a). With regard to (c), no general methods are known. However, letting $v(\mathscr{G})$ denote the minimum number of points possible for a $\mathscr{G}$-graph, Babai [13] has shown that $v(G) \leq 2|G|$, except when G is cyclic of order 3, 4, or 5. (It is known [11], that in these exceptional cases the minimum number of points is 9, 10, 15, respectively; and otherwise the bound $2p$ is attained whenever $\mathscr{G}$ has order p, a prime.) Also it is easy to see that, except for $\mathscr{G} \simeq id$, $(\mathscr{G}, n)$-graphs exist whenever $n \geq v(\mathscr{G})$.

Concerning (b), we make the following observations. An upper bound on n_0 has been given in terms of $s(\mathscr{G})$, v_0 and the size of certain standard asymmetric forests [6]. This bound is rather crude, however, and better results are obtainable. A tempting conjecture is that stability starts as soon as possible, i.e., at the smallest integer n_1 such that $M \cup F_{n-v_0,s}$ exists, for all $n \geq n_1$—note that for $s = 0$, $n_1 = v_0 + 7$ and for $s > 0$, $n_1 = v_0 + v(T_{s+1})$. Indeed, we have $n_0 = n_1$ whenever $\mathscr{G}$ is symmetric, hyperoctahedral, or isomorphic to $\mathscr{C}_3$ or $\mathscr{S}_k \times \mathscr{S}_2$. The conjecture is false, however, as is shown by the results cited earlier for $\mathscr{S}_k \times \mathscr{S}_3$ $(k \geq 3)$, where $n_1 = k + 29$, but $n_0 = k + 47$. Insight into the situation is provided by the following result.

Let $N_j(k) = v(T_{j+1} \cup \cdots \cup T_{j+k})$ and let $s = s(\mathscr{G})$. For each $v_1 \leq v_0$ let $S(v_1)$ be the set of all semireduced $(\mathscr{G}, v_1)$-graphs, and let d_1 be the minimum defect possible for a graph in $S(v_1)$. Let m be the smallest integer ≥ 1 so that $N_{s+m+1}(d_1 - d_0) \geq v_0 - v_1$, whenever $S(v_1) \neq \phi$; when $s = 0$, we require $m \geq 2$. Then stability holds for $n \geq v_0 + N_s(m)$; i.e., $n_0 \leq v_0 + N_s(m)$.

For example, suppose it happens that v_0 is the minimum number of points possible in a $\mathscr{G}$-graph. Then $S(v_1) = \phi$ for $v_1 < v_0$, so that when $s > 0$ we have $m = 1$. This yields $n_0 \leq v_0 + v(T_{s+1}) = n_1$. When $s = 0$, we have $m = 2$ and obtain $n_0 \leq v_0 + v(T_2) = v_0 + 7 = n_1$. This accounts for the behavior in all cases cited above, where $n_0 = n_1$, except when $\mathscr{G} \simeq \mathscr{C}_3$. In this case, $s = 0$, $v_0 = 18$, and there are ten admissible values for v_1; namely, $9 \leq v_1 \leq 18$. We indicate why $m = 2$. For $d_1 > 0$, we have $N_3(d_1) \geq N_3(1) = v(T_4) = 9$, and this last is $\geq 18 - v_1$, except when $v_1 = 8$ or 9. But in these cases, $d_1 \geq 5$ and $N_3(5) = 57 > 18 - v_1$. Thus we have $m = 2$ and obtain $n_0 = n_1 = 25$. The result can also be used to explain the stability behavior when $\mathscr{G} \simeq \mathscr{S}_k \times \mathscr{S}_3$, $k \geq 3$.

Another problem worth commenting on is the explicit computation of $e(\mathscr{G}, n)$. In view of the STABILITY THEOREM, this reduces (for large n) to computation of $c(F_{n-v_0,s})$. This in turn can be reduced to computing the number of components of F_m, for large m. It is not hard to describe $c(F_m)$ in terms of the number a_k of non-isomorphic $\mathscr{id}$-trees on k points [2, 6]. No closed formula is known for a_k, but the generating function (counting series) has been obtained by Harary and Prins [14], and the a_k are readily calculated using a suitable computer algorithm.

Further work on the minimal-line problem remains to be done for other interesting classes of groups. A case of considerable interest is when $\mathscr{G}$ is a complete monomial group $\mathscr{S}_k[\mathscr{A}]$, $k > 1$. When $\mathscr{A}$ has order 2, we obtain the hyperoctahedral groups discussed earlier. Some work has been done [10] on the case when $\mathscr{A}$ has prime order. The case when $\mathscr{A}$ is symmetric would be of particular interest. Another challenging problem, which is still open, is when $\mathscr{G} \simeq \mathscr{A}_k$, the alternating group of degree k ($k \geq 4$). Although the alternating groups have been examined closely in other contexts, not too much is known with regard to the minimal-line problem for $\mathscr{A}_k$-graphs. Examples of ($\mathscr{A}_4$, 16)-graphs exist (one provided by A. J. Schwenk is given in FIGURE 6) and are believed to be point-minimal among $\mathscr{A}_4$-graphs.

In a different direction, one could ask for the largest number $E(\mathscr{G}, n)$ of lines possible for a $(\mathscr{G}, n)$-graph. But this is easily dealt with, since $E(\mathscr{G}, n)$ is attained by the complement of a minimal-line $(\mathscr{G}, n)$-graph. Thus $E(\mathscr{G}, n) = \binom{n}{2} - e(\mathscr{G}, n)$. An interesting problem posed by Hell [15] in this regard is the following: does there exist a $(\mathscr{G}, n)$-graph having e lines for each value of e between $e(\mathscr{G}, n)$ and $E(\mathscr{G}, n)$? An affirmative answer when $\mathscr{G} \simeq \mathscr{id}$ has been obtained by P. Hell and L. V. Quintas, who also noted that the answer is sometimes negative. For example, when $\mathscr{G} \simeq \mathscr{S}_k$ with $k > 2$, we have $e(\mathscr{G}, k) = 0$ and $E(\mathscr{G}, k) = \binom{k}{2}$, but there are no $(\mathscr{G}, k)$-graphs with e lines whenever $0 < e < \binom{k}{2}$. However, it is true that such an intermediate value result does hold for arbitrary $\mathscr{G}$, provided n is sufficiently large.

Another problem, closely related to the minimal-line problem, is that of determining $e_c(\mathscr{G}, n)$ and $E_c(\mathscr{G}, n)$, the least and greatest number of lines realizable by a *connected* $(\mathscr{G}, n)$-graph. This problem has been solved for the identity group [1, 2], symmetric groups [16, 17], cyclic groups of prime order [18, 10], and hyperoctahedral groups [7].

Finally, we note that one can seek minimal-line $\mathscr{G}$-graphs such that the graphs and/or the groups satisfy additional conditions. For example, the graphs can be required to be planar [19] or imbeddable in a surface of given genus [20], or minimality of lines can be sought independent of the number of points [21] or in connection with additional conditions, such as regularity of given degree [22, 23]. With regard to the group, one can explore the vast area obtained when conditions are imposed on the structure of $\mathscr{G}$ as a group of permutations [24, 25, 26]. The possibilities are clearly quite extensive, and describing them would take us well beyond the intent of the present paper. Nevertheless, it is worth pointing out that it is perhaps in these directions that the minimal-line problem has its greatest potential for being placed in the context of interesting applications.

References

1. Erdös, P. & A. Rényi. 1963. Asymmetric Graphs, Acta Math. Acad. Sci. Hung. **14:** 295–315.
2. Quintas, L. V. 1967. Extrema concerning asymmetric graphs. J. Comb. Theory **3:** 57–82.
3. Quintas, L. V. 1968. The least number of edges for graphs having symmetric automorphism group. J. Comb. Theory **5:** 115–125.
4. Frucht. R., A. Gewirtz & L. V. Quintas. 1970. The least number of edges for graphs having automorphism group of order three. *In* Recent Trends in Graph Theory. Proc. Conf., New York, 1970. Lect. Notes Math. Springer-Verlag, New York. **186:** 95–104.
5. Haggard, G. 1973. The least number of edges for graphs having dihedral automorphism group. Discrete Math. **6:** 53–78.
6. McCarthy, D. J. & L. V. Quintas. 1976. A stability theorem for minimum edge graphs with given abstract automorphism group. Trans. Am. Math. Soc. **208:** 27–39.
7. Haggard, G., D. McCarthy & A. Wohlgemuth. 1976. Extremal edge problems for graphs with given hyperoctahedral automorphism group. Discrete Math. **14:** 139–156.
8. Christopher, P. R. Minimum edge graphs with automorphism group having symmetric direct factors. Proc. 8th Southeast. Conf. Combinatorics. Graph Theory Comput., 1977. To appear.
9. Foy, G. 1975. Unicyclic graphs with a cyclic automorphism group. St. John's University. M.A. thesis.
10. McCarthy, D. 1976. Minimal edge problems for graphs with generalized symmetric automorphism group (Abstract). Not. Am. Math. Soc. **23:** A-512, No. 737-05-25.
11. Meriwether, R. L. 1967. Smallest graphs with a given cyclic group (Summary). *In* Math. Rev. **33** (2563); also, Proc. Calgary Intl. Conf. Combinatorial Struct. Appl., 1970. R. Guy *et al.*, Eds. Gordon & Breach, New York, N.Y. pp. 103–109.
12. Arlinghaus, W. C. Wayne State University. Doctoral dissertation (unpublished).
13. Babai, L. 1974. On the minimum order of graphs with given group. Can. Math. Bull. **17:** 467–470.
14. Harary, F. & G. Prins. 1959. The number of homeomorphically irreducible trees, and other species. Acta Math. **101:** 141–162.
15. Hell, P. 1974. Math. Rev. **48** (157).
16. Gewirtz, A. & L. V. Quintas. 1969. Connected extremal edge graphs having symmetric automorphism group. *In* Recent Progress in Combinatorics. W. T. Tutte, Ed. Academic Press, New York, N.Y. pp. 223–227.
17. Gewirtz, A. & L. V. Quintas, 1971. The uniqueness of a certain graph. J. Comb. Theory. **11:** 45–53.
18. Frucht, R., A. Gewirtz & L. V. Quintas. 1973. El número mínimo de líneas para grafos conexos con grupo de automorfismos de orden 3. Scientia. **142:** 72–85.
19. Marušič, D. 1976. Minimal graphs with given cyclic automorphism group. University of Ljubljana, Yugoslavia. Publikacije El. fakulteta, Beograd, 1977. B.Sc. thesis.
20. Babai, L. 1973. Groups of graphs on given surfaces. Acta Math. Acad. Sci. Hung. **24:** 215–221.
21. Bouwer, I. Z. & R. Frucht. 1973. Line-minimal graphs with cyclic group. *In* A Survey of Combinatorial Theory. J. N. Srivastava *et al.*, Eds. North-Holland, Amsterdam, the Netherlands. pp. 53–67.
22. Baron, G. & W. Imrich, 1969. Asymmetrische reguläre graphen. Acta Math. Acad. Sci. Hung. **20:** 135–142.
23. Gewirtz, A., A. Hill & L. V. Quintas. 1969. El número mínimo de puntos para grafos regulares y asimétricos. Scientia. **138:** 103–111.
24. Bouwer, I. Z. 1969. Section graphs for finite permutation groups. *In* The Many Facets of Graph Theory. Proc. Conf. Kalamazoo, Mich., 1968. Lect. Notes Math. Springer-Verlag, New York, N.Y. **110:** 55–61.
25. Nowitz, L. A. & M. E. Watkins. 1972. Graphical regular representations of non-abelian groups I, II. Can. J. Math. **24:** 993–1008, 109–1018.
26. Imrich, W. & M. E. Watkins. 1974. On graphical regular representations of cyclic extensions of groups. Pac. J. Math. **55:** 461–477.
27. McCarthy, D. J. 1979. Extremal problems for graphs with dihedral automorphism group. Proc. Second Int. Conf. Combinatorial Math., 1978. Ann. N.Y. Acad. Sci.

APPENDIX

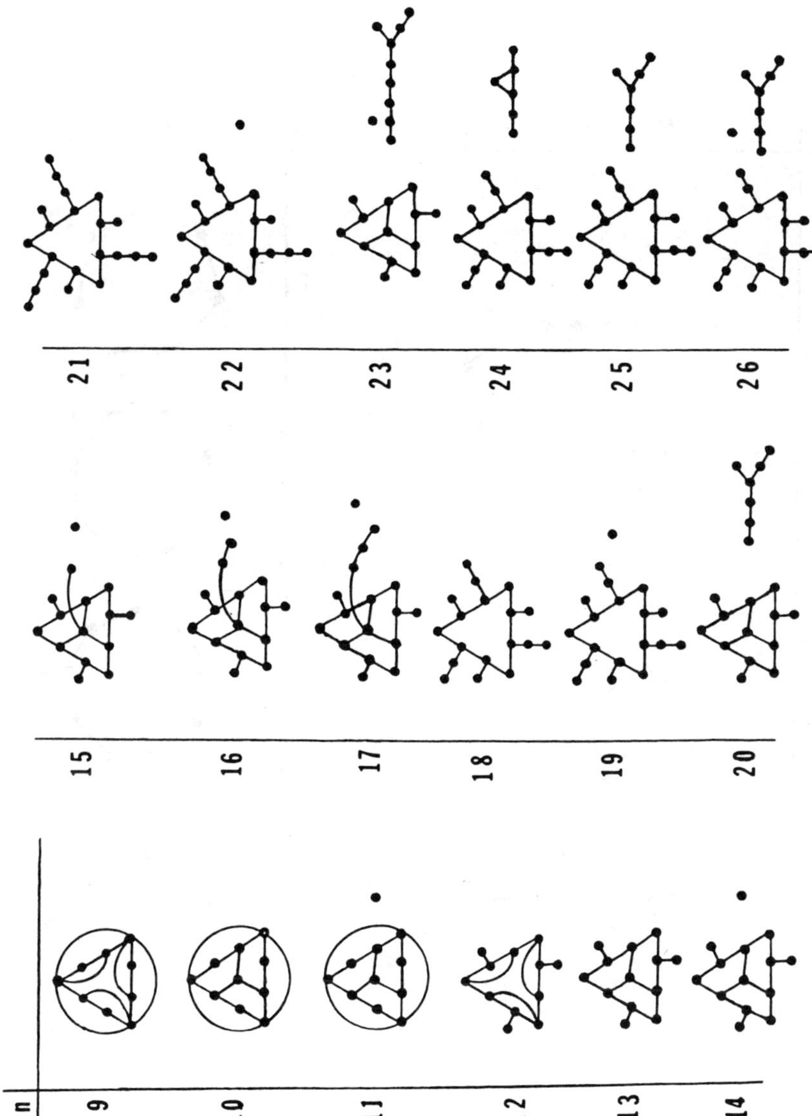

FIGURE 1. Minimal-line $(\mathscr{C}_3, n)$-graphs, $9 \leq n \leq 26$.

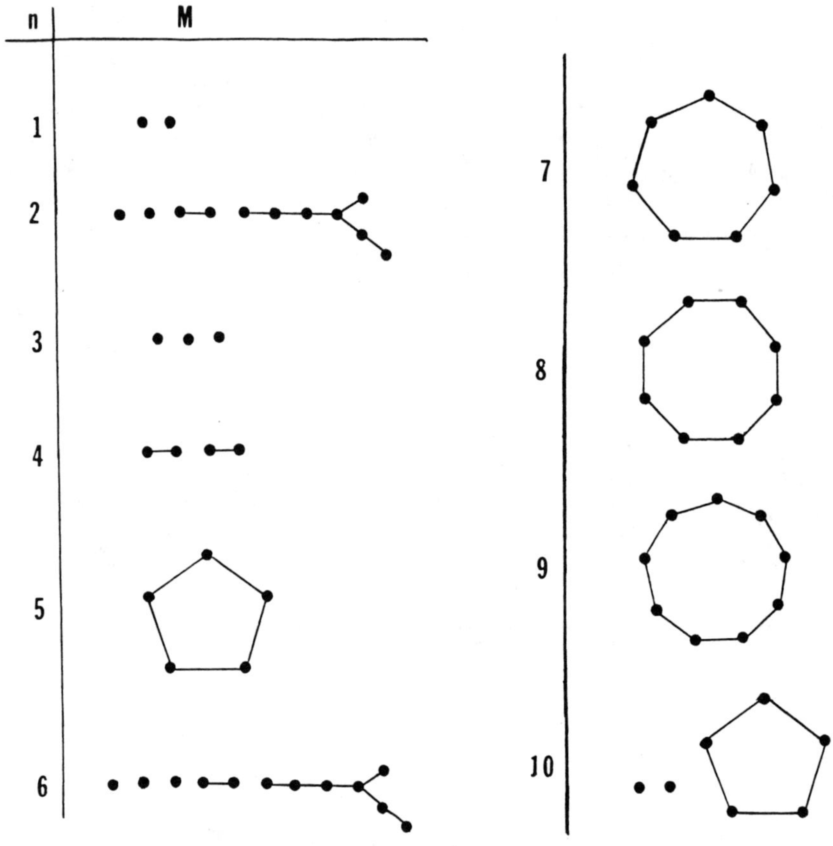

FIGURE 2. Stability graphs for $\mathscr{D}_k$, $1 \leq k \leq 10$.

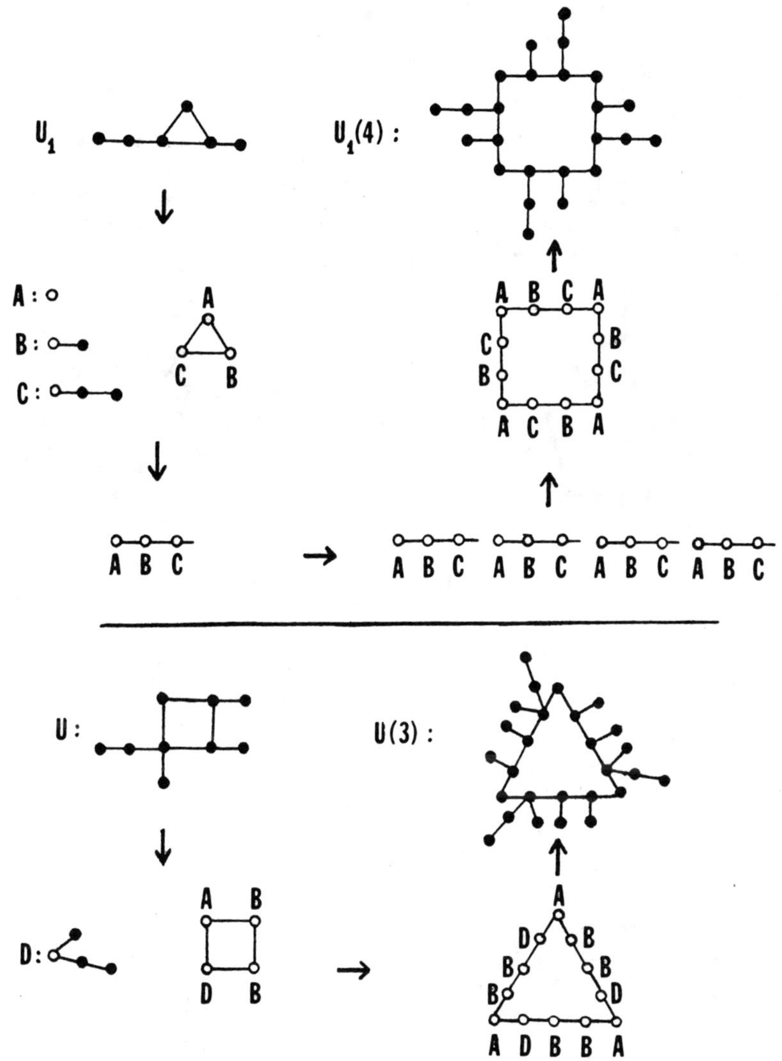

FIGURE 3. Replication of unicyclic graphs.

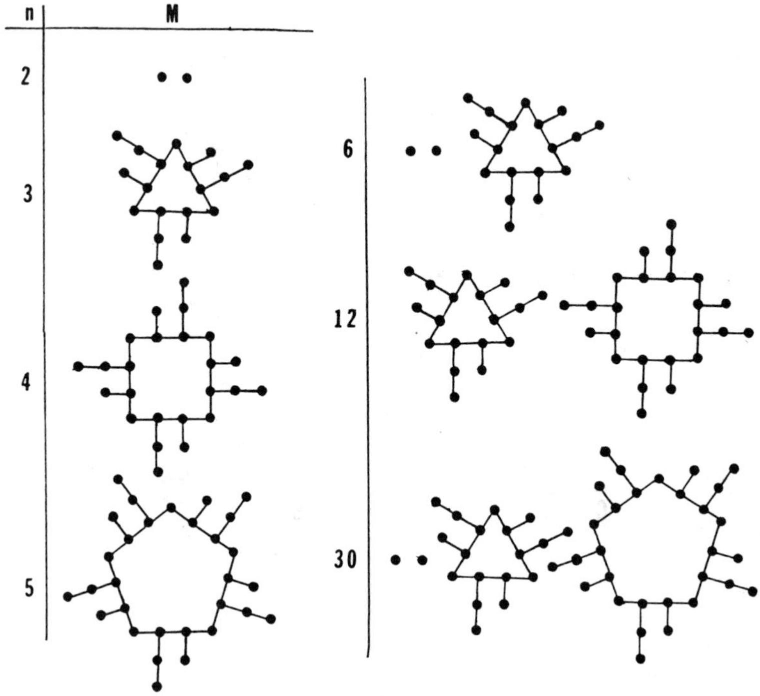

FIGURE 4. Stability graphs for cyclic groups.

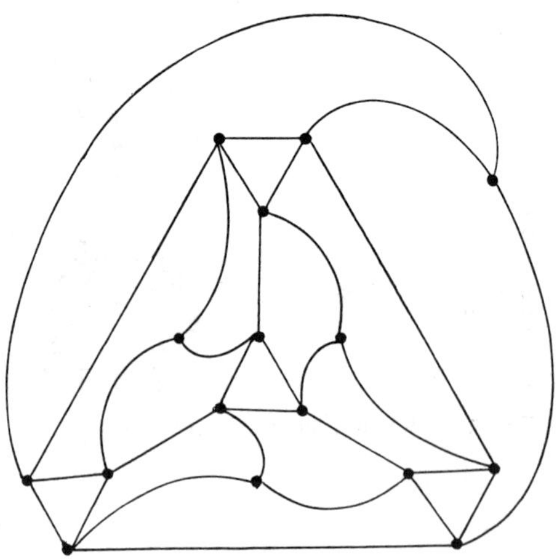

FIGURE 5. An $(\mathscr{A}_4, 16)$-graph.

SOME HAMILTONIAN COUNTEREXAMPLES

John C. Molluzzo

Division of Mathematics and Science
St. John's University
Staten Island, New York 10301

In this article we survey some interesting examples and counterexamples concerning hamiltonian graphs and related concepts. The discussion is divided into three areas: necessary or sufficient conditions, planar graphs, and hamiltonianlike conditions. Throughout we shall use the terminology of Harary [16]. In particular, a graph is hamiltonian if it possesses a cycle which passes through each of its points exactly once. The number of points and lines of a graph are denoted by p and q, respectively. We assume throughout that $p \geq 3$ and that all graphs are connected. The degree of a point v is denoted by $d(v)$ and the minimum degree of the points of G by $\delta(G)$. The degree sequence of a graph is written in nondecreasing order, $d_1 \leq d_2 \leq \cdots \leq d_p$. Let H be a graph of order p which contains the graph G such that $d_H(u) + d_H(v) < p$, for all $uv \notin E(H)$. If H_1 and H_2 are such graphs, then so is $H_1 \cap H_2$. Thus, there exists a unique smallest graph H with the above property. We call this graph the closure of G and denote it by $\mathrm{cl}(G)$. One can thus obtain $\mathrm{cl}(G)$ from G by joining pairs of nonadjacent points of G with degree sum at least p until no such pair remains.

SECTION 1

Most of the sufficient conditions for a graph to be hamiltonian force the graph to have many lines relative to the number of its points. A sequence of successively stronger theorems giving sufficient conditions for a graph to be hamiltonian were developed beginning in 1952. Each of the following is sufficient for a graph to be hamiltonian:

S1. $\delta(G) \geq p/2$ [9].

S2. u nonadjacent to v implies $d(u) + d(v) \geq p$ [27].

S3. For every n, $1 \leq n \leq (p-1)/2$, the number of points of degree not exceeding n is less than n and, for odd p, the number of points of degree at most $(p-1)/2$ does not exceed $(p-1)/2$ [28].

S4. $d_i \leq i$ and $d_j \leq j$ imply $d_i + d_j \geq p$ [1].

S5. $d_i \leq i < p/2$ implies $d_{p-i} \geq p - i$ [6].

S6. If the points of G are $v_1, \ldots, v_p$, let there be no i, j with $i < j$, $v_i v_j \notin E(G)$, $d(v_i) \leq i$, $d(v_j) \leq j - 1$, and $d(v_i) + d(v_j) \leq p - 1$, with $i + j \geq p$ [20].

S7. $\mathrm{cl}(G) = K_p$ [3].

It is interesting to note that none of these conditions is strong enough to prove that the simple 5-cycle C_5 is hamiltonian.

One of the few useful necessary conditions for hamiltonicity may be phrased in terms of the toughness of a graph. If $k(G)$ denotes the number of connected components of the graph G, we define G to be t-tough if $k(G - S) > 1$ implies that $|S| \geq t \cdot k(G - S)$ for all sets S of points of G. Chvátal [7] has proved that every hamiltonian graph is 1-tough. That 1-toughness is not sufficient for hamiltonicity is shown by the graph of FIGURE 1.

If G is not complete, the largest t for which G is t-tough is called the toughness of G and is denoted $t(G)$. Chvátal conjectured that every graph G with $t(G) > 3/2$

0077-8923/79/0328-0157$1.75/1 © 1979, NYAS

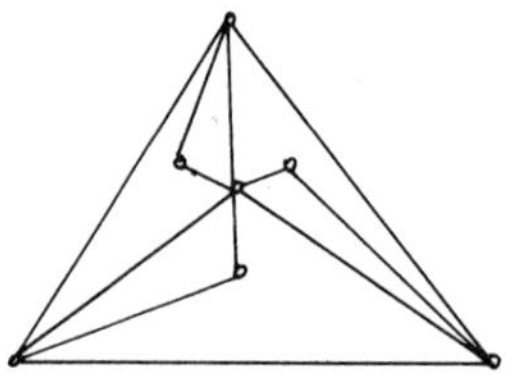

FIGURE 1. A 1-tough nonhamiltonian graph.

is hamiltonian. That this conjecture could not be improved to include $t = 3/2$ is shown by the nonhamiltonian graph of FIGURE 2b. This graph is the inflation of the Petersen graph which is depicted in FIGURE 2a. The inflation G^* of a graph G is the graph whose points are all ordered pairs (v, x), where x is a line of G and v is an endpoint of x, with two points of G^* adjacent if and only if they differ in exactly one coordinate. Although true for planar graphs, since a graph with $t > 3/2$ is 4-connected (see Section 2), Chvátal's conjecture has recently been disproved by C. Thomassen [34].

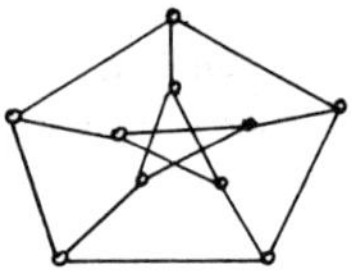

FIGURE 2(a). The Petersen graph.

A graph which is obtained in a manner similar to that of the inflation of the Petersen graph is due to Meredith [25] and appears in FIGURE 3. This graph was obtained as a counterexample to the following conjecture of Nash-Williams [26]: If G is 4-connected and 4-regular, then G is hamiltonian. If one shrinks each of the $K_{3,4}$'s to a point, it is easy to show that the resulting multigraph is nonhamiltonian. This in turn implies that the original graph is nonhamiltonian.

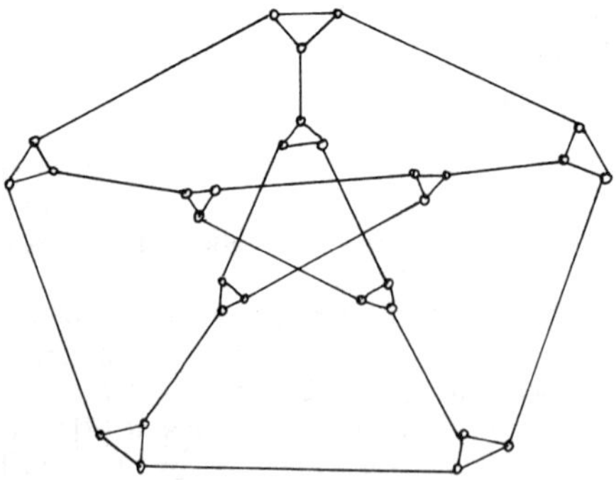

FIGURE 2(b). A nonhamiltonian graph with $t = 3/2$.

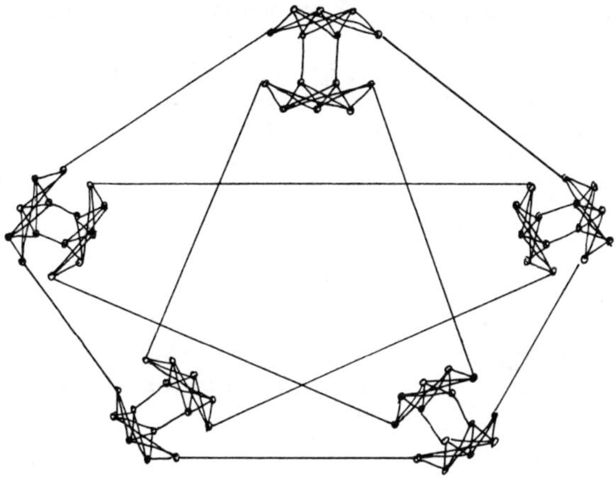

FIGURE 3. A 4-connected, 4-regular, nonhamiltonian graph.

SECTION 2

Planar hamiltonian graphs historically have been the object of intensive investigation because of their relation to the Four Color Problem. More recently they have found application to the structure of organic compounds and to linear programming algorithms.

A cyclic cutset L of a 3-connected graph G is a set of lines of G such that $G - L$ has two components each of which contains a cycle. The cyclic connectivity of G, $c\lambda(G)$, is the minimum cardinality of all the cyclic cutsets of G. The graph G is cyclically n-connected if $c\lambda(G) \geq n$. We define a graph G to be polyhedral if it is cubic, planar, and 3-connected.

A proof of the Four Color Theorem could be based on any one of the following statements, if any were true:

P1. Every polyhedral graph is hamiltonian.

P2. Every polyhedral cyclically 4-connected graph is hamiltonian.

P3. Every polyhedral cyclically 5-connected graph is hamiltonian.

Unfortunately, for the Four Color Problem, they are all false. The counterexamples that have been constructed to prove them false, however, are very interesting graphs.

The first counterexample to P1 was the forty-six-point graph of FIGURE 4 constructed by Tutte [35] and is now known as the Tutte graph. For later reference we

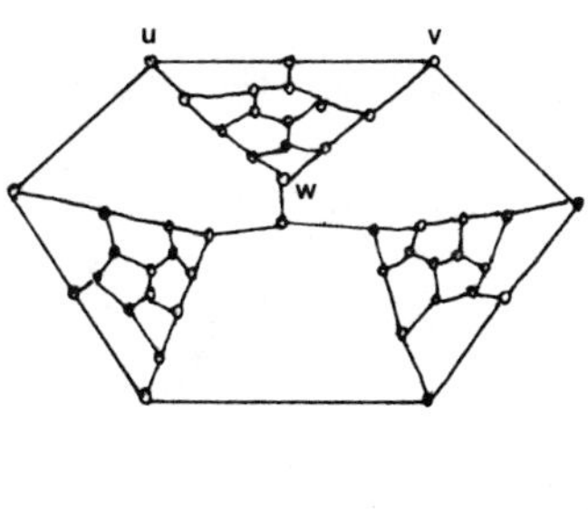

FIGURE 4. The Tutte graph: a nonhamiltonian polyhedral graph.

shall refer to the fifteen-point triangular configuration with vertices *u*, *v*, and *w* as the Tutte configuration.

Since 1946 a great deal of effort has gone into finding the smallest nonhamiltonian polyhedral graph. It has been shown [14] that all polyhedral graphs with up to twenty points are hamiltonian. The smallest known nonhamiltonian polyhedral graph, discovered independently by Lederberg [21], Barnette, and Bosak [5], has thirty-eight points and is depicted in FIGURE 5. Note the use of the Tutte configuration.

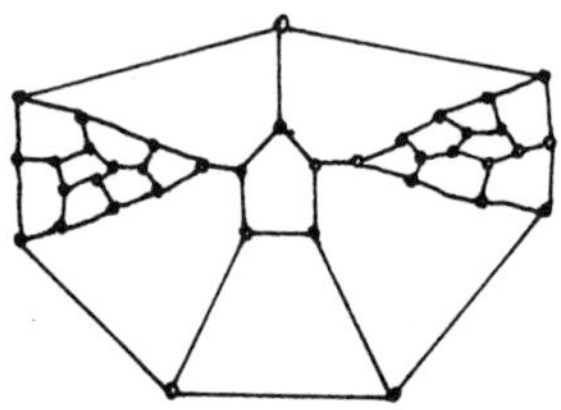

FIGURE 5. The smallest-known nonhamiltonian polyhedral graph.

A variation of statement P1 has given rise to a very interesting graph due to J. D. Horton [4]. Tutte conjectured that every bipartite cubic 3-connected graph is hamiltonian. Horton's graph (FIGURE 6) shows this to be false.

Still other variations on statement P1 may be obtained as follows. Tutte [36] proved that every planar 4-connected graph is hamiltonian. If we now consider 3-connected planar graphs, we can ask what additional condition on these graphs will force them to be hamiltonian? The hypothesis of statement P1 adds the condition that G be regular of degree 3, which is not enough to force G to be hamiltonian. Suppose we require that G be regular of degree 4, or regular of degree 5 (if G were regular of degree 6 or greater, then G would not be planar). Sachs [29] showed that neither condition forces G to be hamiltonian. Recently Zaks [39] simplified Sachs' constructions by finding a planar 3-connected, 4-regular, nonhamiltonian graph with 114 points and a planar 3-connected, 5-regular, nonhamiltonian graph with 228 points. We shall omit these graphs but mention that they are derived from the graph G of FIGURE 5 by replacing the edges of a 1-factor of G by certain configurations.

The first counterexample to statement P2 was constructed by Tutte [37] and has sixty points. The smallest known counterexample to P2, which has forty-two

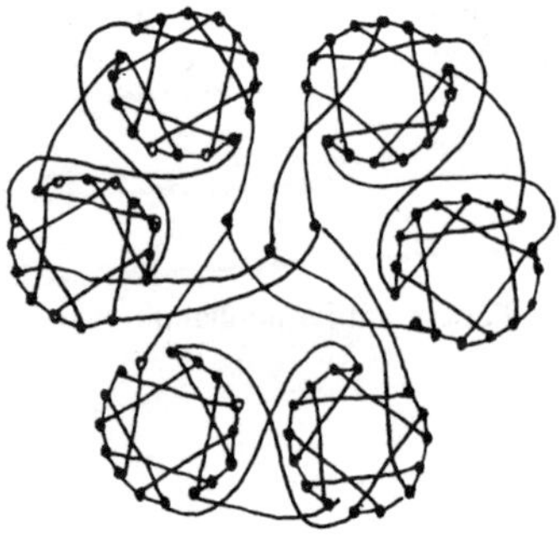

FIGURE 6. A bipartite 3-connected nonhamiltonian graph.

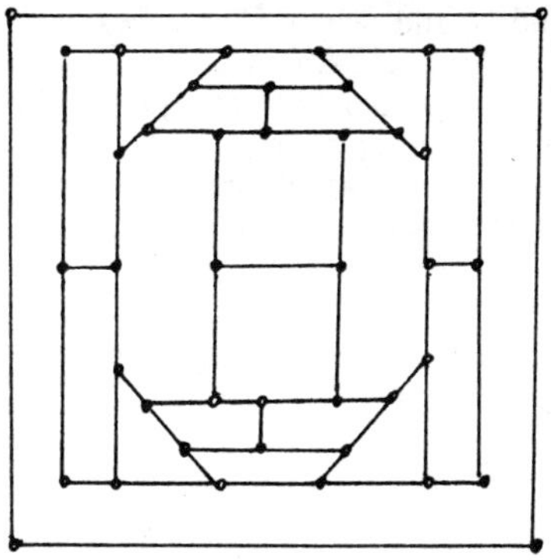

FIGURE 7. The smallest-known polyhedral cyclically 4-connected nonhamiltonian graph.

points, is shown in FIGURE 7 [15]. The first counterexample to statement P3 was constructed by Walther [38] and has 162 points. The graph of FIGURE 8, which appears in [15], is due independently to Tutte and Grinberg, has forty-four points, and is the smallest known counterexample to P3.

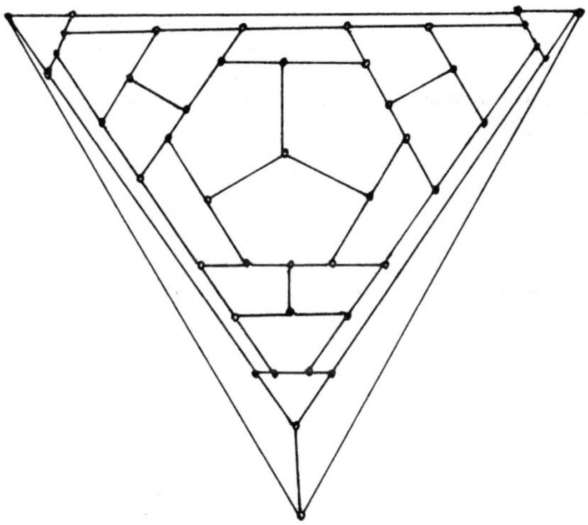

FIGURE 8. The smallest-known polyhedral cyclically 5-connected nonhamiltonian graph.

SECTION 3

We shall first consider some conditions weaker than that of a graph being hamiltonian. The nth power G^n of G has the same set of points as G with two points adjacent if and only if the distance between them in G is n or less. Sekanina [31] and Karaganis [19] have proved that every connected graph G has G^3 hamiltonian. Fleischner [12] then showed that if G is 2-connected, then G^2 is hamiltonian.

(In fact, in this case much more is true of G^2. Faudree and Schelp [11] have shown that G^2 is strongly path-connected, i.e., between each pair of distinct points u and v of G^2 there exists a path in G^2 of every possible length greater than or equal to the distance between u and v in G^2.) Fleischner's theorem cannot be improved. FIGURE 9a shows a tree whose square is nonhamiltonian and FIGURE 9b shows a graph, due to Fleischner and Kronk [13], which is bridgeless and has a nonhamiltonian square.

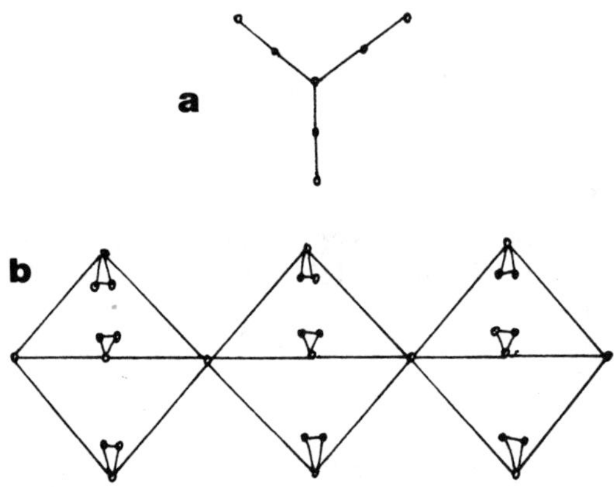

FIGURE 9. (a) A tree whose square is nonhamiltonian. (b) A bridgeless graph whose square is nonhamiltonian.

A graph is traceable if it has a hamiltonian path. Balinski conjectured that every polyhedral graph is traceable. The graph in FIGURE 10, constructed by T. A. Brown [14], shows this to be false. In each triangle T of FIGURE 10 is a copy of the Tutte configuration placed so that the points u, v, and w of FIGURE 4 get mapped to the correspondingly labelled points.

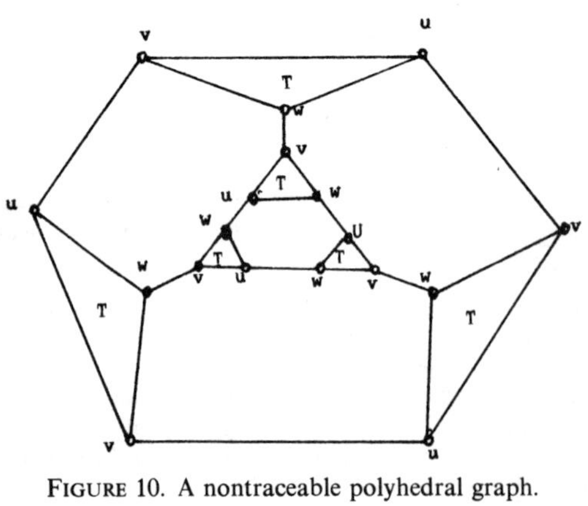

FIGURE 10. A nontraceable polyhedral graph.

A nonhamiltonian (nontraceable) graph G is hypohamiltonian (hypotraceable) if $G - v$ is hamiltonian (traceable) for every point v of G. The following theorem [17, 23, 32, 10, 8] establishes the existence of hypohamiltonian graphs of almost all orders p.

THEOREM. There exist no hypohamiltonian graphs on points $p < 10$, $p = 11, 12$, or 14. For $p = 10$, 13, and $p \geq 15$, except possibly $p = 17$, there exists a hypohamiltonian graph of order p.

The smallest hypohamiltonian graph is the Petersen graph (FIGURE 2a).

Less is known about hypotraceable graphs. We have the following theorem of Thomassen [32, 34].

THEOREM. There exist hypotraceable graphs of order p for $p = 34$, 37 and all $p \geq 39$.

It is not known if hypotraceable graphs of orders 35, 36, or 38 exist, nor is the order of the smallest hypotraceable graph known. The smallest hypotraceable graph constructed to date is depicted in FIGURE 11.

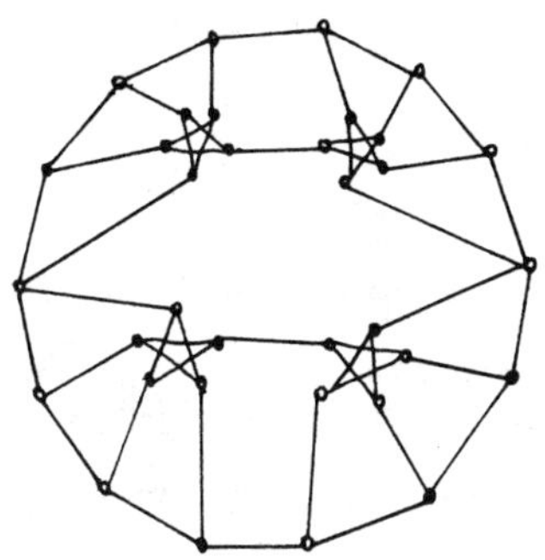

FIGURE 11. The smallest-known hypotraceable graph.

The only condition stronger than hamiltonicity that will be mentioned is that of a graph being pancyclic. The graph G is pancyclic if it contains cycles of all lengths n, $3 \leq n \leq p$. Many of the conditions which are sufficient for a graph to be hamiltonian are also sufficient for a graph to be pancyclic or bipartite. For example, Ore's condition S2 implies that G is pancyclic or bipartite [2]; Chvátal's condition S5 implies that G is pancyclic or bipartite [30]; G 2-connected implies that G^2 is pancyclic [18]. Such results led Bondy to make the metaconjecture that almost all conditions that imply that G is hamiltonian also imply that G is pancyclic or bipartite.

The last examples we present are cases where the metaconjecture is false. As mentioned previously, every planar 4-connected graph is hamiltonian. Malkevitch [24] has constructed the graph of FIGURE 12 which is planar and 4-connected but which is not pancyclic since it does not contain a 4-cycle.

The graph of FIGURE 13, due to C. Thomassen, was communicated to the author by J. A. Bondy. It is an example of a graph with a complete closure but which is not pancyclic since it contains no triangles. Hence, condition S7 does not imply that G is pancyclic.

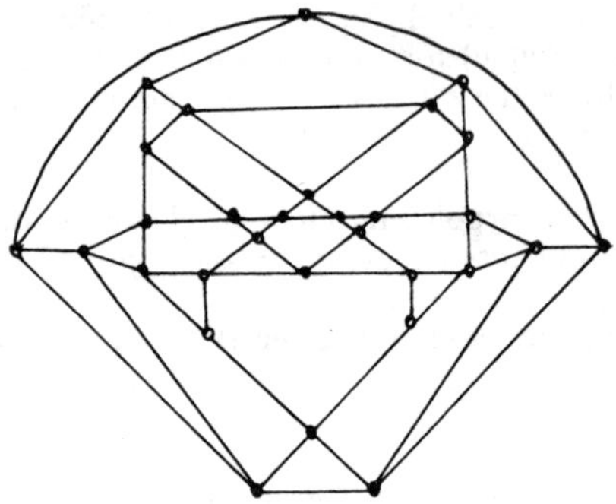

FIGURE 12. A planar 4-connected nonpancyclic graph.

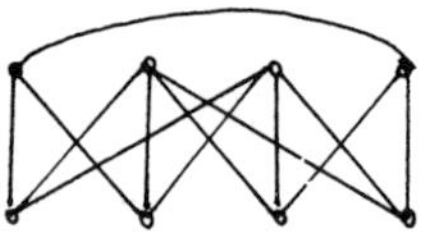

FIGURE 13. A nonpancyclic graph with a complete closure.

REFERENCES

1. BONDY, J. A. 1969. Properties of graphs with constraints on degrees. Stud. Sci. Math. Hung. **4:** 473–475.
2. BONDY, J. A. 1971. Pancyclic graphs. I. J. Comb. Theory. **11:** 80–84.
3. BONDY, J. A. & V. CHVÁTAL. 1976. A method in graph theory. Discrete Math. **15:** 111–135.
4. BONDY, J. A. & U. S. R. MURTY. 1976. Graph Theory with Applications. American Elsevier, New York, N.Y.
5. BOSAK, J. 1967. Hamiltonian lines in cubic graphs. *In* Theory of Graphs. Int. Symp., Rome (1966). P. Rosenstiehl, Ed. Gordon and Breach, New York, N.Y. pp. 35–46.
6. CHVÁTAL, V. 1972. On hamilton's ideals. J. Comb. Theory Ser. B. **12:** 163–168.
7. CHVÁTAL, V. 1973. Tough graphs and hamiltonian circuits. Discrete Math. **5:** 215–228.
8. COLLIER, J. B. & E. F. SCHMEICHEL. To appear.
9. DIRAC, G. A. 1952. Some theorems on abstract graphs. Proc. London Math. Soc. Ser. **3** (2): 69–81.
10. DOYAN, J. & V. VAN DIEST. 1975. New families of hypohamiltonian graphs. Discrete Math. **13:** 225–236.
11. FAUDREE, R. J. & R. H. SCHELP. 1976. The square of a block is strongly path-connected. J. Comb. Theory Ser. B. **20:** 47–61.
12. FLEISCHNER, H. 1974. The square of every two-connected graph is hamiltonian. J. Comb. Theory Ser. B. **16:** 29–34.
13. FLEISCHNER, H. & H. KRONK. 1972. Hamiltonische Linien im Quadrat brückenloser Graphen mit Artikulationen. Monat. Math. **76:** 112–117.
14. GRÜNBAUM, B. 1967. Convex Polytopes. John Wiley & Sons, New York, N.Y.
15. GRÜNBAUM, B. 1970. Polytopes, graphs, and complexes. Bull. Am. Math. Soc. **76:** 1131–1201.
16. HARARY, F. 1969. Graph Theory. Addison-Wesley, Reading, Mass.
17. HERZ, J. C., J. J. DUBY & F. VIGUÉ. 1967. Recherche systématique des graphes hypohamiltoniens. *In* Theory of graphs. Int. Symp. (Rome, 1966). P. Rosenstiehl, Ed. Gordon and Breach, New York, N.Y. pp. 153–159.

18. Hobbs, A. M. 1976. The square of a block is vertex pancyclic. J. Comb. Theory Ser. B. **17:** 22–34.
19. Karaganis, J. J. 1969. On the cube of a graph. Can. Math. Bull. **11:** 295–296.
20. Las Vergnas, M. 1972. University of Paris VI. Doctoral dissertation.
21. Lederberg, J. 1966. Systematics of organic molecules, graph theory and hamiltonian circuits. Instrum. Res. Rep. No. 1040. Stanford University. Stanford, Cal.
22. Lederberg, J. 1967. Hamiltonian circuits of convex trivalent polyhedra (up to 18 vertices). Am. Math. Mon. **74:** 522–527.
23. Lindgren, W. F. 1967. An infinite class of hypohamiltonian graphs. Am. Math. Mon. **74:** 1087–1089.
24. Malkevitch, J. 1971. On the lengths of cycles in planar graphs. *In* Recent trends in graph theory. Springer-Verlag, New York, N.Y. pp. 191–195.
25. Meredith, G. H. J. 1973. Regular *n*-valent *n*-connected non-hamiltonian non-*n*-edge-colorable graphs. J. Comb. Theory Ser. B. **14:** 55–60.
26. Nash-Williams, C. St. J. A. 1971. Possible directions in graph theory. *In* Combinatorial mathematics and its applications. D. J. A. Welsh, Ed. Academic Press, New York, N.Y.
27. Ore, O. 1960. A note on hamiltonian circuits. Am. Math. Mon. **67:** 55.
28. Pósa, L. 1962. A theorem concerning hamiltonian lines. Magy. Tud. **7:** 225–226.
29. Sachs, H. 1967. Construction of non-hamiltonian planar regular graphs of degrees 3, 4, and 5 with the highest possible connectivity. *In* Theory of graphs. Int. Symp. (Rome, 1966). P. Rosenstiehl, Ed. Gordon and Breach. New York, N.Y. pp. 373–382.
30. Schmeichel, E. F. & S. L. Hakimi. 1974. Pancyclic graphs and a conjecture of Bondy and Chvátal. J. Comb. Theory Ser. B. **17:** 22–34.
31. Sekanina, M. 1960. An ordering of the set of vertices of a connected graph. Publ. Fac. Sci. Univ. Brno. **412:** 137–142.
32. Thomassen, C. 1974. Hypohamiltonian and hypotraceable graphs. Discrete Math. **9:** 91–96.
33. Thomassen, C. 1976. Planar and infinite hypohamiltonian and hypotraceable graphs. Discrete Math. **14:** 377–389.
34. Thomassen, C. Private communication.
35. Tutte, W. 1946. On hamiltonian circuits. J. Lond. Math. Soc. **21:** 98–101.
36. Tutte, W. 1956. A theorem on planar graphs. Trans. Am. Math. Soc. **8:** 99–116.
37. Tutte, W. 1960. A non-hamiltonian planar graph. Acta. Math. Ac. Sci. Hung. **11:** 371–375.
38. Walther, H. 1965. Ein kubischer, planaren zyklisch fünffach zusammenhängender Graph, der keinen Hamiltonkreis besitzt. Wiss. Z. Tech. Hochsch., Ilmenau. **11:** 163–166.
39. Zaks, J. 1976. Pairs of hamiltonian circuits in 5-connected planar graphs. J. Comb. Theory Ser. B. **21:** 116–131.

ON PROPERTIES OF A WELL-KNOWN GRAPH
OR
WHAT IS YOUR RAMSEY NUMBER?

Tom Odda

Department of Mathematics
Xanadu University

Introduction

During the past ten years various authors (see e.g., [1, 2, 3, 5]) have investigated properties of a certain graph G, often called the *collaboration graph of mathematicians* (see Figure 1). In this note we explore a number of questions concerning G and suggest numerous new open problems, not only for G but also for a variety of related graphs.

Notation and Definitions

Strictly speaking, the collaboration graph G is really a hypergraph. The *vertex set* $V(G)$ of G is usually taken to be the set of mathematicians. A subset $X \subseteq V(G)$ is an *edge* of G iff there exists a published paper (or book)* which has X as its set of authors. However, for the purposes of this note, we consider G as a graph by replacing each (hyper)edge X by the complete graph $K(X)$ on X. In accordance with the usual conventions for this topic, at most one edge in $K(X)$ will be allowed in any solution of the various extremal problems for G we will consider. We should note that occasionally the restricted graph $G^* \subseteq G$ is considered in which only papers having exactly two authors are allowed to generate edges [1].

We should say a few words about the notation used in the figures. A solid edge between A and B usually indicates that a paper having A and B as authors has appeared or is about to appear. A dotted line indicates that the paper in question will probably appear (perhaps at present is just a technical report). Other various notations will be described in the text.

The Main Result

Some Comments

1. It has long been known that G is nonplanar. This seems to have first been noted by Schinzel (see [1]), who found a $K_{3,3}$ consisting of the two vertex sets {Chowla, Mahler, Schinzel} and {Davenport, Erdös, Lewis}. (In fact, this $K_{3,3} \subseteq G^*$.) However, a brief examination shows that K_5 is also a subgraph of G, the vertex set being {Erdös, Graham, Rothschild, Spencer, Straus}. In forming this K_5, it certainly did not hurt that these five authors are actually part of three 6-tuple papers (each of which however is allowed to contribute only one edge to the K_5).

* Various conventions are possible with regard to abstracts, technical reports, memoranda, lecture notes, correspondence, etc. However, we shall follow none of them.

0077-8923/79/0328-0166$1.75/1 © 1979, NYAS

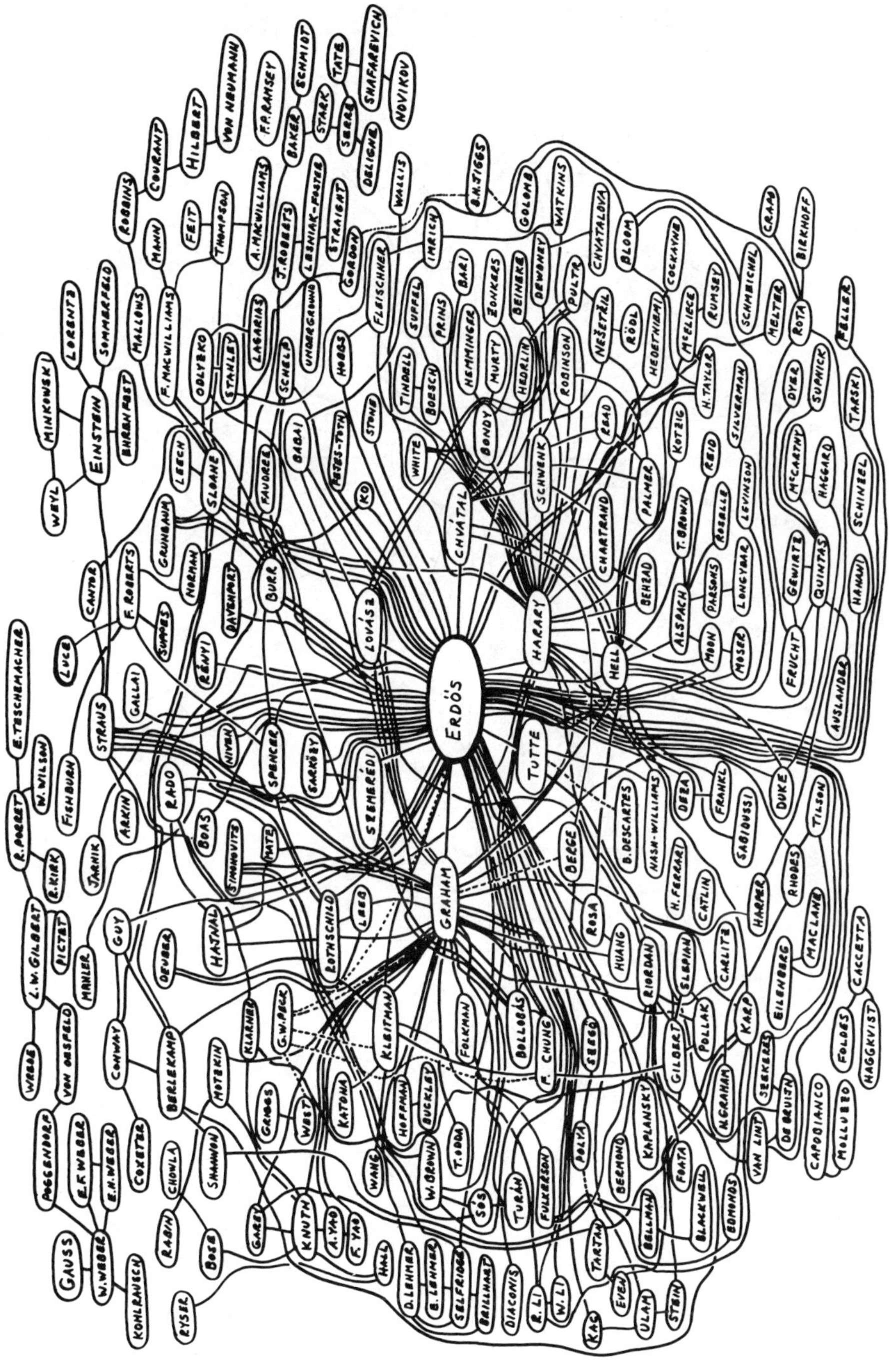
ERDÖS
HARARY
GRAHAM
TUTTE
CHVÁTAL
LOVÁSZ
SZEMERÉDI
EINSTEIN
MINKOWSKI
LORENTZ
SOMMERFELD
WEYL
EHRENFEST
GAUSS
W. WEBER
KOHLRAUSCH
POGGENDORF
E.F. WEBER
E.H. WEBER
VON NEUMANN
HILBERT
COURANT
ROBBINS
F.P. RAMSEY
SCHMIDT
BAKER
STARK
TATE
SHAFAREVICH
NOVIKOV
SERRE
DELIGNE
KNUTH
A. YAO
F. YAO
ULAM
STEIN
EVEN
TARTAN
RYSER
RABIN
CHOWLA
CONWAY
COXETER
BERLEKAMP
SHANNON
DUKE
ROTA
BIRKHOFF
TARSKI
FELLER
STRAUS
CANTOR
LUCE
F. ROBERTS
SUPPES
SLOANE
LEECH
GRUNBAUM
NORMAN
DAVENPORT
BURR
RÉNYI
GALLAI
SPENCER
SÁRKÖZY
NIVEN
RADO
BOAS
HAJNAL
ROTHSCHILD
KLEITMAN
F. CHUNG
BOLLOBÁS
FOLKMAN
KATONA
HOFFMAN
BUCKLEY
W. BROWN
T. ODA
WANG
SÓS
TURÁN
FULKERSON
PÓLYA
SZEGÖ
BERGE
ROSA
HUANG
RIORDAN
GILBERT
POLLAK
KARP
KAPLANSKY
BELLMAN
BLACKWELL
EDMONDS
VAN LINT
DE BRUIJN
SEEKERS
CAPOBIANCO
MOLLUZZO
FOLDES
HAGGKVIST
CACCETTA
MACLANE
RHODES
HARPER
CARLITZ
CATLIN
H. FERRARI
B. DESCARTES
HELL
DEZA
FRANKL
SABIDUSSI
TILSON
QUINTAS
FRUCHT
GEWIRTZ
HANANI
SCHINZEL
HAGGARD
McCARTHY
DYER
SUPNICK
MELTER
SCHMEICHEL
RUMSEY
HEDETNIEMI
McELIECE
H. TAYLOR
SILVERMAN
LEVINSON
REID
KOTZIG
PALMER
CHARTRAND
READ
SCHWENK
BONDY
ROBINSON
HEDRLIN
BEINEKE
ZONKERS
MURTY
HEMMINGER
PRINS
BARI
IMRICH
FLEISCHNER
GOLOMB
WATKINS
DEWDNEY
CHVÁTALOVÁ
BLOOM
COCKAYNE
ERDŐS
WALLIS
GORDON
STRAIGHT
BABAI
HOBBS
STONE
WHITE
TINDELL
KO
FAUDREE
SCHELP
STANLEY
ODLYZKO
F. MACWILLIAMS
MALLOWS
MAHN
FEIT
THOMPSON
A. MACWILLIAMS
J. ROBERTS
LAGARIAS
UNDERGROUND
LESNIAK-FOSTER
BEHZAD
ALSPACH
T. BROWN
ROSELLE
LONGYEAR
PARSONS
MOON
MOSER
AUSLANDER
GOLDMAN
GUY
MAHLER
PICTET
L.W. GILBERT
VON OSSFELD
WEBBE
R. PORRET
E. TESCHEMACHER
W. WILSON
FISHBURN
ARKIN
JARNIK
B. KIRK
MOTZKIN
KLARNER
G.W. PECK
WEST
GRIGGS
GAREY
BOSE
HALL
D. LEHMER
E. LEHMER
SELFRIDGE
BRILLHART
DIACONIS
R. LI
W. LI
KAC
FOATA
BERMOND
CARLITZ
EILENBERG
SIMONOVITS
MATE
LEEB
DEUBER
CRAPO

FIGURE 1.

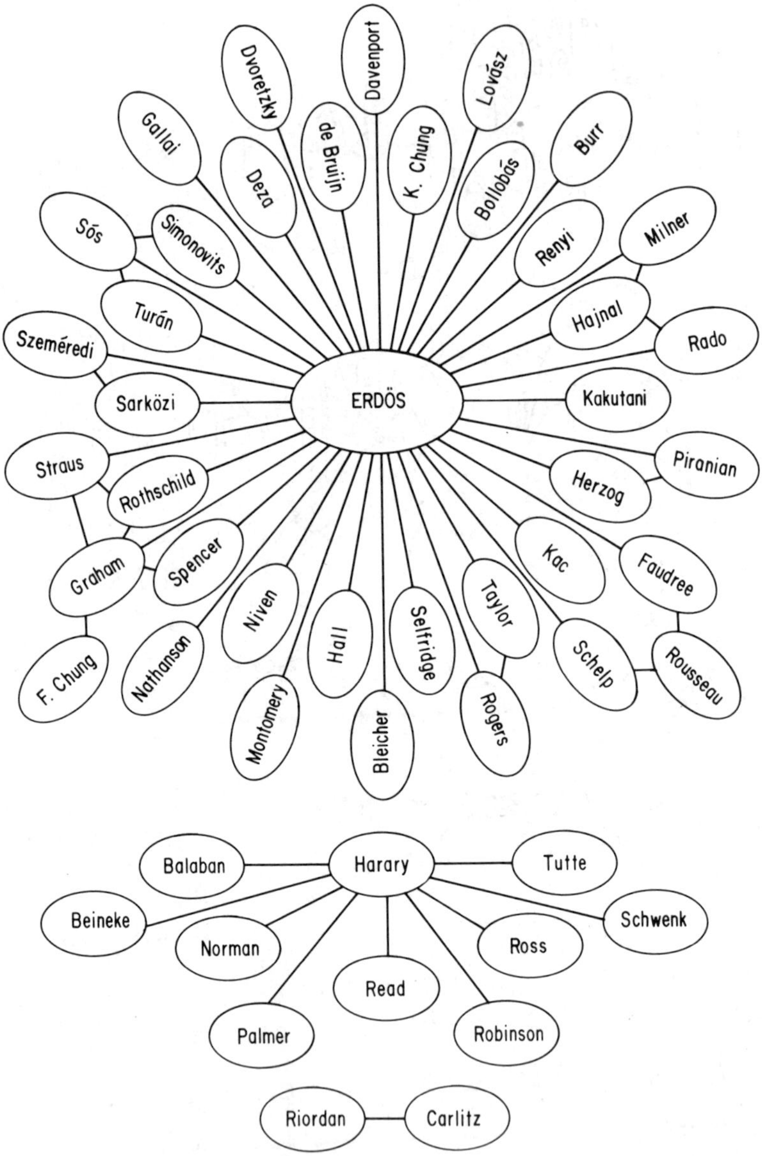

FIGURE 2. A portion of G_5.

2. D. West is the only known person to have a purely *imaginary* Erdös number,† where the Erdös number of a vertex V is defined to be length of a shortest path from V to Erdös (if its exists). This comes about because of the recent paper of G. W. Peck [6]. The actual authors are {Chung, Erdös, Graham, Kleitman, Purdy, West}. This has a precedent in the well-known paper of B. H. Jiggs [4], whose actual authors were {Baumert, Golomb, Gordon, Hales, Jewett, Selfridge}. The i was imaginary (as it should be). As a consequence, West's coauthor, Griggs, has Erdös number $1 + i$.
3. It was pointed out by Erdös [1] that G^* should actually be written as $G^{(t)*}$ since its current state depends on the time t at which it is considered.

   Erdös went on to conjecture that, for any r, there is a time $t(r)$ so that $G^{(t(r))*}$ will contain a complete graph K_r on r vertices. However, there may be some reason for doubting the validity of this conjecture because of the following observations. If a person P writes one (joint) paper per day for (say) eighty years, he/she will still only have written less than 30,000 papers. Of course, he/she may write a greater rate than one paper per day and may be productive for more than eighty years, but in any case, it seems reasonably safe to say that (at least in the foreseeable future) no one will write 100,000 papers, each with a different coauthor. If $G^{(t)*}$ were ever to contain $K_{100,000}$, a complete graph on 100,000 points, we would need 99,999 people doing this! Thus, it appears somewhat unlikely that $K_{100,000}$ will ever be a subgraph of $G^{(t)*}$, for any t.

   However, there is no obvious reason why K_r could not be a subgraph of $G^{\omega} = \bigcup_{t \geq 0} G^{(t)}$ for any fixed value of r. All that is needed for $K_r \subseteq G^{\omega}$ is that we have an r-tuple-authored paper. For trends in this direction (in physics), see [7].
4. The vertex {Thompson} is the only known example of a person connected to both a woman (F. MacWilliams) and her daughter (A. MacWilliams). Curiously, no corresponding example for a father and son is known (to the author).
5. There is still no 4-set of vertices X known in G for which every nonempty subset of X has a joint paper [1]. For the weak solution (every subset of X occurs in the set of authors of a distinct paper), the closest example known is {Erdös, Graham, Rothschild, Straus} for which fourteen of the fifteen required papers exist.
6. An interesting question concerning G which has never been investigated concerns its robustness. That is, suppose we consider the subgraph $G_k \subseteq G$ in which A and B have an edge between them only if they have at least k joint papers. In FIGURE 2 we show a portion of G_5. In FIGURE 3 we show a portion of G_{25}.
7. A related graph is the *joint book graph B* (whose name is self-explanatory). The strict interpretation does not allow jointly edited books or contributed chapters in a book to count. A portion of B is shown in FIGURE 4.

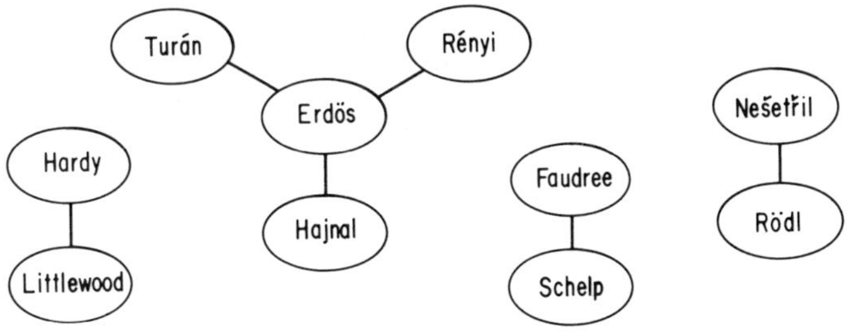

FIGURE 3. A portion of G_{25}.

† This was first noted by Kleitman.

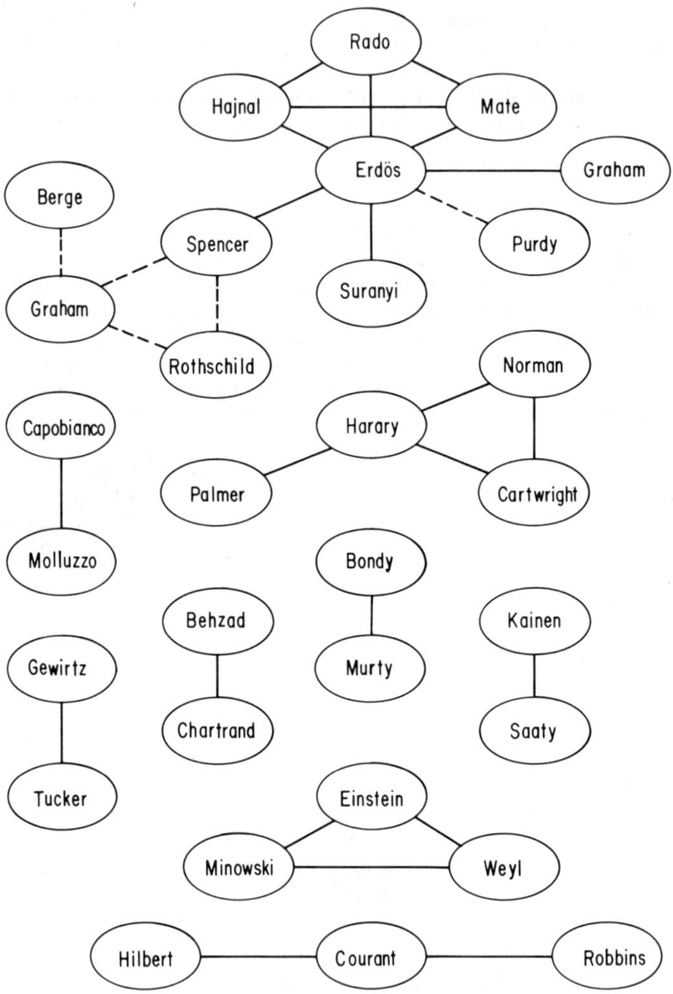

FIGURE 4. A portion of B.

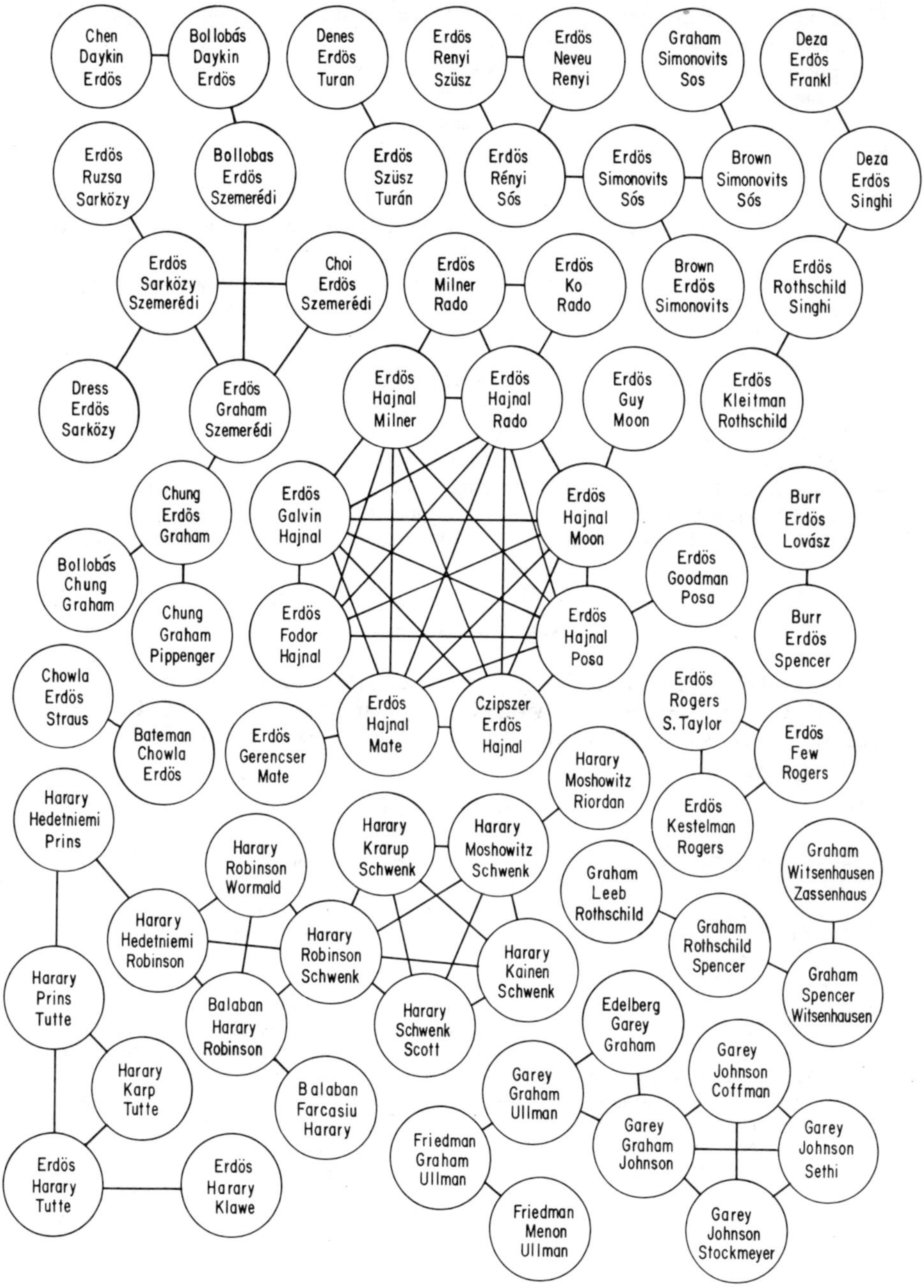

FIGURE 5. A portion of G_{STP}.

8. An interesting variation is to consider the strict triple paper graph G_{STP}. The vertices of G_{STP} are sets of three people which exactly form the set of authors of a published paper. Two such vertices are joined by an edge if they have two people in common. We show a portion of G_{STP} in FIGURE 5. It is actually somewhat surprising that G_{STP} has such large components.
9. Just as in the case of the Erdös number of a vertex V, one can just as well ask about the Hilbert number, the Einstein number, the Gauss number, or even the Ramsey number‡ of V. However, it is not yet known whether Gauss or Ramsey are actually connected to Erdös in G. It may be reassuring to know however that everyone in the largest component of G, shown in FIGURE 1, at least has a Rumsey number. We also note, in passing, that Tarjan has an odd Even number, and an even Odda number.

CONCLUDING REMARKS

We hope we have given the reader a feeling for the main ideas of this subject. Of course, the variety and depth of future results in this area are limited only by the ingenuity and persistence of the investigators.§ We look forward with anticipation to the great strides which will (unfortunately) inevitably occur in the field and hope that this modest effort has made some small contribution to filling a much needed gap in the literature.

ACKNOWLEDGMENTS

The author wishes to acknowledge here the various suggestions of S. A. Burr, F. R. K. Chung, P. Erdös, A. Gewirtz, R. L. Graham, L. V. Quintas, E. G. Straus, and M. Van Ness in spite of whose comments this note was written.

REFERENCES

0. BURR, S. A. 1979. A survey of noncomplete Ramsey theory for graphs. In this volume.
1. ERDÖS, P. 1972. On the fundamental problem of mathematics. Am. Math. Mon. **79:** 149–150.
2. GOFFMAN, CASPER. 1969. And what is your Erdös number? Am. Math. Mon. **76:** 791.
3. HARARY, F. 1971. The collaboration graph of mathematicians and a conjecture of Erdös. J. Rec. Math. **4:** 212–213.
4. JIGGS, B. H. 1963. Recent results on comma-free codes. Can. J. Math. **15:** 178–187.
5. ODDA, TOM. On properties of a well-known graph. In this volume.
6. PECK, G. W. Maximum antichains of rectangular arrays. J. Comb. Theory (A). To appear.
7. WISS, J. E., G. GOLDHABER, G. S. ABRAMS, M. S. ALAM, A. M. BOYARSKI, M. BREIDENBACH, W. C. CARITHERS, S. COOPER, R. G. DEVOE, J. M. DORFAN, G. J. FELDMAN, C. E. FRIEDBERG, D. FRYBERGER, G. HANSON, J. JAROS, A. D. JOHNSON, J. A. KADYK, R. R. LARSEN, D. LÜKE, V. LÜTH, H. L. LYNCH, R. J. MADARAS, C. C. MOREHOUSE, H. K. NGUYEN, J. M. PATERSON, M. L. PERL, I. PERUZZI, M. PICCOLO, F. M. PIERRE, T. P. PUN, P. RAPIDIS, B. RICHTER, B. SADOULET, R. H. SCHINDLER, R. F. SCHWITTERS, J. SIEGRIST, W. TANENBAUM, G. H. TRILLING, F. VANUCCI & J. S. WHITAKER. 1976. Evidence for parity nonconservation in the decays of the narrow states near 1.87 GeV/c^2*. Phys. Rev. Lett. **37:** 1531–1533.

‡ This should not be confused with the generalized Ramsey numbers referred to by Burr in his paper [0] in this Volume. On the other hand, perhaps it should.

§ No doubt other qualities are also helpful.

INDIFFERENCE AND SERIATION

Fred S. Roberts

Department of Mathematics
Rutgers University
New Brunswick, New Jersey 08903

Introduction

Much mathematical work in the social sciences concerns itself with the organization of data, with the attempt to put complex interrelationships into a coherent pattern. In this paper, we shall discuss two problems of data organization in the social sciences which are solved using graph theoretical techniques. The first has to do with measurement, specifically measurement of indifference. The second has to do with seriation, the attempt to put data in a sequence or in serial order. We will use the results on indifference measurement to solve the problem of seriation.

In what follows, we shall denote by (V, A) a directed graph, or digraph, where V is the set of vertices and A the set of arcs, and by (V, E) an (undirected) graph, where V is the set of vertices and E is the set of edges. We shall not allow multiple arcs or edges, but shall assume, unless mentioned otherwise, that in a graph (V, E) each vertex has a loop, an edge from the vertex to itself. Our diagrams of graphs will omit the loops. Finally, we shall assume that V is finite throughout.

Measurement of Indifference

Suppose V is a set of objects or alternatives being compared and, for all x and y in V, we include an arc from x to y in A if and only if x is preferred to y, judged more important than y, judged more qualified than y, etc. (We use the generic term *preference* for all of these relationships.) One way of trying to organize the data encoded by the digraph (V, A) is by techniques of measurement. Following such authors as Scott and Suppes [21], Suppes and Zinnes [22], Krantz *et al.* [11], and Roberts [19], we shall treat measurement as the assignment of numbers preserving certain observed relationships. In this spirit, measurement of preference is the assignment of a number $u(x)$ to each vertex x in V such that x is preferred to y if and only if x gets a higher number than y, i.e., such that, for all x, y in V,

$$(x, y) \in A \quad \text{iff} \quad u(x) > u(y) \tag{1}$$

A major problem of measurement is the existence question: when does there exist such an assignment of numbers? In the case of Equation 1, there exists a function u satisfying this equation if and only if (V, A) defines what is called in the literature of measurement theory a (strict) weak order—this is like a linear order except that ties are allowed.

Related to the judgments of preference are judgments of indifference: we are indifferent between two alternatives if and only if we prefer neither. (*Indifference* is again a generic term, corresponding to judged equally important, judged equally qualified, etc.) Then indifference defines an undirected graph $G = (V, E)$, with an edge between two vertices if and only if they are judged indifferent. It is convenient to think of G as having a loop from each vertex to itself: each alternative

0077-8923/79/0328-0173$1.75/1 © 1979, NYAS

is judged indifferent to itself. If a function u satisfying Equation 1 exists, then, for all x, y in V,

$$\{x, y\} \in E \quad \text{iff} \quad u(x) = u(y) \tag{2}$$

Among other things, the existence of a function u satisfying (**2**) implies that indifference is a transitive relation. Arguments against transitivity of indifference go back to the economist Armstrong [1] in this century and perhaps, as Menger [14] claims, even to Poincaré. One such argument is the following. Certainly you have a preference between a sandwich with ham and one thick slice of cheese and one which has added to the first sandwich five more thick slices of cheese. However, if cheese is added to the first sandwich in slices 1 millimeter in thickness at a time, you are almost certain to be indifferent between successive sandwiches. Hence, if indifference is transitive, you should be indifferent between the first and last sandwich as well.

Examples such as this suggest that indifference corresponds more to closeness than to equality. In the case of preference, they led Luce [12] to ask for an assignment of numbers $u(x)$ satisfying

$$(x, y) \in A \quad \text{iff} \quad u(x) > u(y) + \delta, \tag{3}$$

where δ is a positive number measuring closeness or threshold; we prefer x to y if and only if x is "sufficiently" better than y. In the case of indifference, examples such as the ham and cheese, lead to the assignment

$$\{x, y\} \in E \quad \text{iff} \quad |u(x) - u(y)| \leq \delta \tag{4}$$

The existence of a function u satisfying (**3**) is equivalent to the digraph (V, A) defining a special kind of partial order called a semiorder. Semiorders were first defined by Luce [12]. See Scott and Suppes [21] for a definition and proof of equivalence to the existence of a function satisfying (**3**). We shall describe in detail conditions for the existence of a function satisfying (**4**).

In terms of graphs, the existence of a function satisfying (**4**) is equivalent to the existence of an assignment of numbers to the vertices such that two vertices get numbers which are "close" if and only if the two vertices are joined by an edge. FIGURE 1 shows such an assignment with $\delta = 1$. We shall say that a graph is an *indifference graph* if there is an assignment satisfying (**4**). Thus, the graph of FIGURE 1 is an indifference graph. However, Z_4, the circuit of length 4, is not. For suppose there were an assignment u satisfying (**4**). Then, using the labelling of vertices shown in FIGURE 2, we would have $u(b) \geq u(a)$ without loss of generality. Moreover, since $\{a, b\}$ and $\{b, c\}$ are edges, while $\{a, c\}$ is not, we must have

$$|u(a) - u(b)| \leq \delta$$
$$|u(b) - u(c)| \leq \delta$$
$$|u(a) - u(c)| > \delta$$

This could only happen if the numbers $u(a)$, $u(b)$, and $u(c)$ were as in FIGURE 3. But we must also have

$$|u(a) - u(d)| \leq \delta$$

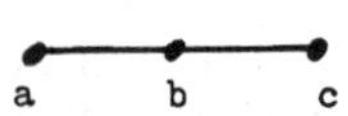

u(a) = 0, u(b) = .6, u(c) = 1.2

FIGURE 1.

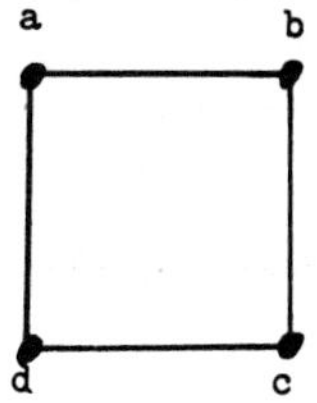

FIGURE 2.

and

$$|u(c) - u(d)| \leq \delta$$

while

$$|u(b) - u(d)| > \delta$$

This is impossible. A similar argument shows that Z_n, $n \geq 4$, is not an indifference graph.

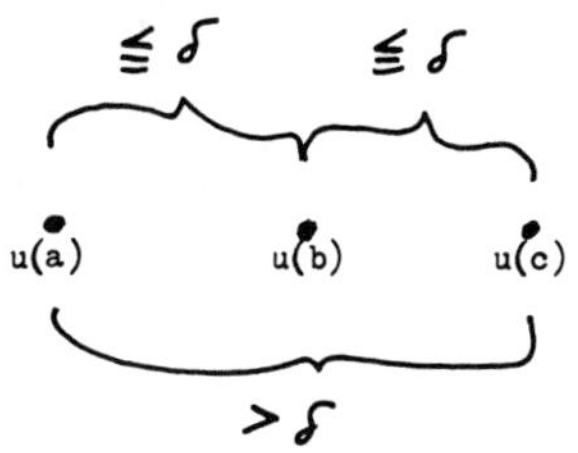

FIGURE 3.

A second example of a graph which is not an indifference graph is the graph $K(1, 3)$ shown in FIGURE 4. For, suppose $K(1, 3)$ were an indifference graph, with an assignment u satisfying Equation 4. Then without loss of generality, $u(a) \leq u(b) \leq u(c)$. Hence, since

$$|u(a) - u(b)| > \delta$$

and

$$|u(b) - u(c)| > \delta$$

it follows that

$$|u(c) - u(a)| > 2\delta$$

Now we also require

$$|u(x) - u(a)| \leq \delta$$

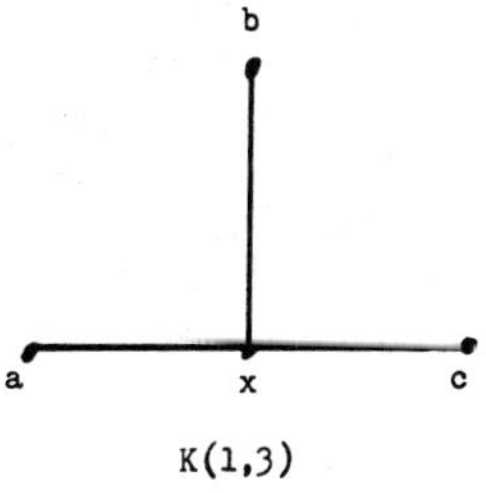

FIGURE 4.

and

$$|u(x) - u(c)| \leq \delta$$

This is impossible. A similar argument shows that the graphs of FIGURES 5 and 6 are not indifference graphs.

The following is a forbidden subgraph characterization of indifference graphs.

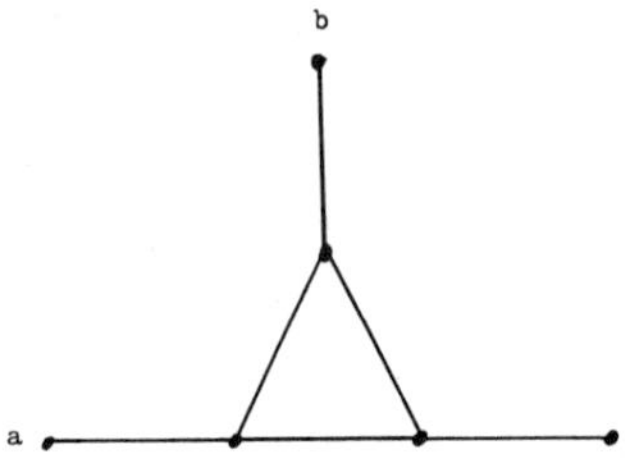

FIGURE 5.

THEOREM 1 (Roberts [16]). A graph (with a loop at each vertex) is an indifference graph if and only if it does not contain as a generated subgraph Z_n, $n \geq 4$, or any of the graphs of FIGURES 4, 5, and 6.

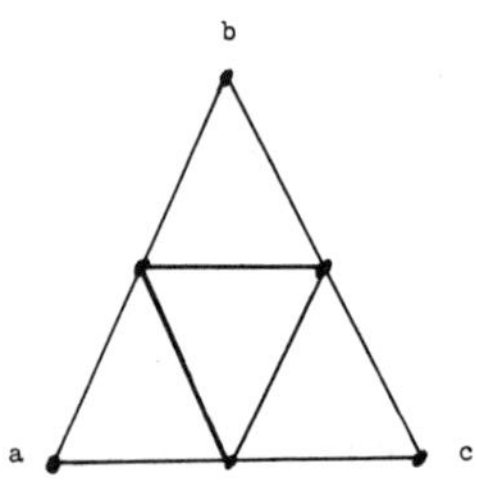

FIGURE 6.

To gain further insight into THEOREM 1, we introduce an equivalence relation $\approx$ on the vertices of a graph $G = (V, E)$ as follows:

$$x \approx y \quad \text{iff} \quad (\forall z \in V)(\{x, z\} \in E \quad \text{iff} \quad \{y, z\} \in E)$$

It is easy to see that $\approx$ is indeed an equivalence relation. Since our graphs have loops at each vertex, equivalence implies adjacency. Thus, in $K(1, 3)$ of FIGURE 4, a and b are not equivalent, for $\{a, a\}$ is an edge (not shown) while $\{b, a\}$ is not. The graph $G/\approx$ is the graph whose vertex set is the set of equivalence classes, and which has an edge between the equivalence classes containing a and b if and only if G has an edge between a and b. Clearly, $G/\approx$ is well defined. We say G is *reduced* if G is isomorphic to $G/\approx$, i.e., if each vertex is equivalent only to itself.

A vertex e in a graph is called an *extreme vertex* if whenever e is adjacent to u and v, then u and v are adjacent to each other and there is a vertex z such that z is adjacent to u and v but not to e. The graph of FIGURE 6 has three extreme vertices, those labelled a, b, and c. Similarly, the vertices labelled a, b, and c are (vacuously) extreme in the graphs of FIGURES 4 and 5. However, the graphs Z_n, $n \geq 4$, have no extreme vertices. It is not hard to show that, at least if G is connected and reduced, then extreme vertices correspond under an assignment u satisfying Equation 4 exactly to the endpoints or extreme points in the set

$$\{u(x)\colon x \in V(G)\},$$

a finite set of points on the line. Now each such finite set of points on the line with more than one point has two endpoints. Hence, we have the following theorem.

THEOREM 2 (Roberts [16]). A graph G (with a loop at each vertex) is an indifference graph if and only if for every connected, generated subgraph H, $H/\approx$ has exactly one vertex or it has exactly two extreme vertices, where $\approx$ is the $\approx$ relation of H.

We see from this theorem that the graphs Z_n (the circuits of length n), $n \geq 4$ and the graphs of FIGURES 4, 5, and 6 could not be indifference graphs. For the Z_n's do not have enough extreme vertices and the others have too many—and all are connected and reduced. It is not good enough to state this theorem for connected components of G, as the example of FIGURE 7 shows. This graph is not an indifference graph, for it has $K(1, 3)$ as the subgraph generated by vertices x, a, b, and c. Yet it is connected and reduced, and has only two extreme vertices, a and c.

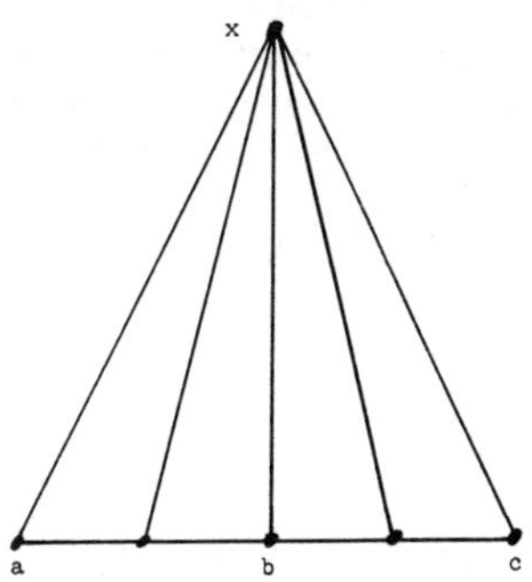

FIGURE 7.

INTERVAL GRAPHS

Suppose $\mathscr{F}$ is a family of sets. The *intersection graph* of $\mathscr{F}$ has as its vertices the sets in $\mathscr{F}$ and it has an edge between two sets if and only if they overlap. Marczewski [13] showed that every graph is (isomorphic to) an intersection graph of some family of sets. A graph is an *interval graph* if it is the intersection graph of a family of real intervals and a *unit interval graph* if it is the intersection graph of a family of closed intervals of unit length. Interval graphs have many applications: to genetics, development psychology, and phasing of traffic lights, to mention a few. (See Roberts [18; § 3.4, 3.5].)

THEOREM 3 (Roberts [16]). A graph G is a unit interval graph if and only if it is an indifference graph.

Proof: There exists a function u satisfying Equation 4 with $\delta > 0$ if and only if there is such a u with $\delta = 1$. Let

$$J(a) = [u(a) - \tfrac{1}{2}, u(a) + \tfrac{1}{2}]$$

Then it is easy to see that u satisfies (4) with $\delta = 1$ if and only if G is the intersection graph of the closed unit intervals $J(a)$. □

Seriation Using Overlap Data

A very common problem of data organization in the social sciences is the problem of seriation—the problem of placing variables in order. We encounter this problem in developmental psychology, where the variables are stages of development. We encounter it in archaeology, where the variables are artifacts and the order is chronological. We encounter it any time we order people by opinions—for example, from least to most enthusiastic, from least to most liberal, or from least to most hawkish.

One approach to the seriation problem is to gather data about whether or not two variables overlap (chronologically, opinionwise, etc.). For example, in archaeology, this is done with artifacts by assuming that their time periods overlap if and only if they were found together in some grave or dig. Then we seek an assignment of an interval of time to each artifact so that two artifacts get overlapping time intervals if and only if they were found together in some grave. Such an assignment of intervals is possible if and only if the graph defined by the "found together is some grave" data is an interval graph. The assignment of intervals gives a possible chronological ordering in several ways. First, use of (left or right) endpoints of the intervals induces a linear ordering. Second, the relation that one interval strictly precedes another induces a partial ordering. Neither of these induced orderings is unique, and indeed the problem of how unique they can be is an interesting one mathematically.

This approach to seriation in archaeology using interval graphs is due to Kendall [6, 7]. A similar procedure leads to possible seriations in developmental psychology (Coombs and Smith [2]), opinion data, etc.

Seriation from Proximity Matrices

A more quantitative approach to seriation starts with judgments of proximity or similarity. Our approach finds its foundations in the work of Kendall [5–10]. Let r_{ij} be the judged similarity between alternatives i and j, with $r_{ij} > r_{kl}$ meaning that i and j are more similar to one another than k and l. We shall assume that $R = (r_{ij})$ is a symmetric matrix and use it to put alternatives in serial order.

Following Hubert [4], let us say that R is in *strong Robinson form* if whenever $i \leq j \leq k \leq l$, then $r_{jk} \geq r_{il}$. This means that, in the ordering of alternatives given by the ordering of rows of the matrix, if two alternatives are between two others, the two between are at least as similar as the two outside. The matrix,

$$R = \begin{pmatrix} 25 & 7 & 8 \\ 7 & 18 & 9 \\ 8 & 9 & 16 \end{pmatrix} \tag{5}$$

is not in strong Robinson form. For $1 \leq 1 \leq 2 \leq 3$, yet $r_{12} < r_{13}$. Even if a matrix is not in strong Robinson form, some permutation applied simultaneously to the rows and columns might put it into this form. Still following Hubert, we can define the goal of seriation in the present context as being to find a permutation which puts a proximity matrix into strong Robinson form. In our example, interchanging the second and third rows and second and third columns will suffice, for it leads to the matrix

$$\begin{pmatrix} 25 & 8 & 7 \\ 8 & 16 & 9 \\ 7 & 9 & 18 \end{pmatrix}$$

We shall ask for conditions under which there is such a permutation.

It is not so easy to see from the definition whether or not a matrix is in strong Robinson form. Fortunately, there is an easy criterion. We say a symmetric matrix R is in *weak Robinson form* if the entries in each row (and hence each column) do not decrease as they approach the main diagonal. This notion of Robinson form was introduced into the seriation literature by the archaeologist W. S. Robinson [20], and is the notion used by Kendall. Following an observation made tacitly in Hubert [4], we have the following.

THEOREM 4. A symmetric matrix R is in strong Robinson form if and only if it is in weak Robinson form.

Proof: Suppose R is in strong Robinson form. Suppose $i \le j \le k$. Then $i \le i \le j \le k$, so $r_{ij} \ge r_{ik}$. Similarly, $i \le j \le k \le k$, so $r_{jk} \ge r_{ik}$ and, by symmetry, $r_{kj} \ge r_{ki}$. It follows that R is in weak Robinson form. Conversely, suppose R is in weak Robinson form, and $i \le j \le k \le l$. Then $i \le k \le l$ implies $r_{ik} \ge r_{il}$ and $i \le j \le k$ implies $r_{kj} \ge r_{ki}$. Hence, by symmetry of R, $r_{jk} \ge r_{il}$. □

Henceforth, we shall use the term *Robinson form* to stand for either strong or weak Robinson form. To give a condition under which an $n \times n$ matrix R is permutable to Robinson form, let us define for every $\varepsilon > 0$ a graph G_ε as follows. The vertices of G_ε are the integers 1, 2, ..., n and there is an edge between i and j if and only if $r_{ij} \ge \varepsilon$. Again following Hubert [4], it is easy to see that if R is in Robinson form, then the following condition holds for every $\varepsilon > 0$: if $i \le j \le k \le l$, and $\{i, l\} \in E(G_\varepsilon)$, then $\{j, k\} \in E(G_\varepsilon)$. Conversely, if this condition holds for every $\varepsilon > 0$, then R is in Robinson form. Generalizing, suppose $\lesssim$ is a (linear) ordering of the vertices of a graph G. We say $\lesssim$ is a *compatible ordering* if whenever $i \lesssim j \lesssim k \lesssim l$ and $\{i, l\} \in E(G)$, then $\{j, k\} \in E(G)$.

THEOREM 5 (Roberts [17]). A graph G (with a loop at every vertex) has a compatible vertex ordering if and only if G is in indifference graph.

Proof: If G is an indifference graph, there is a function u satisfying Equation 4. Without loss of generality, we can show that u is one-to-one. Then it is easy to show that the ordering defined by $x \lesssim y$ iff $u(x) \le u(y)$ is a compatible vertex ordering. Next, it is easy to show that none of the forbidden subgraphs of THEOREM 1 have compatible vertex orderings, and hence the converse follows. □

It is tempting to combine THEOREM 5 and the observation made before the theorem to say that a matrix R is permutable into Robinson form if and only if every G_ε is an indifference graph. This is false, for two reasons. One is that we must have the *same* compatible ordering for each G_ε. The other is that G_ε may not have a loop at each vertex, as is required of indifference graphs. Now it is easy to replace R by a matrix R' with ∞ on each diagonal element. Then so long as the diagonal elements of R maximize their respective rows, R is permutable into Robinson form if and only if R' is. Also, if G'_ε is defined from R' as was G_ε from R, then G'_ε has a loop at each vertex. Hence, we have the following theorem.

THEOREM 6. A symmetric matrix R is permutable into Robinson form if and only if every diagonal element of R is maximal in its row and the family $\{G'_\varepsilon\}$ is a family of indifference graphs each of which has the same compatible vertex ordering.

It should be noted that the condition of THEOREM 6 is testable because there are really only finitely many different G'_ε.

To give another criterion for a symmetric $n \times n$ matrix R to be permutable into Robinson form, let us associate with R the weighted graph $G(R)$ defined as follows. The vertices are the integers 1, 2, ..., n, there is an edge between i and j if and only if $i \neq j$ and $r_{ij} \neq 0$, and the weight r_{ij} appears on the edge $\{i, j\}$. Thus, $G(R)$ has no loops. To give an example, the matrix R of Equation 5 gives rise to the weighted graph $G(R)$ shown in FIGURE 8. The weight associated with a spanning tree of $G(R)$ will be the sum of the weights of its edges, and a *maximal spanning tree* will be one whose weight is maximal.

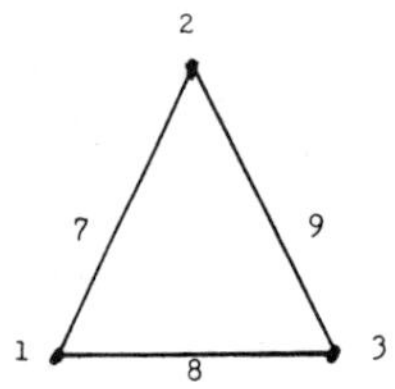

FIGURE 8.

THEOREM 7 (Wilkinson [23]). If a symmetric matrix R is permutable into Robinson form, then some maximal spanning tree of $G(R)$ is a chain. The order of rows permuting R into Robinson form defines such a chain.

Proof: One well-known algorithm for finding a maximal spanning tree starts by including an arbitrary vertex and adding, at each step, an edge which maintains a tree and has as large a weight as possible. Suppose 1, 2, ..., n is a permutation putting R into Robinson form. It is easy to show, by definition of the weak Robinson form, that each step will add an edge of the form $\{i, i + 1\}$, and hence produce the chain 1, 2, ..., n. □

To illustrate this theorem, note that in the graph of FIGURE 8, the maximal spanning tree consisting of the edges $\{1, 3\}$, $\{3, 2\}$ defines such a chain, and it corresponds to the permutation 132 putting the corresponding matrix R into Robinson form. Note that the converse of the theorem is false. To see why, consider the graph $G(R)$ of FIGURE 9. It is easy to see that the unique maximal spanning tree is given by the edges $\{1, 2\}$, $\{2, 3\}$, $\{3, 4\}$, which define a chain. Hence, if the corresponding R is permutable into Robinson form, it must already be in Robinson form. However, it is not, since $r_{24} < r_{14}$.

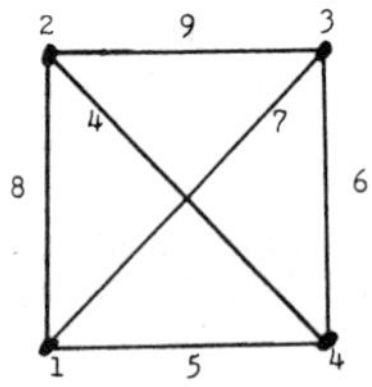

FIGURE 9.

THE CONSECUTIVE 1'S PROPERTY AND ROBINSON FORM

We close by stating a few more results about Robinson form, specifically relating it to the consecutive 1's property which arises in the study of interval graphs and indifference graphs. Fulkerson and Gross [3] prove that a graph G is an interval graph if and only if the maximal cliques of G can be ordered $K_1, K_2, \ldots, K_p$ in such a way that if a vertex belongs to maximal cliques K_r and K_t, and $r < s < t$, then the vertex belongs to K_s as well. This condition can be stated in terms of matrices. A (0, 1)-matrix M is said to have the *consecutive* 1's *property* (for columns) if some permutation of the rows leads to the 1's in each column appearing consecutively. The *maximal clique-vertex incidence matrix* $M = (m_{ij})$ of a graph G has rows corresponding to maximal cliques and columns to vertices, where m_{ij} equals 1 if vertex j belongs to maximal clique i, and otherwise m_{ij} equals 0. The Fulkerson-Gross result can now be stated as follows.

THEOREM 8 (Fulkerson & Gross [3]). A graph G is an interval graph if and only if its maximal clique-vertex incidence matrix has the consecutive 1's property.

FIGURE 10 illustrates the theorem, giving a graph G, its maximal clique-vertex incidence matrix, and a permutation of the rows which leads to consecutive 1's. Thus, G is an interval graph.

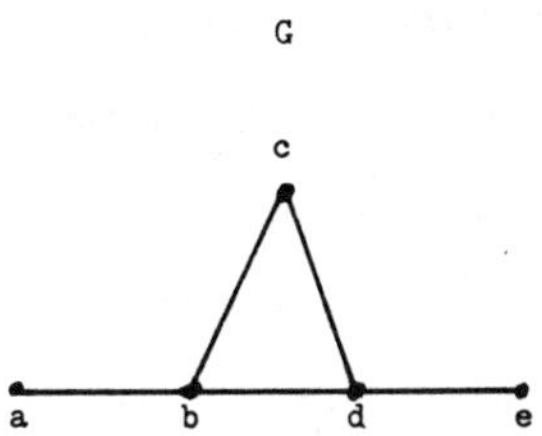

$$\begin{array}{c} \\ ab \\ de \\ bcd \end{array}\begin{array}{c} \begin{array}{ccccc} a & b & c & d & e \end{array} \\ \begin{pmatrix} 1 & 1 & 0 & 0 & 0 \\ 0 & 0 & 0 & 1 & 1 \\ 0 & 1 & 1 & 1 & 0 \end{pmatrix} \end{array} \qquad \begin{array}{c} \\ ab \\ bcd \\ de \end{array}\begin{array}{c} \begin{array}{ccccc} a & b & c & d & e \end{array} \\ \begin{pmatrix} 1 & 1 & 0 & 0 & 0 \\ 0 & 1 & 1 & 1 & 0 \\ 0 & 0 & 0 & 1 & 1 \end{pmatrix} \end{array}$$

FIGURE 10.

The *adjacency matrix* of a graph is the square matrix whose i, j entry is 1 if vertices i and j are adjacent and 0 if they are not.

THEOREM 9 (Roberts [15]). A graph (with a loop at each vertex) is an indifference graph if and only if its adjacency matrix has the consecutive 1's property.

THEOREM 10 (Kendall [6]). Suppose B is a (0, 1)-matrix with the consecutive 1's property. The permutations of the rows of B which produce consecutive 1's correspond exactly to the permutations which, when applied simultaneously to rows and columns, put BB^T into Robinson form.

COROLLARY 1. If M is the maximal clique-vertex incidence matrix of an interval graph, then MM^T is permutable into Robinson form.

COROLLARY 2. If A is the adjacency matrix of an indifference graph, then AA^T is permutable into Robinson form.

FITTING MODELS APPROXIMATELY

Real data rarely fits a model exactly. However, an empirical data matrix might come close to having Robinson form or the consecutive 1's property, or empirical indifference judgments might come close to having an assignment satisfying Equation 4, i.e., indifference measurement with threshold. It is left to define precisely what closeness in this sense means and to find procedures which organize data approximately into a given pattern. Thus, for example, Kendall [6] asks what happens to the Fulkerson and Gross [3] algorithm for obtaining a permutation leading to consecutive 1's if this is only approximately attainable and Hubert [4] discusses the use of the maximal spanning tree in $G(R)$ to obtain an approximate Robinson form. There is room for much more work on these and related questions.

References

1. Armstrong, W. E. 1939. The determinateness of the utility function. Econ. J. **49:** 453–467.
2. Coombs, C. H. & J. E. K. Smith. 1973. On the detection of structure in attitudes and developmental processes. Psych. Rev. **80:** 337–351.
3. Fulkerson, D. R. & O. A. Gross. 1965. Incidence matrices and interval graphs. Pac. J. Math. **15:** 835–855.
4. Hubert, L. 1974. Some applications of graph theory and related non-metric techniques to problems of approximate seriation: The case of symmetric proximity measures. Brit. J. Math. Stat. Psychol. **27:** 133–153.
5. Kendall, D. G. 1963. A statistical approach to Flinders Petrie's sequence dating. Bull. Int. Stat. Inst. **40:** 657–680.
6. Kendall, D. G. 1969. Incidence matrices, interval graphs, and seriation in archaeology. Pac. J. Math. **28:** 565–570.
7. Kendall, D. G. 1969. Some problems and methods in statistical archaeology. World Archaeol. **1:** 61–76.
8. Kendall, D. G. 1971. A mathematical approach to seriation. Phil. Trans. R. Soc. London A. **269:** 125–135.
9. Kendall, D. G. 1971. Abundance matrices and seriation in archaeology. Z. Wahrschein. **17:** 104–112.
10. Kendall, D. G. 1971. Seriation from abundance matrices. *In* Mathematics in the Archaeological and Historical Sciences. F. R. Hodson *et al.*, Eds. Edinburgh University Press, Edinburgh, Scotland.
11. Krantz, D. H., R. D. Luce, P. Suppes & A. Tversky. 1971. Foundations of Measurement. Vol. 1. Academic Press, New York, N.Y.
12. Luce, R. D. 1956. Semiorders and a theory of utility discrimination. Econometrica. **24:** 178–191.
13. Marczewski, E. 1945. Sur deux propriétés des classes d'ensembles. Fund. Math. **33:** 303–307.
14. Menger, K. 1951. Probabilistic theories of relations. Proc. Nat. Acad. Sci. **37:** 178–180.
15. Roberts, F. S. 1968. Representations of indifference relations. Stanford University, Stanford, California. Doctoral dissertation.
16. Roberts, F. S. 1969. Indifference graphs. *In* Proof Techniques in Graph Theory. F. Harary, Ed. Academic Press, New York, N.Y. pp. 139–146.
17. Roberts, F. S. 1971. On the compatibility between a graph and a simple order. J. Comb. Theory. **11:** 28–38.
18. Roberts, F. S. 1976. Discrete Mathematical Models, with Applications to Social, Biological, and Environmental Problems. Prentice-Hall, Englewood Cliffs, N.J.
19. Roberts, F. S. 1979. Measurement Theory, with Applications to Decision-Making, Utility, and the Social Sciences. Addison-Wesley, Reading, Mass. To appear.
20. Robinson, W. S. 1951. A method for chronologically ordering archaeological deposits. Am. Antiq. **16:** 293–301.
21. Scott, D. & P. Suppes. 1958. Foundational aspects of theories of measurement. J. Symb. Logic. **23:** 113–128.
22. Suppes, P. & J. Zinnes. 1963. Basic measurement theory. *In* Handbook of Mathematical Psychology. Vol. 1. R. D. Luce, R. R. Bush & E. Galanter, Eds. John Wiley & Sons, New York, N.Y. pp. 1–76.
23. Wilkinson, E. M. 1971. Archaeological seriation and the traveling salesman problem. *In* Mathematics in the Archeological and Historical Sciences. F. R. Hodson *et al.*, Eds. Edinburgh University Press, Edinburgh, Scotland.

SPECTRAL RECONSTRUCTION PROBLEMS*

Allen J. Schwenk

Department of Mathematics
U.S. Naval Academy
Annapolis, Maryland 21402

INTRODUCTION

In the standard *reconstruction problem* (in the form introduced by Harary [6]), we are given a deck of cards each bearing a graph providing some information about a hidden graph. There is one card for each vertex of the hidden graph, and the graph on the card is obtained by removing that particular vertex from the hidden graph. The problem is to reconstruct the original graph from this *deck of vertex deleted subgraphs*. Of course the subgraphs are unlabelled and redrawn in an arbitrary manner, for otherwise the reconstruction would be trivial.

Bondy and Hemminger [2] have compiled a lengthy survey of graphical reconstruction, including a bibliography containing about 125 entries. The subtle difficulty of the original problem has led to publications which indicated partial information about the hidden graph which can be obtained even if we can't recover the graph itself. Another strategy has been to restrict the graphs under consideration to obtain reconstruction theorems for certain classes of graphs. For example, Kelly [8] reconstructed trees. Most papers containing partial results of either of these two types would be subsumed by a solution of the original problem.

Analogous problems have also been studied. We are asked to reconstruct the graph from its edge-deleted subgraphs. Lovász [9] has made elegant progress on this problem. Similarly, we may be asked to reconstruct digraphs, or tournaments, from vertex-deleted (or edge-deleted) subgraphs. In this area Stockmeyer [15] has found one of the most striking negative results in the field by constructing infinite families of nonreconstructible tournaments. This result casts a whole new light on the subject. No longer does it appear that we, in frail human fashion, are nibbling away at a grand yet almost certain truth. Instead, we can now recognize that reconstructibility is necessarily limited, and we can proceed with our newly obtained perspective to try to map out those limits.

SPECTRAL RECONSTRUCTION

In this article, we intend to consider reconstruction in conjunction with the spectrum of a graph. The *characteristic polynomial* of the adjacency matrix A of the graph G is denoted

$$\phi(G; x) = \det(xI - A) = \sum_{n=0}^{p} a_n x^{p-n} = \prod_{j=1}^{p} (x - \lambda_j)$$

The *eigenvalues* $\lambda_1, \lambda_2, \ldots, \lambda_p$ are called the *spectrum* of G. Another useful set of variables, the *moments* $M_n = \sum_{j=1}^{p} \lambda_j^n$, with $1 \leq n \leq p$, may be interpreted as the number of closed walks of length n. This may be computed as the trace of A^n.

* Research supported by a grant from the Naval Academy Research Council.

0077-8923/79/0328-0183$1.75/1 © 1979, NYAS

In principle it has long been known how to use one of these sets of variables (coefficients, eigenvalues, or moments) to compute the others (e.g., see David and Barton [4]). Although the practical aspects of this computation almost seem to be a lost art, we shall presume that knowledge of any of these sets is equivalent to knowing the spectrum.

We consider four reconstruction problems: We may be given a deck of subgraphs, $G_i = G - v_i$, or perhaps, we are only given a deck of polynomials $\phi(G_i;\ x)$. We are asked to reconstruct the graph G, or perhaps, we settle for the less ambitious goal of reconstructing $\phi(G;\ x)$. These four problems are depicted in TABLE 1. The double arrows indicate that a successful reconstruction theorem for certain problems would imply *a fortiori* reconstructibility in other cases. We omit proofs of these evident relations. Note that the standard problem, labelled A, will not be solved in this article.

We now present a useful little theorem which is a special case of a more general result from matrix theory.

TABLE 1
FOUR SPECTRAL RECONSTRUCTION PROBLEMS

Reconstruct \ Given	Deck of G_i's		Deck of $\phi(G_i)$'s
G	Problem A ⇓	⇐	Problem B ⇓
$\phi(G; x)$	Problem C	⇐	Problem D

THEOREM 1. The derivative of the characteristic polynomial is given by

$$\phi'(G;\ x) = \sum_{n=1}^{p} \phi(G_i;\ x)$$

Proof: In $\phi(G;\ x) = \det\,(xI - A)$, replace the x's by variables $z_1, z_2, \ldots, z_p$. We differentiate using the multivariable form of the chain rule from calculus:

$$\phi'(G;\ x) = \sum_{i=1}^{p} \frac{\partial}{\partial z_i} [\det\,(\text{diag}\,(z_1, \ldots, z_p) - A)] \frac{dz_i}{dx}$$

But every $z_i = x$, so we drop the factor $dz_i/dx = 1$. By expanding the determinant along row i, we see that the partial with respect to z_i evaluated at $z_1 = \cdots = z_p = x$ is simply $\phi(G_i;\ x)$. Thus, the proof is complete. □

This theorem clearly demonstrates that by integration of $\phi'(G;\ x)$ we may derive from the subspectra all but the constant term in $\phi(G;\ x)$. This much has been obtained independently by this author [11], Vance Faber (in work not published), Pouzet [10], and Gutman and Cvetković [5]. The first three noted that finding the constant term a_p is equivalent to determining h, the number of hamiltonian cycles in G. This is seen by first using *Newton's recurrence* (see [4])

$$-ka_k = \sum_{n=1}^{k} M_n a_{k-n}$$

to obtain the first $p - 1$ moments from the coefficients $a_0, a_1, \ldots, a_{p-1}$. Next, we observe that a closed walk W of length p having $|V(W)|$ vertices is counted exactly $p - |V(W)|$ times among the closed p-walks in the subgraphs G_i. Thus,

$$M_p = 2ph + \sum_{i=1}^{p} \sum_{W \subset G_i} \frac{1}{p - |V(W)|}$$

Finally, Newton's recurrence is used to recover a_p.

From another point of view, if any single eigenvalue λ_k can be evaluated, the polynomial $\psi(x) = \phi(G; x)/(x - \lambda_k)$ has moments $N_n = M_n - \lambda_k^n$. Again applying Newton's recurrence, we evaluate all $p - 1$ coefficients in $\psi(x)$ from which $\phi(G; x)$ is obtained by multiplying $\psi(x)(x - \lambda_k)$. The interlacing theorem guarantees that, for every j,

$$\max_{1 \le i \le p} \lambda_j(G_i) \le \lambda_j(G) \le \min_{1 \le i \le p} \lambda_{j-1}(G_i)$$

If any choice of j gives equality throughout, then $\lambda_j(G)$ is known. For example, this must occur if any G_i happens to have a repeated eigenvalue.

Gutman and Cvetković [5] considered a variety of circumstances in which the reconstruction of $\phi(G; x)$ could be completed. Nevertheless, Problem C remained open in general until Tutte [16] provided the *coup de grace* with his development of "musters and clusters."

We now provide a nonreconstructible pair of graphs for Problem B. Let $G = K_4 \times K_4$ be the cartesian product of two complete graphs, and let H be the triangulation of the torus depicted in FIGURE 1. Certainly G and H are distinct graphs, for each vertex in G has the neighborhood $2K_3$ while the common neighborhood in H is C_6. The spectrum of G is found by using the formula for a cartesian product [12]; that of H is obtained by brute force, and we find

$$\phi(G; x) = \phi(H; x) = (x - 6)(x - 2)^6(x + 2)^9$$

Not only are G and H cospectral, but also, as already noted [11], they possess a collection of common eigenvectors as displayed in TABLE 2. Four of these eigenvectors have opposite eigenvalues in the two graphs. We have shown previously [11] that two graphs with common eigenvectors and common *corresponding* eigenvalues must be isomorphic.

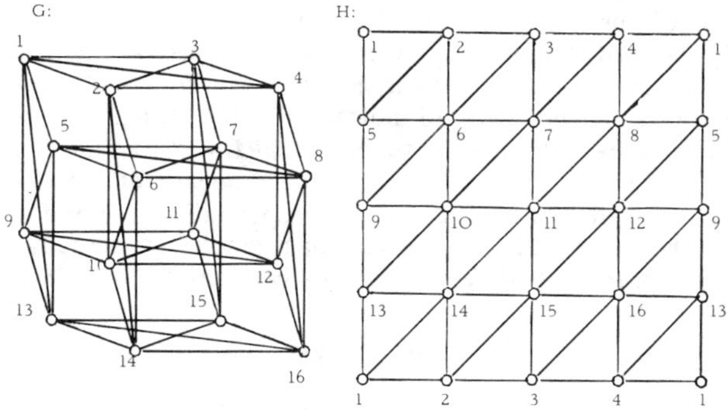

FIGURE 1. Cospectral graphs with cospectral vertex-deleted subgraphs and with cospectral edge-deleted subgraphs.

TABLE 2
THE SIXTEEN COMMON EIGENVECTORS FOR THE GRAPHS OF FIGURE 1

Eigenvector	Eigenvalue in G	Eigenvalue in H
$(++++\ ++++\ ++++\ ++++)$	6	6
$(++++\ ++++\ ----\ ----)$	2	2
$(++++\ ----\ ++++\ ----)$	2	−2
$(++++\ ----\ ----\ ++++)$	2	2
$(++--\ ++--\ ++--\ ++--)$	2	2
$(+-+-\ +-+-\ +-+-\ +-+-)$	2	−2
$(+--+\ +--+\ +--+\ +--+)$	2	2
$(++--\ +--+\ --++\ -++-)$	−2	2
$(+--+\ --++\ -++-\ ++--)$	−2	2
$(++--\ -++-\ --++\ +--+)$	−2	−2
$(+--+\ ++--\ -++-\ --++)$	−2	−2
$(+-+-\ -+-+\ +-+-\ -+-+)$	−2	−2
$(+-+-\ 0000\ -+-+\ 0000)$	−2	−2
$(0000\ +-+-\ 0000\ +-+-)$	−2	−2
$(+0-0\ -0+0\ +0-0\ -0+0)$	−2	−2
$(0+0-\ 0-0+\ 0+0-\ 0-0+)$	−2	−2

The easiest way to obtain the characteristic polynomials for the vertex deleted subgraphs is to recognize that both graphs are vertex symmetric. Thus, in light of THEOREM 1, for each i,

$$\phi(G_i; x) = \tfrac{1}{16}\phi'(G; x) = (x^2 - 6x + 2)(x - 2)^5(x + 2)^8 = \phi(H_i; x)$$

Thus, if we were given a deck of sixteen cards each possessing this polynomial, it would be impossible to decide whether the original graph was G or H. That is, conjecture B is false.

This pair of strongly regular graphs has been examined in several different contexts. The graph H is the unique exception to the proposition that the graphs $K_m \times K_n$ are characterized by their spectra (see [14, 13]). As we have just seen, H is cospectral with $K_4 \times K_4$.

The graphs $K_n^3 = K_n \times K_n \times K_n$ have been called *cubic lattice graphs.* Aigner [1] found that for $n \neq 4$ these graphs are characterized by a list of four properties, and when $n = 4$, there is just one exceptional 64-point graph L with the stated properties. Cvetković [3] observed that the list of properties were spectrally determined, and so the graphs K_n^3 are characterized by their spectra except for $n = 4$, in which case K_4^3 is cospectral with L. Aigner gave no indication of the structure of L, but merely presented its adjacency matrix. We have found that $L = K_4 \times H$. Consequently, the cospectrality of L and K_4^3 is really a direct consequence of the formula for the spectrum of a cartesian product [12] and the cospectrality of H and $G = K_4 \times K_4$. Thus we have shown that these two exceptional graphs are actually variations of the same exception, the only difference being a cartesian factor of K_4.

EDGE RECONSTRUCTION

The problem of reconstructing G from the edge deleted subgraphs $G - e_i$ can be modified in the same manner as the vertex problem was in TABLE 1. Problem C is much easier than Tutte's result, and its solution has been long known by this author. We note that G is a cycle C_p if and only if every $G - e_i$ is a path. For any other graph, we let $|E(W)|$ be the number of edges used in the closed n-walk W. Since $|E(W)| < |E(G)|$, we may determine that for each $n \le p$,

$$M_n = \sum_{G - e_i} \sum_{W \subset G - e_i} \frac{1}{|E(G)| - |E(W)|}.$$

This and Newton's recurrence provides $\phi(G; x)$.

Surprisingly, the same pair of graphs serves as a counterexample to version B. This time there is no evident way to compute $\phi(G - e_i; x)$, but we use the computer to evaluate successive moments and then appeal to Newton to get

$$\phi(G - e_i; x) = \phi(H - e_i; x) = (x^3 - 5x^2 - 9x + 22)(x^2 - x - 3)(x - 2)^4(x + 2)^7$$

The multiplicities of the eigenvalues ± 2 illustrate the following simple theorem.

THEOREM 2. If G has an eigenvalue λ with multiplicity k, then m, the multiplicity of λ in $G - e_i$, is bounded by $k - 2 \le m \le k + 2$.

Proof: Let $\alpha_1, \alpha_2, \ldots, \alpha_k$ be eigenvectors for λ in G. By suitable linear combinations, we may obtain $k - 2$ independent eigenvectors with coordinates equal to zero at both endpoints of edge e_i. But now the presence of edge e_i is immaterial for these eigenvectors, so $m \ge k - 2$. Similarly, we could start with the eigenvectors $\beta_1, \beta_2, \ldots, \beta_m$ in $G - e_i$ and conclude that $k \ge m - 2$. □

Just as in the vertex case, Problems A and D remain open.

CONCLUDING REMARKS

Another nonreconstructible pair of graphs for vertex Problem A is provided by the smallest known pair of cospectral regular graphs [7]. One is a planar line graph $G = L(K_{4,4} - 4K_2)$ and the other H is a toroidal graph. These graphs, shown in FIGURE 2, have the common spectrum

$$\phi(G; x) = \phi(H; x) = (x - 4)(x - 2)^3 x^3 (x + 2)^5$$

Since H is not vertex symmetric, it is not automatically true that the vertex deleted subgraphs are cospectral. Nevertheless, a computer based computation verifies that,

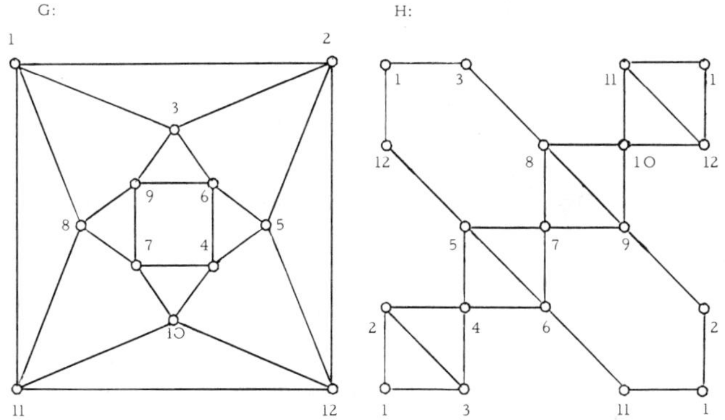

FIGURE 2. Cospectral graphs with cospectral vertex-deleted subgraphs, but with distinct edge-deleted spectra.

for every i,

$$\phi(G_i) = \phi(H_i) = (x^3 - 4x^2 + 4)(x - 2)^2x^2(x + 2)^4$$

However, this pair fails to be edge nonreconstructible. For every edge in G, we find

$$\phi(G - e_i; x) = (x^4 - 3x^3 - 7x^2 + 13x + 4)(x^3 - x^2 - 3x + 1)(x - 2)x(x + 2)^3$$

But H has three similarity classes of edges. The sixteen edges similar to the edge joining 1 and 2 give this same polynomial. The four edges similar to the edge joining 2 and 3 produce

$$\phi(G - e_i; x) = (x^3 - 3x^2 - 7x + 14)(x^2 + x - 1)(x - 2)^2x^2(x + 2)^3$$

The four remaining edges yield

$$\phi(G - e_i; x) = (x^4 - 3x^3 - 7x^2 + 12x + 8)(x^2 - x - 3)(x - 2)x^2(x + 2)^3$$

To summarize, we repeat that identical conclusions have been obtained for vertex and edge spectral reconstruction problems. The weakest conjecture (Problem C), reconstructing the spectrum from the subgraphs, is true. The strongest possible conjecture, Problem B, is false. The original Problem A, and the problem of reconstructing the spectrum from either vertex-deleted or edge-deleted subspectra remain open. This author suspects that the type D conjectures are false, but counterexamples will be difficult to find. We feel that the negative result to Problem B adds weight to Stockmeyer's negative result [15], mentioned previously. Increased uncertainty for the standard conjectures, however slight it may be, expands the appeal of studying the spectrum of reconstruction problems.

Summary

Consideration of spectra led to four variations of the usual vertex-deletion reconstruction conjecture. We may be given a deck of point-deleted subgraphs, or possibly just the characteristic polynomials of these subgraphs, and then we are challenged to reconstruct the graph, or possibly just its characteristic polynomial. Clearly the graph-graph problem is just the standard problem, which, of course, remains unsolved. The graph-polynomial variation has a positive resolution due to Tutte. The polynomial-graph version has been shown to be nonreconstructible. The fourth variation remains open. Corresponding problems with analogous results have been found for the edge deletion reconstruction conjecture.

References

1. Aigner, M. 1969. The uniqueness of the cubic lattice graph. J. Comb. Theory **6:** 282–297.
2. Bondy, J. A. & R. L. Hemminger. 1977. Graph reconstruction—A survey. J. Graph Theory. **1:** 227–268.
3. Cvetković, D. M. 1970. New characterizations of the cubic lattice graph. Publ. Inst. Mat. (Beograd). **10**(24): 195–198.
4. David, F. N. & D. E. Barton. 1962. Combinatorial Chance. Latimer, Trend, Plymouth, England.
5. Gutman, I. & D. M. Cvetković. 1975. The reconstruction problem for characteristic polynomials of graphs. University of Beograd Publication. Electrotehn. Fak. Ser. Mat. Fiz. **498–541:** 45–48.

6. HARARY, F. 1964. On the reconstruction of a graph from a collection of subgraphs. *In* Theory of Graphs and Its Applications. M. Fiedler, Ed. Czechoslovak Academy of Sciences, Prague, reprinted, Academic Press, New York, N.Y. pp. 47–52.
7. HOFFMAN, A. J. & D. K. RAY-CHAUDHURI. 1965. On the line graph of a symmetric balanced incomplete block design. Trans. Am. Math. Soc. **116:** 238–252.
8. KELLY, P. J. 1957. A congruence theorem for trees. Pac. J. Math. **7:** 961–968.
9. LOVÁSZ, L. 1972. A note on the line reconstruction problem. J. Comb. Theory (B). **13:** 309–310.
10. POUZET, M. 1977. Note sur le problème de Ulam. To appear.
11. SCHWENK, A. J. 1973. The spectrum of a graph. University of Michigan. Doctoral dissertation.
12. SCHWENK, A. J. 1974. Computing the characteristic polynomial of a graph. *In* Graphs and Combinatorics. R. Bari & F. Harary, Eds. Springer-Verlag, Berlin, Federal Republic of Germany. pp. 153–172.
13. SEIDEL, J. J. 1969. Strongly regular graphs. *In* Recent Progress in Combinatorics. W. T. Tutte, Ed. Academic Press, New York, N.Y. pp. 185–198.
14. SHRINKHANDE, S. S. 1959. On a characterization of the triangular association scheme. Ann. Math. Stat. **30:** 39–47.
15. STOCKMEYER, P. K. 1977. The falsity of the reconstruction conjecture for tournaments. J. Graph Theory. **1:** 19–26.
16. TUTTE, W. T. 1977. All the king's horses. To appear.

PACKING TREES OF DIFFERENT SIZE INTO THE COMPLETE GRAPH

H. Joseph Straight

Department of Mathematics
Western Michigan University
Kalamazoo, Michigan 49008

Department of Mathematics
College at Fredonia
State University of New York
Fredonia, New York 14063

In this paper a summary is presented of work that has been done concerning the following conjecture.

TREE PACKING CONJECTURE (TPC). Let n be a positive integer, $n \geq 2$. For $1 \leq i \leq n-1$, choose T_i to be any tree on i edges. Then there exists a partition of the edge set of the complete graph K_n say $E_1 \cup E_2 \cup \cdots \cup E_{n-1}$, such that the subgraph of K_n induced by E_i equals T_i for $1 \leq i \leq n-1$.

Such a partition of the edge set of K_n will be called a *packing* of $T_1, T_2, \ldots, T_{n-1}$ into K_n.

Let S_i and P_i denote the star and path on i edges, respectively. Let L_i be the tree on i edges shown in FIGURE 1. Also let $\Delta(T)$ denote the maximum degree of the tree T.

Before considering what has been done concerning the tree packing conjecture, let us look at an example. Consider K_6 and the sequence of trees S_1, S_2, P_3, P_4, L_5. In FIGURE 2 we show one possible packing of these trees into K_6.

It is easy to argue that the sequence of stars $S_1, S_2, \ldots, S_{n-1}$ can always be packed into K_n. Fink and Straight [1] have shown that the paths $P_1, P_2, \ldots, P_{n-1}$ can be packed into K_n. A result which combines and strengthens these two has been found by Gyárfás and Lehel [2], and independently by Zaks and Liu [3], who give a very nice proof. This result states that $T_1, T_2, \ldots, T_{n-1}$ can be packed into K_n if each T_i is a path or a star. Gyárfás and Lehel also show in their paper that a packing is possible if each T_i, with at most two exceptions, is a star. The author has recently been able to further extend these results.

The first theorem is shown using the proof technique of Zaks and Liu, which focuses on the adjacency matrix of K_n.

THEOREM 1. Let n be a positive integer, $n \geq 2$. For $1 \leq i \leq n-1$, let $T_i \in \{S_i, P_i, L_i\}$. Then $T_1, T_2, \ldots, T_{n-1}$ can be packed into K_n.

Since $T_i \in \{S_i, P_i, L_i\}$, for $i \leq 4$, it follows from THEOREM 1 that the TPC holds for $n \leq 5$.

Our next two results generalize that of Gyárfás and Lehel.

0077-8923/79/0328-0190\$1.75/1 © 1979, NYAS

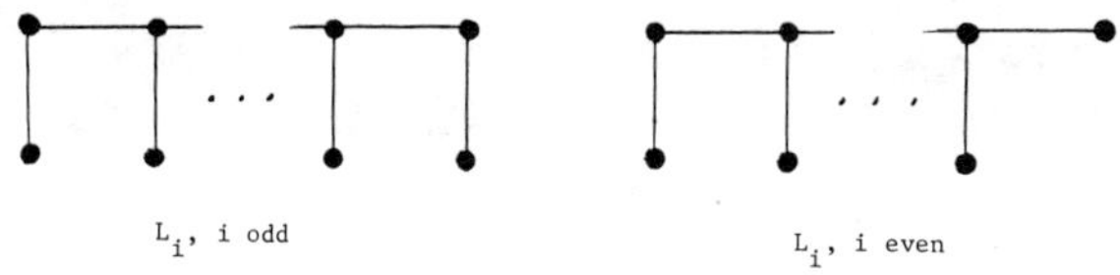

FIGURE 1.

THEOREM 2. Let n be a positive integer, $n \geq 2$. Suppose that for each T_i, $1 \leq i \leq n-1$, with at most two exceptions, $\Delta(T_i) \geq i-1$. Then $T_1, T_2, \ldots, T_{n-1}$ can be packed into K_n.

Since $\Delta(T_i) \geq i-1$ holds for $i \leq 3$, it follows from THEOREM 2 that the TPC is true for $n = 6$.

THEOREM 3. Let n be a positive integer, $n \geq 2$. Suppose that for each T_i, $1 \leq i \leq n-1$, with at most one exception, $\Delta(T_i) \geq i-2$. Then $T_1, T_2, \ldots, T_{n-1}$ can be packed into K_n.

Let us now see what can be said about the TPC in the case $n = 7$. Let $T_1, T_2, \ldots, T_6$ be a given sequence. If $T_4 \neq P_4$, then $\Delta(T_4) \geq 3$, and thus the sequence is packable, by THEOREM 2. So assume T_4 equals P_4. If $T_5 \neq P_5$, then $\Delta(T_5) \geq 3$, and thus the sequence can be packed into K_7, by THEOREM 3. Hence we may assume that T_5 equals P_5. Now if $\Delta(T_6) \geq 4$, then, again, we can apply THEOREM 3. If $T_6 \in \{S_6, P_6, L_6\}$, then the sequence can be packed into K_7, by THEOREM 1. We are thus left with eight sequences where the theorems do not apply, namely, when T_3 equals S_3 or P_3 and T_6 is one of the four trees of size 6 having maximum degree 3 which are not isomorphic with L_6. These four trees are shown in FIGURE 3.

In each of these eight remaining cases one can exhibit a packing of the sequence into K_7. Therefore, the TPC does hold when $n = 7$. In FIGURE 4 we show such a packing for the sequence $S_1, S_2, S_3, P_4, P_5, D$.

It is interesting to note that if the conjecture fails to hold for some integer N, then it fails to hold for all integers n satisfying $n \geq N$. For suppose $T_1, T_2, \ldots, T_{n-1}$

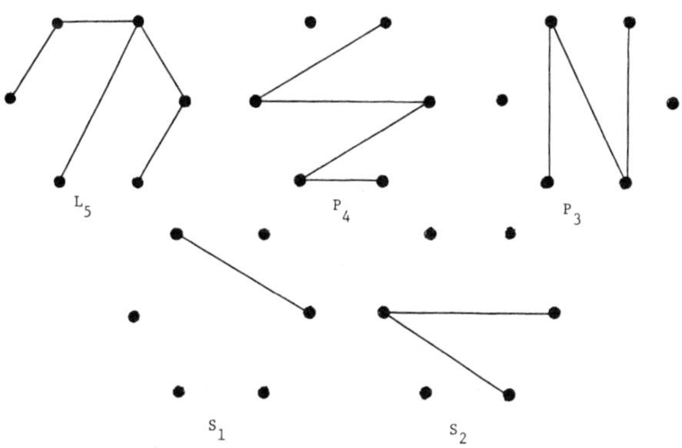

FIGURE 2.

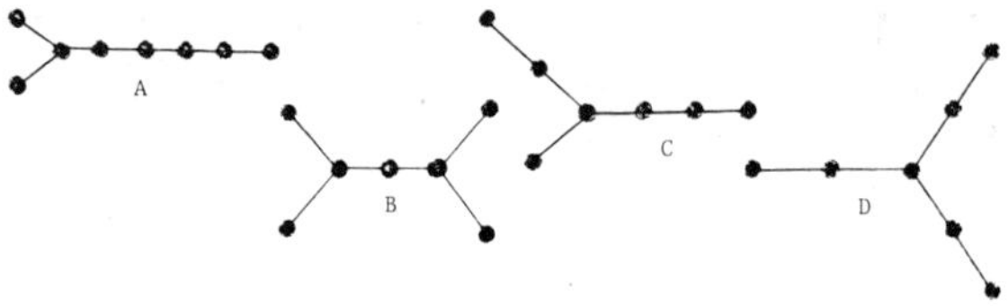

FIGURE 3.

cannot be packed into K_n. Then $T_1, T_2, \ldots, T_{n-1}$ cannot be packed into K_{n+1}. It follows that either the TPC holds for all the integers, or it holds for the first N_0 consecutive integers (other than 1), where $N_0 \geq 7$.

Finally we mention a generalization of the tree packing problem. Let $K_{n(m)}$ denote the complete n-partite graph having m vertices in each partite set. This graph has $[m^2n(n-1)]/2$ edges. For each i and j, $1 \leq i \leq n-1$, $1 \leq j \leq m^2$, let T_{ij} be a tree on i edges. We may then ask if it is possible to partition the edges of $K_{n(m)}$ into

$$\begin{array}{ccccccc} E_{1,1} & \cup & E_{1,2} & \cup \cdots \cup & E_{1,m^2} & \cup \\ E_{2,1} & \cup & E_{2,2} & \cup \cdots \cup & E_{2,m^2} & \cup \\ \vdots & & \vdots & \vdots & \vdots & \vdots \\ E_{n-1,1} & \cup & E_{n-1,2} & \cup \cdots \cup & E_{n-1,m^2} & \end{array}$$

such that the subgraph of $K_{n(m)}$ induced by $E_{i,j}$ equals T_{ij}, $1 \leq i \leq n-1$, $1 \leq j \leq m^2$. Here again it is not difficult to argue that a packing is possible if each tree is a star.

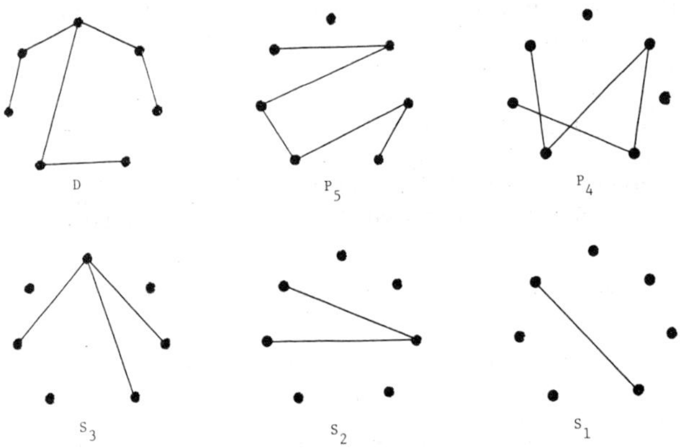

FIGURE 4.

REFERENCES

1. FINK, J. F. & J. STRAIGHT. Path-perfect graphs. To appear.
2. GYÁRFÁS, A. & T. LEHEL. Packing trees of different order into K_n. Proc. 1976 Kenthely Colloq. To be published.
3. ZAKS, S. & C. L. LIU. Decompositions of graphs into trees. Proc. Eighth Southeast. Conf. Combinatorics, Graph Theory, Comput., Baton Rouge, Louisiana (Feb. 28–Mar. 3, 1977). To be published.

THE SPANNING SUBGRAPHS OF EULERIAN GRAPHS*

C. L. Suffel
Department of Pure and Applied Mathematics
Stevens Institute of Technology
Hoboken, New Jersey 07030

INTRODUCTION

In 1736 Euler settled the Königsberg bridge puzzle in the negative by solving a class of graph theoretic problems abstracted from the puzzle [3] and thereby discovered the first theorem of graph theory:†

THEOREM 1. A (pseudo) graph G has an eulerian trail if and only if G is connected and each point has even degree.

Of course most connected graphs have some odd-degree points so that in many cases it is impossible to traverse the lines, each one exactly once, and return to the starting point. In 1962 Mei-ko [6] posed the problem of finding a shortest closed covering walk, i.e., a closed walk including all the lines of a connected graph G. Because he originally formulated the problem in terms of the shortest distance a postman must walk to complete his tour, the problem is usually referred to as the "Chinese Postman Problem." A myriad of variations exist including applications to garbage collection, security patrolling, and rail-line inspection. Several authors including Mei-ko [6], Edmonds and Johnson [2], and Goodman and Hedetniemi [4] have solved the problem using somewhat different approaches. Mei-ko characterized the minimum closed covering walks in terms of a cycle property they possess. Edmonds and Johnson used matching theory and obtained a sophisticated algorithm for determining the minimum-length tour, while Goodman and Hedetniemi facilitated the solution with the aid of a "minimum pairing" of the odd-degree points of the graph—a notion we shall discuss later.

So that we can easily pose the problems that interest us let us first reformulate the postman problem. Suppose W: $v_0, x_1, v_1, x_2, v_2, \ldots, v_{n-1}, x_n, v_n = v_0$ is a closed covering walk of a graph G. For each repetition of a line x_i in W add a copy of x_i to G. The resulting multigraph is clearly eulerian. Thus another way to pose the postman problem is to ask for the minimum number of copies of lines from G that have to be added to G to obtain an eulerian multigraph. Other questions concerned with the addition of lines to a given graph, multigraph, or pseudograph to produce eulerian super structures arise quite naturally. We shall consider the following problems in the sequel:

(1) Given a (possibly disconnected) multigraph, when is it possible to obtain an eulerian spanning supermultigraph by the addition of lines and what is the minimum number of required lines?
(2) Given a (possibly disconnected) graph, when is it possible to obtain an eulerian spanning supergraph by the addition of lines (necessarily from the complement of the graph) and what is the minimum number of required lines?

* This paper is an expanded and illustrated version of an article by Boesch *et al.* [1].
† The notation and terminology used here follows that used by Harary [5].

0077-8923/79/0328-0193$1.75/1 © 1979, NYAS

Although the first problem is significantly easier to solve, its solution does provide insight into the second question.

After answering the first question we treat both problem (2) for connected graphs and the postman problem in a unified manner using techniques similar to those employed by Goodman and Heditniemi. In the final section disconnected graphs are considered and an answer to the minimum line question is given for any graph (connected or not) which spans an eulerian supergraph.

Henceforth a noneulerian graph (multigraph) G is *subeulerian* if it spans an eulerian supergraph (supermultigraph). The minimum number of additional lines required to produce an eulerian super structure is called the *euler completion number* and is denoted by $\mathrm{eu}^+(G)$.

Subeulerian Multigraphs

It is evident that any noneulerian multigraph M on p points is subeulerian since it spans the multigraph obtained from K_p by replacing each line of K_p with $2n$ lines for sufficiently large n. If M is connected, then it is only necessary to change the parity of the odd-degree points of M. As there are evenly many odd-degree points, and, unlike the postman problem, we are not constrained to choose the lines from M, the odd-degree points may be "paired off" and each pair joined with a line. Thus $\mathrm{eu}^+(M) = p_0/2$, where p_0 denotes the number of odd-degree points. In the disconnected case it is necessary to connect the components of M together in addition to changing the parity of the odd-degree points. Since a means of entering and leaving each of the $k(M)$ components must be provided, we add $k(M)$ lines according to the following scheme: when the original components are collapsed to points, the new lines form a cycle. The two new lines incident to a component of M having only even-degree points are to have a common endpoint in that component and the two new lines incident to a component containing at least two odd-degree points are to be attached at distinct odd-degree points of that component. Since there are evenly many odd-degree points in the resulting connected multigraph, it remains to

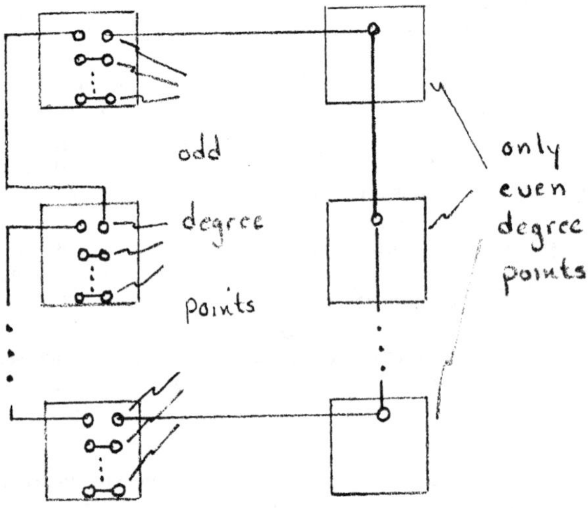

FIGURE 1.

pair off and join them with lines (see FIGURE 1). It is not difficult to see that this construction uses a minimum number of additional lines and, denoting the number of components having only even-degree points by $k_e(M)$, $\mathrm{eu}^+(M) = p_0/2 + k_e(M)$—a formula which subsumes the connected case.

SUBEULERIAN CONNECTED GRAPHS

The reader has probably already realized that the question of when a graph is subeulerian is trivial if the graph has an odd number p of points since it spans the eulerian graph K_p. However it is easily verified that the complete bipartite graph $K_{1,3}$ spans no eulerian graph. In fact, as we shall presently see, the essence of the structure of graphs which are not subeulerian is captured in this example.

Let G be a connected graph. Recall that for a connected multigraph it is enough to pair off the odd-degree points and join each pair with a line. Since, in the present case, we are constrained to use lines from $\bar{G}$, the complement of G, we cannot hope to mimic the multigraph case exactly. The idea of changing the parity of odd-degree points in pairs can be retained however if we observe that the addition of a path to a graph only affects the parity of the endpoints of the path. Thus if the odd-degree points $\theta(G)$ of G can be joined in pairs with paths taken from $\bar{G}$, then G is subeulerian.

Conversely, suppose G is a subeulerian, connected graph so that G spans an eulerian supergraph E. If the graph $E \cap \bar{G}$ contains a cycle, the lines which constitute the cycle are superfluous since removal of those lines from E leaves a supergraph of G which is necessarily connected and has only even-degree points. Thus there is a minimal spanning eulerian graph F trapped between G and E such that $F \cap \bar{G}$ is a forest. But any such forest can be decomposed into a union of paths, no two of which share an endpoint. Since the degree of a point in F is the sum of its degrees in G and $F \cap \bar{G}$, the set of endpoints of the paths coincides with the set $\theta(G)$ of odd-degree points of G.

The above conclusions are succinctly summarized with the aid of the following definition and proposition.

DEFINITION 1. If S is a set of points of a graph L, then a *pairing of S on L* is a collection $\mathscr{P} = \{P_1, P_2, \ldots, P_k\}$ of paths in L with $S = \bigcup_{i=1}^{k} \partial P_i$ and $\partial P_i \cap \partial P_j = \phi$, when $i \neq j$; here ∂P denotes the set of endpoints of P. If there is at least one pairing of S on L, we denote the minimum number of edges in $\bigcup_{i=1}^{k} P_i$ over all pairings by $m(S; L)$; a pairing that realizes this value is called a *minimum pairing.*

PROPOSITION 1. A connected graph G is subeulerian if and only if there is a pairing of $\theta(G)$ on $\bar{G}$. Furthermore, $\mathrm{eu}^+(G) = m(\theta(G), \bar{G})$.

It is appropriate at this time to comment on the Chinese Postman Problem. Whereas for the subeulerian case the additional lines must be taken from $\bar{G}$, the conditions of the postman problem dictate that the lines be selected from the connected graph G. Since pairings of $\theta(G)$ on G always exist, a spanning supermultigraph can be obtained from G by the addition of the lines of a minimum pairing of $\theta(G)$ on G. Conversely, just as in the subeulerian case, any multigraph $G + M$, where the lines of M are copies of lines of G, may be reduced to one for which M is a forest and, therefore, consists of the lines of a minimum pairing of $\theta(G)$ on G. This argument not only proves that $q + m(\theta(G), G)$ is the minimum length of a closed covering walk but also that any minimum closed covering walk *never* includes more than one additional copy of any line of G.

Returning to the subeulerian problem, we consider next the question of when pairings of $\theta(G)$ on $\bar{G}$ exist. It is readily seen that, in general, a pairing of S on L exists if and only if S meets each component of L in evenly many points. Thus pairings of $\theta(G)$ on $\bar{G}$ exist if and only if $\theta(G)$ meets each component of $\bar{G}$ in evenly many points. Now if G is defined on an odd number of points—a case in which G must be subeulerian—then the odd-degree points of G coincide with the odd-degree points of $\bar{G}$, and each component of $\bar{G}$ contains evenly many points of $\theta(G)$. On the other hand, if G is defined on an even number of points, the odd-degree points of G meet each component of $\bar{G}$ in the even-degree points of that component. Hence each component of $\bar{G}$ meets $\theta(G)$ in evenly many points if and only if each component of $\bar{G}$ is defined on evenly many points. But this is the case if and only if G is *not* spanned by a complete bipartite graph $K_{2m+1,2n+1}$. Of course, the latter condition is trivially satisfied when the number of points of G is odd and we may state our main theorem:

THEOREM 2. A connected graph G is subeulerian if and only if G is not spanned by $K_{2m+1,2n+1}$, for any value of m. Furthermore, if G is a connected subeulerian graph then $\mathrm{eu}^+(G) = m(\theta(G), \bar{G})$.

SUBEULERIAN DISCONNECTED GRAPHS

Let us begin this discussion with the assumption that G is a disconnected graph which is not subeulerian and see where it leads us. If G has more than two components, additional lines may be added to G to obtain a spanning supergraph H with exactly two components. Then, since G is not subeulerian, each of the spanning supergraphs $H + x$, where x is a bridge joining the components of H, fails to be subeulerian, as well. It follows from THEOREM 2 and the fact that each graph $H + x$ has line connectivity equal to one that one of the components of H is K_1. But this means that K_1 is a component of G and, since a graph which is not subeulerian must be defined on an even number of points, G is a spanning subgraph of $K_1 \cup K_{2n+1}$ for the appropriate value of n. Now it is readily seen that $K_1 \cup (K_{2n+1} - y)$ is subeulerian whenever y is an edge of K_{2n+1}, so that all the lines of K_{2n+1} belong to G, i.e., $G = K_1 \cup K_{2n+1}$.

Conversely, suppose $G = K_1 \cup K_{2n+1}$. If F is any connected spanning supergraph of G, then the line set of F must include a bridge from K_1 to K_{2n+1}. Thus F is spanned by $K_{1,2n+1}$. Summarizing the foregoing discussion, we have:

THEOREM 3. A disconnected graph G is subeulerian if and only if G is not $K_1 \cup K_{2n+1}$.

Finally we determine $\mathrm{eu}^+(G)$ for a subeulerian disconnected graph G. Since any spanning eulerian supergraph of G may be considered as an eulerian multigraph, $\mathrm{eu}^+(G) \geq p_0/2 + k_e$, where p_0 is the number of points having odd degree and k_e is the number of components having only even-degree points. It is clear that if there is a pairing of $\theta(G)$ using lines (i.e., paths of length 1) then the construction given in the first section for disconnected multigraphs can be applied here to yield a spanning eulerian supergraph of G *provided* $p_0 > 0$ or $p_0 = 0$ and $k_e \geq 3$. On the other hand, if $p_0 = 2$ and $k_e = 2$, i.e., G is the union of two eulerian components, four additional lines are needed if both components are complete; otherwise only three are required (see FIGURE 2).

Thus, with the exception of the case where G is the union of two eulerian

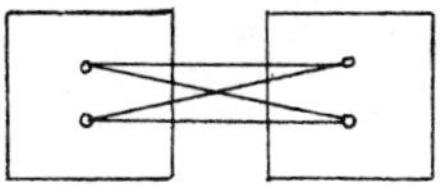

FIGURE 2a. Both components are complete.

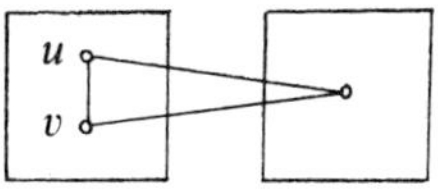

FIGURE 2b. Endpoints u, v are not adjacent in G.

components, if $m(\theta(G), \bar{G}) = p_0/2$, $\mathrm{eu}^+(G) = p_0/2 + k_e$. It is not intuitively displeasing that more than just $m(\theta(G), \bar{G}) = p_0/2$ lines are needed in this case. In fact, it is probably reasonable to suspect that for any disconnected subeulerian G more than $m(\theta(G), \bar{G})$ lines should be necessary to "complete" G to an eulerian graph. This is *not* the case however as we shall now see.

As a first step it is necessary to establish the fact that *pairings of $\theta(G)$ on $\bar{G}$ do exist for any disconnected subeulerian G*. The reader will recall that this was done for connected G, and it suffices to note that essentially the same argument applies here.

Secondly we observe that if $m(\theta(G), \bar{G}) > p_0/2$ and $\mathscr{P} = \{P_1, P_2, \ldots, P_k\}$ is a minimum pairing of $\theta(G)$ on $\bar{G}$, then each path in $\mathscr{P}$ has length 1 or 2 with at least one of length 2. Furthermore, *all the endpoints of the paths of length 2 and at least one point of each path of length 1 lie within a single component K of G*. As a first step in establishing this claim suppose $P \in \mathscr{P}$ has endpoints u and v which belong to separate components of G. Then, since the paths have length no greater than 2 and any pair of paths have distinct endpoints, the edge uv belongs to no path of $\mathscr{P}$, and P may be replaced by uv, thereby obtaining a pairing with fewer lines than $\mathscr{P}$. The same kind of contradiction follows from each of the assumptions that there are two paths of length 2 with endpoints in different components of G or that there is a path of length 1 with both endpoints outside the component containing the endpoints of the paths of length 2.

Thus, each component of G different from K and containing odd-degree points is connected to K in $L = G \cup \bigcup_{i=1}^{k} P_i$ via at least two paths of length 1. Hence if the internal point of a path of length 2 lies within such a component, removal of the path from L will not disconnect the graph.

With these facts in mind consider the graph $L = G \cup \bigcup_{i=1}^{k} P_i$. The entire set $\theta(G)$ lies within a single component H of L and the remaining components, if any, consist entirely of even-degree points of G. Some of the components of G having only even-degree points, the number of which we denote by s ($s \geq 0$), contain internal points of paths from $\mathscr{P}$ having length 2, and are contained in H. The remaining $k_e - s$ components of G with only even-degree points are disjoint from H. Unless $s = k_e$, we may assume that each of the aforementioned s components contains only one internal point of a path of length 2. Indeed, if $s < k_e$ and there is an eulerian component of G containing the internal points of two or more paths of length 2 from $\mathscr{P}$ then we may replace any one of these paths by a path having the same endpoints and an internal point in one of the $k_e - s$ components of G which are disjoint from H. Using this "surgical technique" repeatedly we arrive at the desired conclusion after a finite number of steps.

If $s = k_e$, then L is eulerian, and so $\mathrm{eu}^+(G) = m(\theta(G), \bar{G})$. If $s < k_e$, designate points, one from each of the eulerian components of G disjoint from H, by $w_1, \ldots, w_{k_e - s}$. Let r denote the number of paths in $\mathscr{P}$ of length 2, with internal points *not* in eulerian components of G. If $r > 0$ let $P_1, \ldots, P_r$, denote those paths and u_i, v_i the elements of ∂P_i, for $i = 1, \ldots, r$.

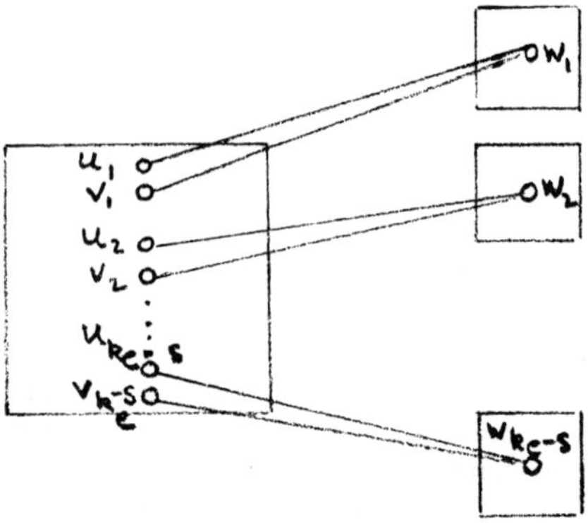

FIGURE 3.

Two separate possibilities must be considered: namely, (1) $0 < k_e - s \leq r$ and (2) $k_e - s > r$. In the first case, we may replace the paths $P_1, P_2, \ldots, P_{k_e-s}$ by paths $u_1 w_1 v_1, u_2 w_2 v_2, \ldots, u_{k_e-s} w_{k_e-s} v_{k_e-s}$ so that the resulting graph is connected and therefore eulerian (see FIGURE 3). In this instance $\mathrm{eu}^+(G) = m(\theta(G), \bar{G})$. In the second case, if $r \geq 1$, replace $P_1, \ldots, P_{r-1}$, as before, and P_r by $u_r w_r w_{r+1}, \ldots, w_{k_e} v_r$, thereby obtaining an eulerian graph (see FIGURE 4). Now the number of lines in excess of $m(\theta(G), \bar{G})$ necessary for this construction is just $k_e - s - r$. But $m(\theta(G), \bar{G}) = 2r + 2s + (p_0/2 - r - s) = p_0/2 + r + s)$ so that $k_e - r - s = k_e - (m(\theta(G), \bar{G}) - p_0/2)$. Thus the total number of additional lines is just $k_e + p_0/2$. Finally, if $r = 0$, then let P: uwv be a path of length 2 with internal point w in an eulerian component E of G. Replace P by the path $uww_1, \ldots, w_{k_e-s}v$ (see FIGURE 5) with the result being an eulerian graph. The calculations used in the previous case apply here to show that a total of $p_0/2 + k_e$ additional lines are required.

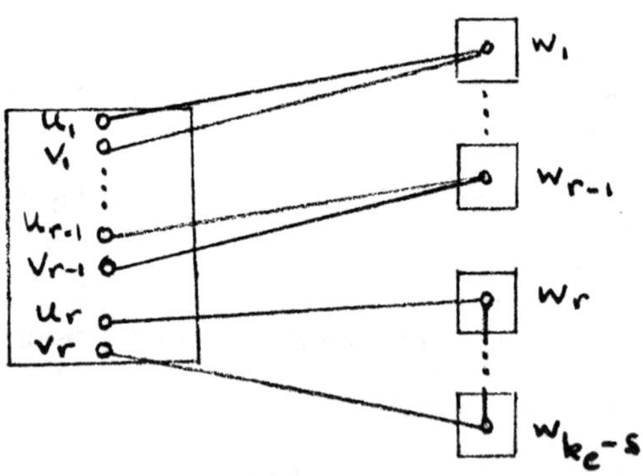

FIGURE 4.

Summarizing, we have seen that for any subeulerian, disconnected graph G which is not the union of two eulerian components $\mathrm{eu}^+(G) = \max(m(\theta(G), \bar{G}), p_0/2 + k_e)$, with $p_0/2 + k_e$ being the answer when there are fewer paths of length 2 than components of H having only even-degree points. Of course, if G is a connected subeulerian graph then $k_e = 0$ and, therefore, $m(\theta(G), \bar{G}) \geq p_0/2 = p_0/2 + k_e$. Hence we have:

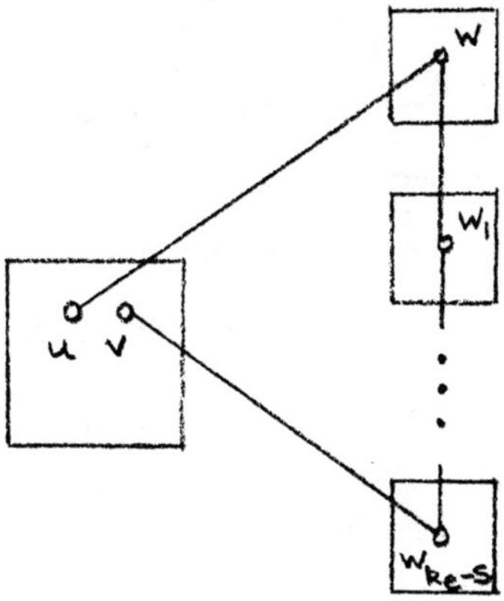

FIGURE 5.

THEOREM 4. If G is a subeulerian graph (connected or not) which is not the union of two eulerian components $\text{eu}^+(G) = \max(m(\theta(G), \bar{G}), k_e + p_0/2)$. If $G = E_1 \cup E_2$, where E_1 and E_2 are eulerian, then $\text{eu}^+(G) = 4$ if both E_1 and E_2 are complete; 3 otherwise.

REMARK. Some related open questions are: (1) characterizing those digraphs (with at most one line between any pair of points) which are spanned by eulerian digraphs, and (2) characterizing those graphs which are spanned by eulerian subgraphs.

It is interesting to note that the class of graphs mentioned in (2) contains the class of hamiltonian graphs.

SUMMARY

In this paper we are concerned with the problem of adding lines to a non-eulerian structure to produce a spanning eulerian superstructure. Three cases to be considered are determined by the nature of the original structure and by the origin of the additional lines as follows:

(1) The original structure is a connected graph G and the additional lines are chosen from G so that the superstructure is a multigraph (the Chinese Postman Problem).
(2) The original structure is a graph G and the additional lines are taken from the complement of G so that the superstructure is a graph as well.
(3) The original structure is a multigraph and no restriction on the origin of the additional lines is imposed.

The idea of "pairing the odd-degree points" unifies the solutions of all three problems. In cases (1) and (3) a spanning eulerian superstructure always exists, while in case (2), we find that a connected noneulerian graph G spans an eulerian graph if and only if G is not spanned by an odd complete bigraph $K(2m + 1, 2n + 1)$. If G is disconnected, a spanning eulerian supergraph of G exists if and only if G is not of the form $K_1 \cup K_{2n+1}$. Exact formulas for the minimum number of necessary additional lines are obtained in all three cases.

References

1. Boesch, F., C. Suffel & R. Tindell. 1977. The spanning subgraphs of Eulerian graphs. J. Graph Theory. **1:** 69–75.
2. Edmonds, J. & E. L. Johnson. 1973. Matching, Euler tours and the Chinese postman. Math. Programming. **5:** 88–124. North-Holland, Dordrecht, the Netherlands.
3. Euler, L. 1736. Solutio problematis ad geometrian situs pertinentis. Commentarii Academiae Petropolitanae **8:** 128–140.
4. Goodman, S. & S. Hedetniemi. 1973. Eulerian walks in graphs. SIAM J. Comput. **2**(1): 16–27.
5. Harary, F. 1969. Graph Theory. Addison-Wesley, Reading, Mass.
6. Mei-ko, K. 1962. Graphic programming using odd or even points. Chin. Math. **1:** 273–277.

COVERING DIGRAPHS AND STRUCTURED PROGRAM SCHEMES

Ralph Tindell

Department of Pure and Applied Mathematics
Stevens Institute of Technology
Hoboken, New Jersey 07030

1. Introduction

This paper provides a brief excursion into an area of theoretical computer science in which certain labelled directed graphs (flowchart schemes) serve as models in the study of classes of flowchart algorithms. The particular concepts and results presented here were chosen for their highly geometric character and the fact that, while sophisticated results from "mainstream" graph theory are rarely drawn on, the approach and techniques used are clearly in the spirit of the subject.

2. Covering Homomorphisms

Before proceeding to the topic of schemes we pause to lift a concept from elementary algebraic topology. The property of interest here is one of the important properties possessed by covering maps. While we shall not define the notion of topological covering map, we point out two well-known examples: the complex exponential map $x \mapsto e^{i2\pi x}$ mapping the real line onto the circle S of complex numbers of modulus one, and its extension to produce a covering of the torus $S \times S$ by the plane. The property possessed by any topological covering map f: $X \to Y$ germane to our interests is the Unique Path Lifting Property:

For each path λ in Y initiating at $y \in Y$, and each $x \in X$ with $f(x) = y$, there is a unique path $\bar{\lambda}$ in X initiating at x and covering λ—that is, with $f\bar{\lambda} = \lambda$.

If we translate this property into the category of graphs—so that "map" becomes "graph homomorphism," "path" becomes "walk,"* etc.—the resulting "covering homomorphisms" are nothing more than graph homomorphisms that happen to be topological covering maps on the underlying topological spaces. However, translation into the category of directed graphs results in something new. Let us call a digraph homomorphism a *covering homomorphism*, or just a *covering*, if it possesses the Unique (directed) Walk Lifting Property. If there is a covering f: $H \to G$, we say that H *covers* G, and that H is a *covering digraph* of G. It is a simple matter to verify the following "local" characterization of coverings.

Proposition 2.1. A digraph homomorphism f: $H \to G$ is covering iff f is onto and maps the outedges of each vertex $x \in X$ bijectively onto those of $f(x)$.

The notion of covering can be seen to be "inverse" to the combinatorially defined notion of "partial unfolding" via "node splitting," as encountered in the study of

* As an ex-topologist I sometimes inadvertently use "path" where Harary's term *walk* is appropriate; to succumb in the present publication would indeed be unseemly!

0077-8923/79/0328-0201$1.75/1 © 1979, NYAS

flowgraphs and of schemes. The following process is described as *node splitting*: the inedges of some node v are partitioned into sets A, A'; those in A are left undisturbed while those in A' have their targets redirected to a *new* node v'; for each outedge e of v a copy e' is introduced with the same target but with source v'. The resultant digraph G' is said to have resulted from "splitting" v to v, v'. FIGURE 1 gives an example. A digraph H is said to be a *partial unfolding* of a digraph G if $H = G$, or H is obtained from G by a finite sequence of node splittings. In proving certain theorems (or in finding a proof!) the above combinatorial definition of partial unfolding often presents technical difficulties. It should be obvious that "merging" the nodes resulting from splitting v produces a covering homomorphism, so that any partial unfolding of G covers G.

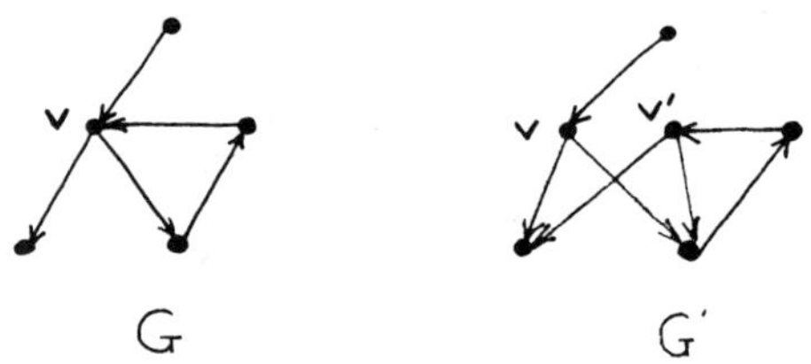

FIGURE 1. Node splitting.

3. FLOWCHART SCHEMES

We now come to the definition of "scheme." The definition may seem cumbersome but all its features are necessary; in fact, it is simpler than most because we consider only "monadic" schemes—that is, ones with a single variable which is never explicitly mentioned. A scheme is merely an abstraction of a flowchart in which the actual functions and tests have been replaced by uninterpreted symbols. Thus we begin with finite sets Γ_1, Γ_2, ... of symbols; the elements in Γ_1 are function symbols; those in Γ_2, predicate symbols; those in Γ_3 symbols for 3-outcome tests; etc. Now the underlying object of a scheme (or a flowchart for that matter) is a directed pseudograph. That is, we allow multiple edges and edges from a point to itself. We then label the modes of positive outdegree by appropriate symbols—if v has outdegree n its label must be in Γ_n. We think of an n-outcome test as having as its possible "outcomes" the integers 1, 2, ..., n; one of the results of the test ought to be an instruction as to where to go next. This is accomplished by assigning to the outedges of an outdegree n vertex distinct labels from the set $[n] = \{1, 2, \ldots, n\}$. Thus, if the outcome is i, we are to "go to the target of the ith outedge." Next we designate one of the nodes as the *begin* node, the point where the initial input is placed. What of the outdegree 0 nodes? We consider them to be the *exits* of the scheme; we often wish to distinguish *where* the output was produced, so we enumerate these exits—e_1, ..., e_p—and say that the scheme is a scheme with a single begin and p exits, or more succinctly, that it is a scheme from 1 to p. To indicate the restriction to the sets Γ_n, we set $\Gamma = \{\Gamma_1, \Gamma_2, \ldots)$ and say our scheme is a Γ-*scheme*. For each $\gamma \in \Gamma_n$, there is a special *atomic* scheme, also called γ, with begin labelled γ and such that the ith outedge of the begin has

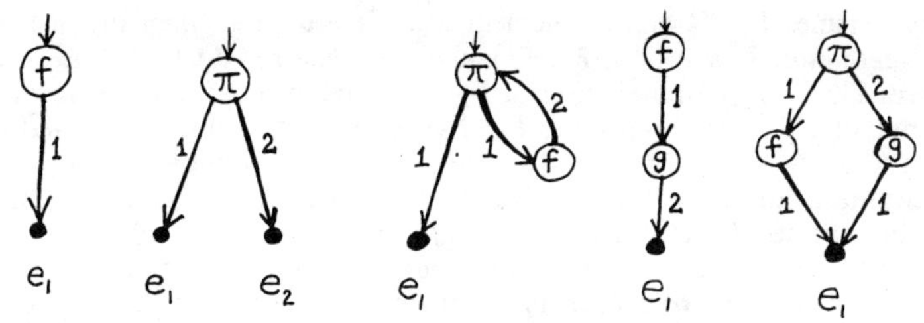

FIGURE 2.

as target the ith exit; there are no other nodes. In FIGURE 2 several schemes are depicted; the first two are the atomic schemes f and π, respectively, where $f \in \Gamma_1$ and $\pi \in \Gamma_2$. In the figure, the sourceless arrow points to the begin.

4. STRUCTURED PROGRAM SCHEMES

We now come to the notions of structured programming in the scheme setting. When a freewheeling programmer constructs flowcharts, he builds without restriction and his final result often looks like spaghetti. In reaction to some of these excesses, which defy subsequent modification or comprehension, many people proposed restrictions on the "operations" allowed in constructing flowchart algorithms. One of the first and most restrictive was that the programmer use only if-then-else, composition, and while-do constructions. These constructions may be formulated as *operations* on schemes.

For the study of "D-schemes," we restrict our attention to schemes with a single exit and assume that $\Gamma_n = \phi$ for $n > 2$. Rather than present formal definitions of the three operations, we indicate the constructions in FIGURE 3. It is assumed that

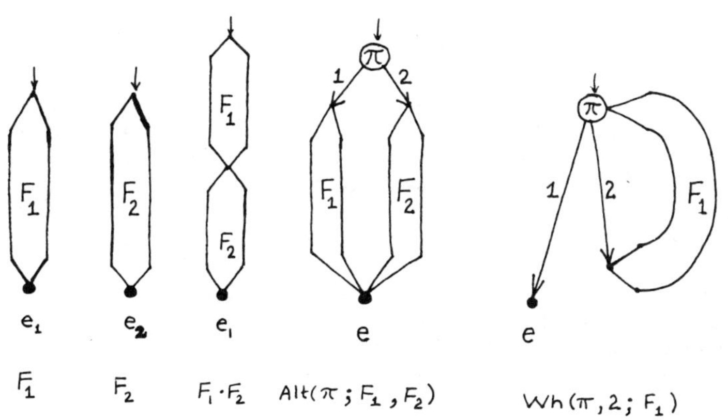

FIGURE 3.

arbitrary schemes F_1, F_2 and an element π of Γ_2 are given; then the *composition* $F_1 \cdot F_2$, *alternation by* π Alt (π; F_1, F_2), and the *while* π, *i do* F_1 Wh (π, i; F) are illustrated. The computational effect of $F_1 \cdot F_2$ is "perform F_1, then perform F_2 on the output of F_1"; that of Alt (π; F_1, F_2) is "if π has outcome i, perform F_i ($i = 1, 2$)"; the effect of Wh (π, i; F_1) is: "test by π; if the result is i, do F_1, and test again; continue until the result of π is not i, then exit." Referring to FIGURE 2, the nonatomic schemes depicted are Wh (π, 2; f), $f \cdot g$, and Alt (π; f, g).

Note that the while-do operation is the only one to introduce new loops. It will be convenient to have a *trivial* scheme T having no effect on its input. Thus T has no edges, and one node that is both the begin and the exit. The schemes Wh (π, 1; T), Alt (π; f, T), and Alt (π; T, f) are pictured in FIGURE 4. Note that T acts as the identity for composition.

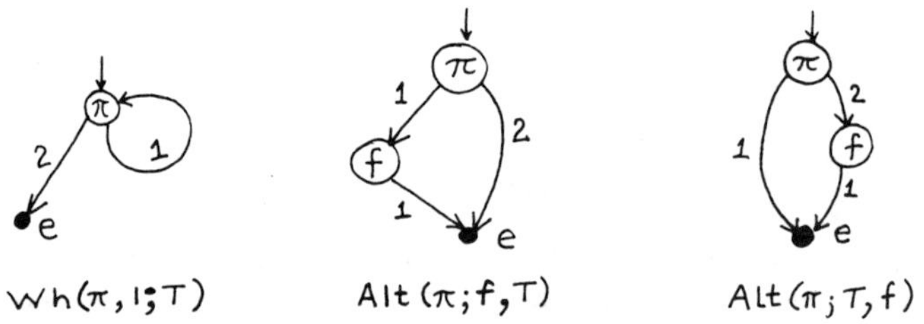

FIGURE 4.

We are now prepared to give the algebraic description of the class D of D-schemes† as the least class containing T, containing all atomic schemes f with $f \in \Gamma_1$ and closed under composition and the operations Alt (π; —, —), Wh (π, i; —) for all $\pi \in \Gamma_2$ and $i \in \{1, 2\}$. Equivalently, the D-schemes are those built from $\{T\} \cup \Gamma_1$ using finitely many applications of the three operations defined above.

Elgot [2] gives a unified description of a variety of classes of structured schemes. This paper serves as our basic reference.

5. Geometric Characterizations

Inspired by the theorem of Elgot and Shepherdson [2, 3], S. L. Bloom and I [1] developed global, geometric conditions on schemes that may be viewed as qualitative measures of "structuredness." We then used these conditions to characterize various algebraically defined classes of structured schemes. In this section, we develop the notions used and state the characterization theorem for D-schemes.

An *initial entry walk* to a subgraph H of a scheme F is a directed walk P from the begin of F to a point u of H such that no vertex preceding u on P lies in H. A *final exit walk* is a directed walk Q from a point v of H to the exit such that no point following v on Q lies in H. In the above instances, u is called an *initial entry node* and v a *final exit node* for H. A scheme F is *reducible* if every directed cycle in F has a unique initial entry node; it is *coreducible* if every directed cycle has a

† Named for E. Dijkstra, the instigator of the discussion of structured programming.

unique final exit node. F is *bireducible* if it is reducible, coreducible, and if, for each directed cycle C, the initial entry and final exit nodes of C coincide. In addition to directed cycles one must consider cycles with a single change of direction: a *bipath* from u to v is the union of two directed paths from u to v, intersecting only at u and v. Finally, we say that F is *biaccessible* if every vertex of F lies on some directed walk from the begin to the exit.

THEOREM 5.1 [1]. A scheme F from 1 to 1 is a D-scheme iff F is bireducible, biaccessible, and for any bipath B from u to v, v is the final exit node of B.

By applying the criteria in THEOREM 5.1, the schemes G_1, G_2, G_3 in FIGURE 5 can be seen *not* to be D-schemes.

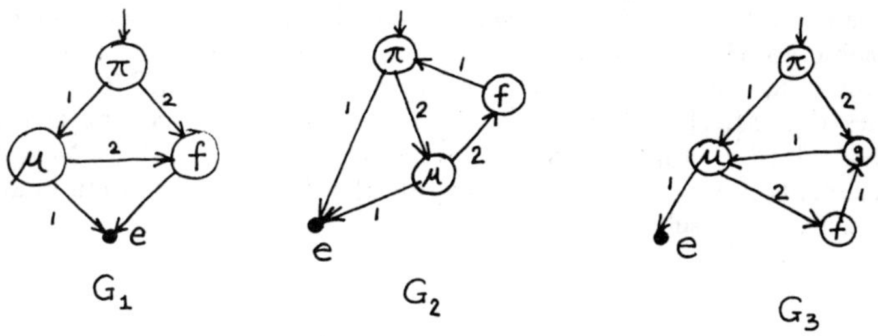

FIGURE 5.

The reasons are: G_1 has a bipath with two final exit nodes, G_2 is not coreducible, and G_3 is not reducible.

6. MORPHISMS BETWEEN SCHEMES

It should be fairly clear what the appropriate notion of scheme (homo) morphism is: a digraph homomorphism preserving begins, labels, and exit enumerations. One interesting fact stands out: every scheme morphism is a digraph covering homomorphism. This follows from the "local" characterization of covers (PROPOSITION 2.1), the preservation of edge labels, and the fact that the outedges of a vertex are *enumerated* by their labels. One may also deduce from this "local ordering" that if there is a morphism from F to G, there is *exactly* one.

Let us consider a slight generalization of the notion of scheme obtained by relaxing the requirement that the vertex set be finite and by requiring of a scheme from 1 to p not that its exits be enumerated but that they be *labelled* by elements of $[p] = \{1, 2, \ldots, p\}$. There may be any number of such exits. In this setting we have the following observation. An *accessible* scheme is one in which each vertex is accessible from the begin.

PROPOSITION 6.1. For each accessible (generalized) scheme F there is a generalized scheme ΠF covering F such that ΠF covers any (generalized) scheme covering F.

Thus the generalized scheme ΠF is a *maximal* covering scheme of F, and is called the *tree* of F, or the *strong behavior* of F. In fact, ΠF is the (possibly infinite) rooted tree whose vertices are the directed walks initiating at the begin b of F. The edges of ΠF are the pairs (W, e), where W is a vertex of ΠF and e is an edge of F with source the terminal point of W; W is the source of (W, e) and its target is the path obtained by extending W by e. The label of W is that of its terminal point in F, and the integer labelling the edge (W, e) is the label of e in F. The constant path b is the begin of ΠF.

The geometric view of F is that it is the "*total* unfolding" of F. ΠF is exactly the set of all formal "computational tracks" through F so that schemes with the same tree will compute the same function no matter how the symbols are interpreted [2]. Such schemes are said to be *strong equivalent* or *tree equivalent*. If F is not accessible, ΠF is still defined but covers only the "accessible part" of F. We may utilize this in the following construction. Define $\Pi_v F$, *the behavior of a vertex v in F*, to be the tree of the scheme obtained from F by changing the designation of the begin to be v. Two vertices are *equivalent* if they have the same behavior (i.e., $\Pi_u F$ is isomorphic to $\Pi_v F$). Reducing F by this relation produces a scheme $M(F)$ which is the unique *minimal* scheme tree equivalent to F in that ΠG isomorphic to ΠF implies that $M(G)$ is isomorphic to $M(F)$. Note that F covers $M(F)$, and G tree equivalent to F implies that G covers $M(F)$. Thus we have *maximal* and *minimal* elements in the class of all (generalized) schemes, and the minimal scheme of a finite scheme is itself finite.

7. Tree Equivalence and D-Schemes

In studying the tree equivalence classes of D-schemes, the first thing one notices is that the minimal scheme $M(F)$ of a D-scheme F is not necessarily a D-scheme. In Figure 6, schemes F, G, $M(F)$, and $M(G)$ are depicted. Note that $F = \text{Alt}\ (\pi;\ \text{Alt}\ (\mu;\ T, f)$ and $G = f \cdot \text{Wh}\ (\pi, 2; f)$ are D-schemes, but their minimal schemes are not.

These two examples reveal that the "bipath exit" property and bireducibility are not preserved by scheme morphisms. In fact, the only one of the four properties (excluding biaccessibility) that is preserved by morphisms is coreducibility. Therefore the minimal scheme of a D-scheme, and hence of any scheme tree equivalent to a D-scheme, is coreducible. The converse to this statement is (a slightly strengthened

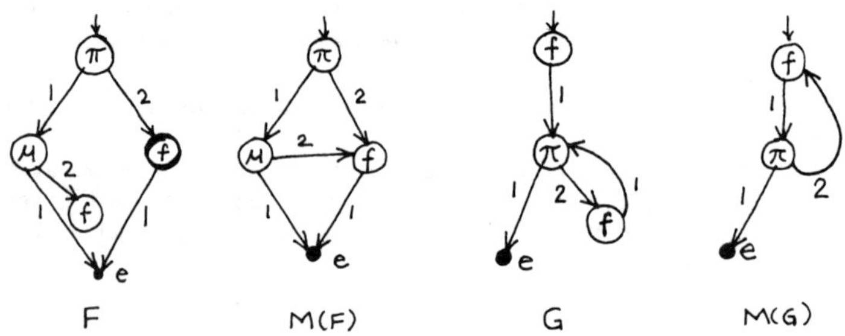

Figure 6.

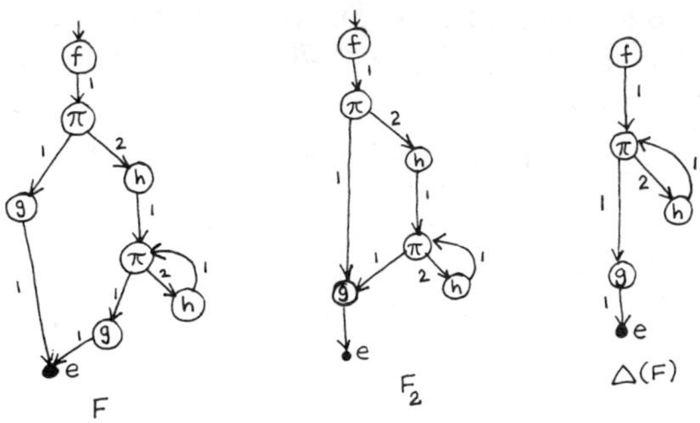

FIGURE 7.

version of) a theorem due to Kasai [4]. Kasai's theorem is a consequence of the following result [4], whose proof is essentially the same as that used by Kasai. Since D-schemes are biaccessible, we shall, for the remainder of this section, assume all schemes considered to be biaccessible.

THEOREM 7.1. A scheme F is covered by a D-scheme iff F is coreducible.

Since the minimal scheme $M(F)$ of a D-scheme F is not necessarily a D-scheme, we might ask whether there are unique minimal schemes within the class of D-schemes. The strongest formulation of this notion would be that for any D-scheme F there is a D-scheme $\Delta(F)$ such that (1) F covers $\Delta(F)$ and (2) any D-scheme G tree equivalent to F also covers $\Delta(F)$. It would then follow in the situation of (2), that $\Delta(G)$ is isomorphic to $\Delta(F)$. That minimal D-schemes exist in this sense follows from the next theorem, due to the present author [5].

THEOREM 7.2. Let G be a coreducible scheme. Then there is a D-scheme $\mathbb{D}(G)$ covering G such that any other D-scheme covering G also covers $\mathbb{D}(G)$.

To obtain the existence of the unique minimal D-scheme $\Delta(F)$, as described above, for a D-scheme F, merely set $\Delta(F) = \mathbb{D}(M(F))$. A remarkable fact that emerged in the development of THEOREM 7.2 was that any morphism between D-schemes is a "combination" [5] of two basic types: one maps Alt $(\pi;\, F_1 \cdot G,\, F_2 \cdot G)$ onto Alt $(\pi;\, F_1,\, F_2) \cdot G$, the other maps Alt $(\pi;\, T,\, G \cdot \text{Wh}\,(\pi, 2;\, G))$ onto Wh $(\pi, 2;\, G)$ (or, alternatively, Alt $(G \cdot \text{Wh}\,(\pi, 1;\, G),\, T)$ onto Wh $(\pi, 1;\, G)$). Thus, by applying a sequence of "reductions" of this type to a D-scheme F, one arrives at $\Delta(F)$. This is the basis for a nice algorithm for constructing $\Delta(F)$ from F. For example, the reduction of $F = f \cdot \text{Alt}\,(\pi;\, g,\, h \cdot \text{Wh}\,(\pi, 2;\, h) \cdot g)$ would proceed as follows $F \to F_2 = f \cdot \text{Alt}\,(\pi;\, T,\, h \cdot \text{Wh}\,(\pi, 2;\, h)) \cdot g \to f \cdot \text{Wh}\,(\pi, 2;\, h) \cdot g = \Delta(F)$. These three schemes are pictured in FIGURE 7.

We now consider the question of the *cost*, in terms of size, in passing from arbitrary schemes to tree equivalent D-schemes. By THEOREM 7.1, we must restrict our attention to those schemes F for which $M(F)$ is coreducible, and the result of THEOREM 7.2 tells us that we should compare the size of $M(F)$ to that of $\mathbb{D}(M(F))$. We close our little excursion into the subject of "schematology" with an answer to this question, due to the present author [5]. Let $\beta(n)$ be the unique function on the integers satisfying $\beta(0) = 0$, $\beta(1) = 1$, and $\beta(n + 2) = 2\beta(n) + 2$. To be explicit, $\beta(n) = 2^{[(n+1)/2]} + 2^{[n/2]} - 2$.

THEOREM 7.3. If G is a coreducible scheme with n edges, then $\mathbb{D}(G)$ has at most $\beta(n)$ edges.

THEOREM 7.4. For each $n \geq 1$, there is a coreducible scheme $G_n = M(G_n)$ having exactly n edges and such that $\mathbb{D}(G_n)$ has exactly $\beta(n)$ edges.

SUMMARY

Monadic flowchart schemes are special labelled directed graphs used as models in the study of classes of flowchart algorithms. The purpose of the present paper is to provide a brief introduction to the topic of flowchart schemes and to show how a geometric concept—that of a covering homomorphism—proves useful in studying the relationship between arbitrary schemes and restricted classes of more highly structured schemes. In particular, recent results are presented that deal with the class of Dijkstra schemes arising from the discipline of structured programming.

REFERENCES

1. BLOOM, S. L. & R. TINDELL. 1979. Algebraic and graph theoretic characterizations of structured flowchart schemes. Theor. Comp. Sci. To appear.
2. ELGOT, C. C. 1976. Structured programming with and without GO-TO statements. IEEE Trans. Software Eng. Vol. SE-2, No. 1 (March). pp. 41–53; Erratum and Corrigendum, ibid. (September).
3. ELGOT, C. C. & J. C. SHEPHERDSON. 1977. A semantically meaningful characterization of reducible flowchart schemes. IBM Res. Rep. 6656 (July).
4. KASAI, T. 1974. Translatability of flowcharts into while programs. J. Comput. Syst. Sci. **9:** 177–195.
5. TINDELL, R. 1979. Unique minimal *D*-schemes. Theor. Comput. Sci. To appear.